Developmental Biology and Musculoskeletal Tissue Engineering

Developmental Biology and Musculoskeletal Tissue Engineering

Principles and Applications

Edited by

Martin J. Stoddart

April M. Craft

Girish Pattappa

Oliver F.W. Gardner

ACADEMIC PRESS

An imprint of Elsevier

Academic Press is an imprint of Elsevier
125 London Wall, London EC2Y 5AS, United Kingdom
525 B Street, Suite 1800, San Diego, CA 92101-4495, United States
50 Hampshire Street, 5th Floor, Cambridge, MA 02139, United States
The Boulevard, Langford Lane, Kidlington, Oxford OX5 1GB, United Kingdom

Notices
Knowledge and best practice in this field are constantly changing. As new research and
experience broaden our understanding, changes in research methods, professional practices,
or medical treatment may become necessary.

Practitioners and researchers must always rely on their own experience and knowledge in
evaluating and using any information, methods, compounds, or experiments described
herein. In using such information or methods they should be mindful of their own safety and
the safety of others, including parties for whom they have a professional responsibility.

To the fullest extent of the law, neither the Publisher nor the authors, contributors, or editors,
assume any liability for any injury and/or damage to persons or property as a matter of
products liability, negligence or otherwise, or from any use or operation of any methods,
products, instructions, or ideas contained in the material herein.

Library of Congress Cataloging-in-Publication Data
A catalog record for this book is available from the Library of Congress

British Library Cataloguing-in-Publication Data
A catalogue record for this book is available from the British Library

ISBN: 978-0-12-811467-4

For information on all Academic Press publications visit our website at
https://www.elsevier.com/books-and-journals

Working together
to grow libraries in
developing countries

www.elsevier.com • www.bookaid.org

Publisher: John Fedor
Acquisition Editor: Mica Haley
Editorial Project Manager: Tracy Tufaga
Production Project Manager: Punithavathy Govindaradjane
Cover Designer: Miles Hitchen

Typeset by TNQ Books and Journals

Contents

10. **Clinical Translation of Cartilage Tissue Engineering, From Embryonic Development to a Promising Long-Term Solution**
Diego Correa and Annie C. Bowles

List of Contributors

Annie C. Bowles, University of Miami, Miami, FL, United States

Soraia P. Caetano-Silva, The Royal Veterinary College, London, United Kingdom

Pauline Colombier, University of California, San Francisco (UCSF), San Francisco, CA, United States

Diego Correa, University of Miami, Miami, FL, United States

April M. Craft, Boston Children's Hospital, Boston, MA, United States; Harvard Stem Cell Institute, Cambridge, MA, United States

Rebekah S. Decker, The Children's Hospital of Philadelphia, Philadelphia, PA, United States

Eric Farrell, Erasmus MC, University Medical Centre Rotterdam, Rotterdam, The Netherlands

Jenna L. Galloway, Massachusetts General Hospital, Harvard Medical School, Boston, MA, United States; Harvard Stem Cell Institute, Cambridge, MA, United States

Mor Grinstein, Massachusetts General Hospital, Harvard Medical School, Boston, MA, United States; Harvard Stem Cell Institute, Cambridge, MA, United States

Behzad Javaheri, The Royal Veterinary College, London, United Kingdom

Caoimhe Kiernan, Erasmus MC, University Medical Centre Rotterdam, Rotterdam, The Netherlands

Callie Knuth, Erasmus MC, University Medical Centre Rotterdam, Rotterdam, The Netherlands

Eiki Koyama, The Children's Hospital of Philadelphia, Philadelphia, PA, United States

Henry M. Kronenberg, Massachusetts General Hospital and Harvard Medical School, Boston, MA, United States

Roberto Narcisi, Erasmus MC, Rotterdam, The Netherlands

Astrid Novicky, The Royal Veterinary College, London, United Kingdom

Noriaki Ono, University of Michigan School of Dentistry, Ann Arbor, MI, United States

Maurizio Pacifici, The Children's Hospital of Philadelphia, Philadelphia, PA, United States

Andrew A. Pitsillides, The Royal Veterinary College, London, United Kingdom

Makarand V. Risbud, Thomas Jefferson University, Philadelphia, PA, United States

Linda J. Sandell, Washington University School of Medicine, St. Louis, MO, United States; Washington University School of Engineering and Applied Science, St. Louis, MO, United States

Nailah M. Seale, University of California, San Diego, La Jolla, CA, United States

Shyni Varghese, University of California, San Diego, La Jolla, CA, United States; Duke University, Durham, NC, United States

Yuze Zeng, Duke University, Durham, NC, United States

Preface

The concept for this book came from a preconference workshop that the editors ran together at the 2015 TERMIS World Congress in Boston, USA. The inspiration for the workshop, and this book, came from the simple idea of bringing together people who work at the interface of musculoskeletal development and tissue engineering to encourage discussion between the disciplines. We sought the opinion of developmental biologists, tissue engineers, and clinicians with experience in these fields to share their work and expertise. In doing so, we hope the insight into the groundbreaking research being conducted in these fields can provide answers or inspiration for those working in both developmental biology and tissue engineering.

To lay the foundations for tissue engineers who have not had previous experience in developmental biology, the initial chapters provide an introduction to musculoskeletal development in health and disease, before specialized chapters focusing on various musculoskeletal tissues. These latter chapters illustrate the trail-blazing work being done by tissue engineers using developmental principles and the efforts being made to translate pioneering scientific work from bench to bedside.

Our motivation in formulating this book is the fascination we share in the truly remarkable processes responsible for embryonic development. We hope that the secrets that these processes hold can be unlocked to pave the way for the therapies of the future. We hope that this book will provide intrigue and foster creativity for those working in both these fields at all levels, whether it is to open up the world of developmental biology to students with an engineering background or to sow the seed of a new avenue of enquiry for those in more senior positions. Our belief is that discussion and collaboration between the fields holds the key toward developing improved clinical solutions in the future.

Acknowledgments

We would like to thank the staff at Elsevier, in particular Mica Haley, Lisa Eppich, and Tracy Tufaga, without whom it would have not been possible to put this book together.

Chapter 1

Developmental Biology of Musculoskeletal Tissues for Tissue Engineers

Noriaki Ono[1], Henry M. Kronenberg[2]
[1]*University of Michigan School of Dentistry, Ann Arbor, MI, United States;* [2]*Massachusetts General Hospital and Harvard Medical School, Boston, MA, United States*

INTRODUCTION

The musculoskeletal system is a highly complex unit in which there are five major elements—bones, muscles, tendons that connect the former two elements, cartilages, and menisci—that function together to achieve locomotion. Although these elements are equally important, bones have been historically regarded as a central component of the musculoskeletal system. Bones have strong and rigid structures owing to mineralized matrix, yet they grow explosively in early life and maintain their strength throughout life. Contrary to their inert appearance, bones are strikingly multifunctional. The primary functions of bones are to protect vital organs and act as levers whereby muscle contraction leads to body movement. In addition, bone cells support hematopoiesis in the adjacent marrow space and secrete hormones that regulate carbohydrate and mineral ion metabolism, as well as fertility and brain function.

Because of their primary functions, bones, muscles, and tendons are by far the most commonly injured tissues in the body. These tissues possess amazing capabilities to repair various degrees of damage incurred on them, ranging from microscopic to substantial damages that disrupt tissue continuity. Tissue engineering has provided great promise in musculoskeletal tissue regeneration. The current approaches are intended to enhance innate regenerative capabilities by supplementing appropriate cells, signals, and scaffolds. Despite some success in many settings, existing approaches have certain limitations in their applicability to musculoskeletal regeneration. Generally speaking, tissue engineers can rebuild the lost component only when there are sufficient pre-existing structures. In other words, they still cannot build skeletal components from nothing. Therefore, the extent that tissue engineers can regenerate now is

Developmental Biology and Musculoskeletal Tissue Engineering. https://doi.org/10.1016/B978-0-12-811467-4.00001-2

1

therefore still at an infancy stage. There has been great progress in developing prostheses, but they do not have the same biological functions as living bones and critically lack many important aspects such as growth and regeneration. As many young and old patients suffer from substantial loss of important musculoskeletal tissues such as digits, limbs, face, skull, or dental structures, more efficient ways for functional regeneration are highly desirable. More specifically, tissue engineers will need to develop a more comprehensive approach to recapitulate the process of development. Thus, it is important to learn how stem cells and signals are used to orchestrate the development of musculoskeletal tissues. In this chapter, we will review the fundamental process of musculoskeletal development that can be instrumental for tissue engineering.

DEVELOPMENTAL ORIGIN OF MUSCULOSKELETAL TISSUES

Developmentally, most musculoskeletal tissues are derived from the mesoderm, except those of the craniofacial complex that are derived from the neural crest, or otherwise known as ectomesenchyme. The mesoderm is formed between the ectoderm and the endoderm as a result of gastrulation. Of its subdomains, the paraxial and lateral plate mesoderm are particularly relevant to the formation of musculoskeletal tissues. The paraxial mesoderm first organizes into somites that give rise to the myotome, sclerotome, and dermatome. The myotome gives rise to skeletal muscles, and the sclerotome gives rise to bones and tendons of the vertebrae (reviewed in Refs. [1,2]).

Limb development undergoes more complex and dynamic processes requiring heterotopic interactions between the ectoderm and the underlying mesoderm. The limb bud, a structure formed in early development, is established by proliferation of mesenchymal cells that originate from the lateral plate mesoderm and the myotome. These limb mesenchymal cells stimulate formation of a signaling center in the ectoderm, termed the apical ectodermal ridge (AER). The AER expresses fibroblast growth factors (FGFs) to communicate with its underlying mesenchyme to pattern the proximal to distal axis by antagonizing retinoic acid signaling present in the proximal limb. They simultaneously stimulate formation of the zone of polarizing activities within the mesenchyme, which expresses sonic hedgehog that patterns the anterior—posterior axis (reviewed in Refs. [3—6]). These signaling centers orchestrate proper development of the limb by recruiting mesenchymal cells from the lateral plate mesoderm to form bones and tendons, also from the myotome to form skeletal muscles.

Development of the craniofacial structure is even more complex than limb development, requiring mesenchymal cells with two distinct origins of the neural crest and the mesoderm (reviewed in Refs. [7,8]). In the face, pharyngeal arches, a series of bulges located laterally, develop through complex interactions among all the primary germ layers and the neural crest (reviewed in Ref. [9]). Of these contributing cell types, neural crest cells give rise to the connective and skeletal element of each arch. Although most of facial skeletal structures are derived purely from the neural crest cells, mesodermal and neural crest cells are intricately combined in the cranial structures, particularly in the calvaria and cranial base, illustrating the complexity of craniofacial development.

FORMATION OF MESENCHYMAL CONDENSATIONS

Bones assume very different shapes in different parts of the human body, but they are formed through only two common mechanisms: intramembranous and endochondral bone formation. Intramembranous bone formation is a straightforward process in which undifferentiated mesenchymal cells directly differentiate into osteoblasts that lay down the mineralized matrix. Intramembranous bones (or dermal bones) evolve earlier in the early fish and comprise part of the skull in mammals regardless of developmental origins of mesenchymal cells in the neural crest or the mesoderm. By contrast, endochondral bone formation is a complex process in which initial cartilaginous templates are later replaced by bones.

In both mechanisms, a primordial structure called mesenchymal condensations frames the future domain of bones. In this process, mesenchymal cells in a specific domain of the embryonic tissue temporarily stop proliferating, then align together to form cell clusters that exclude blood vessels. How self-organization of mesenchymal cells is induced remains unknown. It has been suggested that signaling pathways induced by transforming growth factor β, bone morphogenetic proteins (BMPs), and FGFs regulate formation of mesenchymal condensations. In addition, intrinsic and extrinsic cellular changes are considered to modify the responsiveness of condensing cells to various signals, including reorganization of the cytoskeleton and intercellular adhesions, extracellular matrix milieu, and hypoxic conditions [10]. Formation of mesenchymal condensations is thus a critical step to initiate subsequent steps of differentiation.

ENDOCHONDRAL BONE FORMATION

Most bones in mammals are formed through endochondral bone formation (Fig. 1.1). This process is highly sequential, thus represents one of the best examples of organogenesis requiring heterotypic cellular interactions [11]. In this process, mesenchymal cells in condensations further differentiate into two distinct but closely related cells types, chondrocytes and perichondrial cells. Chondrocytes develop in the vasculature-free central portion of condensations, whereas perichondrial cells develop in the highly vascularized outer layer of condensations. This process is initiated when condensing mesenchymal cells start to express the transcription factor Sox9, a master regulator of chondrogenesis

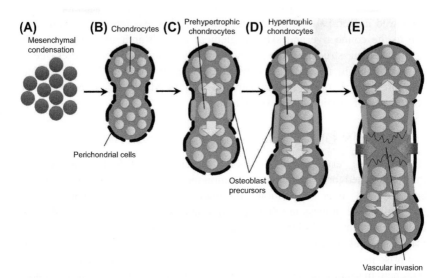

FIGURE 1.1   The process of endochondral bone formation. (A) Bone anlage is formed by condensing mesenchymal cells. (B) Mesenchymal cells differentiate into chondrocytes, and their surrounding cells become perichondrial cells. (C) Chondrocytes continue to proliferate; those in the center stop proliferating and become hypetrophic chondrocytes. (D) First osteoblast precursors appear in the perichondrium adjacent to hypertrophic chondrocytes. (E) Osteoblast precursors invade into the cartilage template along with blood vessels.

[12,13]. Indeed, Sox9 is absolutely required for these mesenchymal cells to stay organized within condensation. Sox9 can directly bind to regulatory elements of major cartilage matrix genes, including those encoding collagens (such as type II, IX, and XI collagen) and proteoglycans (e.g., aggrecan). As a result, mesenchymal cells in condensations become programmed as chondrocytes or perichondrial cells, which surround the chondrocytes.

Formation of the Fetal Growth Plate

Chondrocytes restart proliferation in a relatively uniform manner within the cartilaginous template. As the cartilage enlarges, chondrocytes stop proliferating in the center, drastically change their cell morphology, and become special chondrocytes termed hypertrophic chondrocytes. How this initial hypertrophy is triggered is largely unknown. In the limb, initially continuous mesenchymal condensations undergo the processes termed segmentation and cavitation to give rise to individual skeletal elements. The interzone, a localized dense region in condensations, at a later stage becomes the joint. Secreted proteins, Wnt9a and Gdf5, regulate early joint formation, and the transcription factor c-Jun plays important roles in joint cell specification [14].

When chondrocytes in the center undergo initial hypertrophy, chondrocytes in the flanking regions of the cartilage differentiate and the structure termed

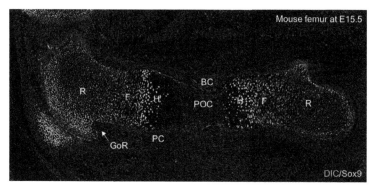

FIGURE 1.2   Endochondral bone in mouse fetus. Shown is mouse femur at embryonic day 15.5 stained with anti-Sox9 antibody. *BC*, bone collar; *blue*, Sox9; *F*, flat chondrocytes; *GoR*, perichondrial groove of Ranvier; *Gray*, differential interference contrast; *H*, hypertrophic chondrocyte; *PC*, perichondrium; *POC*, primary ossification center; *R*, round chondrocyte.

the growth plate is established (reviewed in Ref. [15]). The fetal growth plate is primarily composed of three distinct types of chondrocyte types: round, flat, and hypertrophic chondrocytes (Fig. 1.2). The majority of chondrocytes located in the "head" portion of the cartilage are round chondrocytes, exhibiting a moderate rate for proliferation. A group of round chondrocytes at the end of the cartilage is termed periarticular chondrocytes and forms the joint surface. Another group of round chondrocytes in the "shaft" portion closer to hypertrophic chondrocytes starts to change their cell morphology and becomes stacked up one another like pancakes with distinct cell polarity. They are now flat chondrocytes that continue to proliferate to generate columnar chondrocytes. Chondrocytes divide laterally perpendicular to the long axis of the columns, and their daughter cells crawl over each other to regain the alignment of the original column of the parental cells. This process might be regulated by Wnt/planar cell polarity pathways [16,17]. These chondrocytes eventually stop proliferating, start to change their cell morphology again, and differentiate into prehypertrophic and hypertrophic chondrocytes.

The resultant hypertrophic chondrocytes are postmitotic cells that drastically increase their cellular volume by a combination of true hypertrophy and cell swelling [18]. Most remarkably, three distinct phases underlie hypertrophic cell enlargement: these chondrocytes first undergo true hypertrophy maintaining constant density through an increase in the number of macromolecules and organelles, then undergo massive cell swelling through disproportionate water intake, and finally undergo true hypertrophy again through proportional dry mass increase at a low density. This final phase is controlled by insulin-like growth factor, an important regulator of bone elongation. Although cell proliferation and matrix deposition by round and flat chondrocytes are important for bone lengthening, it is this massive volume enlargement of hypertrophic chondrocytes that makes the greatest contribution to lengthening of the cartilage template that drives longitudinal bone growth during the fetal period.

Cell enlargement is not the only important function of hypertrophic chondrocytes. These cells abundantly express matrix proteins including type X collagen and direct mineralization of their surrounding matrix. More importantly, hypertrophic chondrocytes secrete critical paracrine factors, such as vascular endothelial growth factor (VEGF) that induce invasion of blood vessels from the perichondrium, and Indian hedgehog (Ihh) that regulate proliferation and differentiation of chondrocytes and directs perichondrial cells to become osteoblasts. Thus, they are the master regulator of endochondral bone formation.

Ihh executes a number of important functions through its interactions with its receptor, patched 1 (Ptch1); Ptch1 inhibits smoothened (Smo) and blocks the action of the transcription factors of the Gli family until Ihh binds to Ptch1. Ihh stimulates differentiation of round chondrocytes into flat proliferating chondrocytes and formation of columns [19]. Ihh also promotes proliferation of flat chondrocytes and their differentiation into hypertrophic chondrocytes [20]. At the same time, Ihh acts laterally on the perichondrium and stimulates formation of the bone collar by promoting osteoblast differentiation of perichondrial progenitors. In addition, Ihh acts further on periarticular chondrocytes at the end of the cartilage and therefore promotes the production of parathyroid hormone−related peptide (PTHrP) [21]. PTHrP keeps flat chondrocytes in the proliferative pool and delays their differentiation into prehypertrophic chondrocytes expressing Ihh and into hypertrophic chondrocytes through its interaction with its receptor, the PTH/PTHrP receptor [22]. These series of interactions establish the PTHrP-Ihh feedback loop that is essential to maintaining the growth plate structure [23,24] (Fig. 1.3).

Formation of the Perichondrium

As chondrocytes organize the growth plate in their vasculature-free environment, surrounding perichondrial cells also develop into distinct cell types in their vasculature-rich environment. Unlike chondrocytes, perichondrial cells generally assume flat and elongated fibroblastic cell morphology without any conspicuous morphological differences. The heterogeneity and function of perichondrial cells is not well defined. Perichondrial cells and chondrocytes within the cartilage communicate with each other by sending and receiving signals reciprocally, particularly through FGFs (especially FGF9 and 18) and BMPs (reviewed in Ref. [25]). In addition, some perichondrial cells become osteoblasts and populate the cortical and trabecular bone at a later stage, whereas others may become chondrocytes within the cartilage template [26].

Anatomically, there are three functionally distinct portions of the perichondrium corresponding to the three layers of growth plate chondrocytes: the periround layer in the "head" portion, the periflat layer in the "base" portion, and the perihypertrophic layer in the "shaft" portion of the cartilage. The first

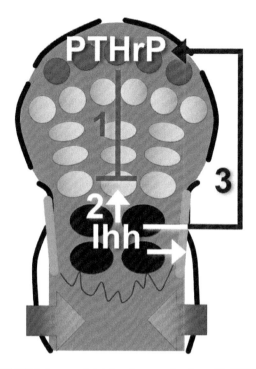

FIGURE 1.3 PTHrP-Ihh loop maintaining the growth plate. (1) PTHrP is expressed by periarticular round chondrocytes and perichondrial cells, as well as maintains chondrocyte proliferation and delays hypertrophy. (2) Ihh is produced by hypertrophic chondrocytes and facilitates proliferation and differentiation of adjacent chondrocytes and osteoblasts in the perichondrium. (3) Ihh stimulates PTHrP expression by round chondrocytes.

group, perichondrial cells of the round layer, generally has ambiguous borders with their underlying chondrocytes in early development. The demarcation becomes clear when the joint and the articular cartilage are fully formed at a postnatal stage. Some of these perichondrial cells express Sox9 in the mesenchymal condensation stage and probably become the source of other perichondrial cells. These perichondrial cells are also involved with formation of the insertion site for tendons. A group of cells coexpressing Sox9 and scleraxis in the perichondrium forms bone eminences in which tendons are inserted [27]. Perichondrial cells in the tendon insertion site also express PTHrP often in a pattern that is contiguous with their underlying periarticular chondrocytes [28]. The second group, perichondrial cells of the flat layer, also possessed unique functions. These cells form a distinct notch undercutting the round layer of the growth plate, termed the perichondrial groove of Ranvier. This structure may house precursors for perichondrial cells and also works as a signaling center [29]. In addition, a special group of chondrocytes termed "borderline" chondrocytes lies immediately beneath the perichondrium [30].

Interestingly, these cells line up parallel to perichondrial cells but perpendicular to other growth plate chondrocytes. Their possible fate is described below. Perichondrial cells adjacent to the prehypertrophic and hypertrophic layers, especially those on the innermost layer of the perichondrium, become committed to the osteoblast lineage largely because of the actions of Ihh released from the adjacent prehypertrophic chondrocytes. This portion of the perichondrium is sometimes described as the osteogenic perichondrium and is indeed the first place that osteoblast precursors appear in endochondral bone formation. This first bone that replaces a layer of perichondrium is termed the bone collar. These cells enter the program governed by the transcription factors Runx2 and osterix (Osx), in which Osx acts at a level genetically downstream of Runx2 [31]. The precise mechanisms underlying this process are unknown. These cells further differentiate into mature osteoblasts and abundantly produce bone matrix proteins including type I collagen, osteopontin, and osteocalcin to make the bone collar and, eventually, the cortical bones. Other perichondrial cells become cells in the periosteum, when they are found adjacent to newly formed cortical bone. These cells are composed of multiple layers of mesenchymal cells with diverse functions.

Formation of the Primary Ossification Center and the Bone Marrow Cavity

Soon after the growth plate and its perichondrium are formed, the primitive bone marrow cavity is formed within the cartilage template by attracting blood vessels. This step is essential to establish bone marrow hematopoiesis by recruiting hematopoietic stem cells (HSCs) toward the end of fetal life. Hypertrophic chondrocytes play central roles in this process by producing VEGFs, one of the most powerful mediators of angiogenesis [32]. Mesenchymal cells coinvade into the marrow cavity closely associated with invading blood vessels and give rise to the mesenchymal stromal compartment of the marrow space. There are at least three sources of these mesenchymal cells. The first source is the adjacent perichondrium. Cells committed to the osteoblast lineage on the innermost layer of the osteogenic perichondrium translocate into the nascent marrow cavity in a pericyte-like fashion before they become mature osteoblasts [33]. The second source is the hypertrophic layer of the growth plate. Some hypertrophic chondrocytes do not die from apoptosis but transform into cells that eventually become osteoblasts in an area right beneath the growth plate [34]. How hypertrophic chondrocytes with a large size can turn into very compact mesenchymal cells in the ossification center is unknown. The third source is the "borderland" between the growth plate and the perichondrium. Borderline chondrocytes (mentioned above) may translocate to the marrow cavity [30], although proof of this concept awaits a formal lineage-tracing experiment to establish it. These mesenchymal cells of heterogeneous origins proliferate and differentiate into osteoblasts and stromal cells within a highly vascularized environment called the primary spongiosa. Whether different cellular origins denote functional differences of these mesenchymal cells is unknown. Interestingly, Runx2 and Osx are also expressed by prehypertrophic and hypertrophic chondrocytes. The partly overlapping gene programs of hypertrophic chondrocytes and osteoblastic cells suggest that mesenchymal cells arriving at the marrow cavity might already have some epigenetic signatures tuned to the osteoblast lineage.

The primary ossification center is established with rich networks of blood vessels and trabecular bones. The bone marrow vasculature has a unique structure in which anastomoses are located in close proximity to the endosteal or trabecular surfaces [35]. Bone marrow arteries run in the central marrow cavity, branch out, and become small arteries (arterioles) as they approach bone surfaces. These vessels transition into sinusoids near bone surfaces and finally collect into central veins back in the central marrow cavity. Bone marrow stromal cells generally assume perivascular locations and bear distinct functions depending on their locations. Bone marrow stromal cells surrounding sinusoids and arterioles express important cytokines such as C-X-C motif chemokine 12 (CXCL12, also known as stromal cell—derived factor 1) [36] and stem cell factor (also known as KIT ligand) [37] to attract and retain HSCs in the marrow cavity. This unique microenvironment provides a favorable site for HSCs to settle, initiate, and maintain active hematopoiesis throughout life.

As the growth plate continues to grow further, the marrow cavity and the primary ossification center continue to enlarge. In this process, osteoblasts and stromal cells need to be replenished constantly. The source of continuous generation of these cells has not been clarified. There are two possible mechanisms to meet the increasing demand for these cells. The first mechanism is that perichondrial cells and growth plate chondrocytes continue to feed into the growing marrow cavity. This is likely to happen because continuous mitotic activities in the flat proliferating layer of the growth plate give rise to an ample number of hypertrophic chondrocytes in the end, some of which may transform into these mesenchymal cells. Moreover, vigorous proliferation in the groove of Ranvier and the osteogenic perichondrium provides a sufficient number of perichondrial cells that can eventually feed into the marrow space. The second mechanism is that precursors for osteoblasts and stromal cells replicate themselves within the marrow space for a prolonged period. This is also likely to happen because bone marrow stromal cells are believed to include "mesenchymal stem cells (MSCs)" with the important properties of potential to differentiate into multiple lineages and ability to self-renew. These cells have long been assessed by their ability to form colonies in vitro with varying capabilities. Roles of such mesenchymal stromal precursor populations in vivo have not been clearly demonstrated because of the complexity of the stromal cell lineage. Further studies are needed to clarify the sources of mesenchymal cells in the marrow space during this rapid phase of bone growth and their physiological fates.

Formation of the Secondary Ossification Center and the Postnatal Growth Plate

The next critical step is the formation of the secondary ossification center within the epiphyseal cartilage remaining on both ends of the bone. This occurs at a postnatal stage and is essential to the formation of the postnatal growth plate and the articular cartilage. Round chondrocytes are mostly uniform in appearance at the ends of the fetal growth plate. However, during an early postnatal phase, a group of round chondrocytes in the center of the cartilage undergoes hypertrophy possibly in response to a hypoxia signal, and invasion of blood vessels and mesenchymal cells occurs in that region. The secondary ossification center is similar to the primary ossification center in that it is composed of a highly vascularized marrow space enriched with trabecular bones. However, the secondary ossification center is surrounded by chondrocytes instead of an immediate perichondrium and associated bone collar. The secondary ossification center is encased within two important cartilaginous tissues; the postnatal growth plate and the articular cartilage. This also indicates that the secondary ossification center provides important signals that differentiate these tissues from the preexisting round layer of chondrocytes. The articular cartilage is a permanent cartilage composed of four distinct layers: nonmineralized superficial, middle, deep zones, and mineralized subchondral bones. Superficial articular chondrocytes are considered to be the source of all other chondrocytes in the articular cartilage [38].

The postnatal growth plate is formed as a disk between the primary and secondary ossification centers with the characteristic columns of chondrocytes. It is similar to the fetal growth plate, but there are certain differences. Most notably, chondrocytes at the top of the postnatal growth plate are slowly dividing and are called resting or reserve chondrocytes. These cells probably serve as precursors for flat proliferating columnar chondrocytes and support continued mitotic activities within the growth plate during the postnatal phase. Indeed, the resting zone has been suggested to contain stem-like cells that give rise to clones of proliferating chondrocytes and produce cytokines that orient columns parallel to the long axis of the bone [39]. The postnatal growth plate continues to work as the principal driving force for bone lengthening well into adulthood in rodents. How these resting chondrocytes maintain themselves long term within the postnatal growth plate is largely unknown. In certain mammals, for example, rabbits and humans, growth plates disappear at the time of puberty through processes that have not been well characterized, except that they depend of the actions of estradiol (in both males and females). When bones continue to grow postnatally, bone superstructures that serve as tendon and ligament insertion sites need to be also moved. A specific balance of growth rates between proximal and distal growth plates of each long bone maintains the relative position of its superstructures, through a process termed isometric scaling [40]. Therefore, the postnatal growth plate is also responsible for maintaining the overall morphology of bones during active bone growth.

Establishment of the Adult Bone Marrow Stroma

In humans, the growth plate disappears during adolescence after the pubertal growth spurt, whereas it does not disappear in mice, although its activity decreases significantly in later life. Even after bone growth slows or stops, bones need constant maintenance to sustain their structures and function throughout the lifespan. As the growth plate activity slows down toward adulthood, the contribution of chondrocytes and perichondrial cells to the marrow cavity is likely to become negligible. It is probably during this adult stage that bone marrow stromal cells make significant contribution to diverse skeletal cell types. It has been demonstrated that a small number of bone marrow stromal cells isolated from humans and mice have the capability to form colonies when cultured in vitro (i.e., colony-forming unit fibroblasts, CFU-Fs), and these cells can make bone ossicles when ectopically transplanted into immunodeficient mice in vivo. Because matrix-producing osteoblasts are relatively short-lived, it is necessary to supply new osteoblasts continually to maintain the bone structure. It is generally believed that bone marrow stromal cells can differentiate into osteoblasts during normal adult skeletal homeostasis, although this hypothesis has not been rigorously tested. Bone marrow stromal cells surrounding sinusoids (perisinusoidal cells) and arterioles (periarteriolar cells) both have the similar capability to make colonies when cultured in vitro and are likely to contribute to the formation of osteoblasts in adulthood. Interestingly, both CXCL12-abundant reticular (CAR) perisinusoidal stromal cells and Nestin-positive periarteriolar stromal cells are known to express transcription factors essential to osteoblast differentiation, such as Runx2 and Osx [41,42]. It is therefore possible that these adult bone marrow stromal cells already have the potential to become osteoblasts. Although significant conceptual understanding of skeletal stem cells (SSCs) or MSCs has been achieved by using ex vivo bone marrow cell culture systems and CFU-Fs, what these "stem cells" actually do in vivo in the physiological and pathological context is largely unknown. It is most likely that bone structure is maintained throughout the lifespan by recruiting cells from diverse sources to make osteoblasts. A small subset of bone marrow stromal cells that have special characteristics as "SSCs" or "MSCs" may or may not play significant roles, and other more differentiated cells (such as "bone-lining cells") may also play important roles in this process.

In Vivo Identification of Skeletal/Mesenchymal Stem Cells in Adult Bone Marrow

Cell surface markers, cell sorting, and mice engineered to express easily assayable proteins (such as green fluorescent protein [GFP] and *Escherichia coli* β-galactosidase [LacZ]) are useful tools for identifying putative stem cell populations in their native environment. If one can identify and purify such putative stem cells properly, tissue engineers may be able to take advantage of these markers and prospectively isolate these cells for regeneration purposes. Thus far, a number of marker combinations have been tested to enrich putative stem cells. Many of these markers are expressed in proximity to the bone marrow vasculature, such as platelet-derived growth factor receptor-α (PDGFRα), stem cell antigen-1 (Sca1), CXCL12 and α-smooth muscle actin (SMA). PDGFRα$^+$Sca1$^+$ nonhematopoietic cells ("PαS cells") reside in the perivascular space in vivo and are enriched for CFU-Fs [43]. These cells, if uncultured and cotransplanted with HSCs can engraft into irradiated recipients and become osteoblasts, stromal cells, adipocytes and, more importantly, PαS cells themselves, suggesting their self-renewal capability in vivo. Interestingly, cultured PαS cells do not have such capability [43].

Recent studies using transgenic mice have identified novel cell types with unique properties; when the human equivalents are identified in a way that allows purification, they may prove to have useful and varied properties. CXCL12 is known as stromal cell−derived factor and a critical regulator of HSCs. In the mouse, perivascular reticular cells adjacent to sinusoids or endosteal surfaces are particularly abundant for *Cxcl12*-GFP expression, termed CAR cells. Individual CAR cells express high level of osteogenic transcription factors, Runx2 and Osx, and adipogenic transcription factor PPARγ, and behave as osteoadipogenic progenitors in vitro [41]. *Leptin receptor-cre* marks these cells and has been used in lineage-tracing experiments to argue that these cells populate the osteoblast compartment in adult mouse bones [44]. Nestin is an intermediate filament protein and a marker for neural stem cells. *Nestin*-GFP is highly expressed in pericytes of bone marrow arterioles (small arteries), and *Nestin*-GFP-positive cells include all CFU-F activities and form self-renewable "mesenspheres" that can pass serial heterotopic transplantations [42]. α-Smooth muscle actin (αSMA) is a marker of pericytes of bone marrow arteries. Pericytes that are associated with vasculature in bone marrow and periosteum are marked by αSMA-GFP/mCherry transgenic expression and exhibit trilineage differentiation potential in vitro [45]. Therefore, comprehensive approaches combining sophisticated cell surface and transgenic markers and more traditional CFU-F and transplantation assays have proven to be useful for understanding SSCs/MSCs in adult bone marrow and revealed an emerging complexity of such cell populations.

INTRAMEMBRANOUS BONE FORMATION

Some bones in mammals, especially those in the skull, are formed through intramembranous bone formation. In this process, mesenchymal cells in condensations directly differentiate into osteoblasts that lay down mineralized matrix without intermediate cartilaginous templates. The mechanism of intramembranous bone formation is less defined than that of endochondral bone formation. In addition, some intramembranous bones such as those in the posterior cranial vault and the mandible have transient cartilages, but these cartilages do not provide matrix templates for osteogenesis; they may provide signaling molecules that aid the adjacent bone formation. This possible function of adjacent cartilage needs further exploration. There are a few possible mechanisms whereby intramembranous bone primordia grow in the absence of growing cartilage templates. One possible explanation is that mesenchymal cells in the adjacent connective tissue are constantly recruited to mesenchymal condensations and expand the "osteogenic island" where active osteoblast differentiation takes place through apposition. Another possible explanation is that mesenchymal cells within the original primordium proliferate and expand the "osteogenic island" by intrinsic growth. These hypotheses have not been rigorously tested thus far, although some evidence supports the latter view [46]. The suture is formed when growth of each bone primordium results in collision of two adjacent bone compartments. The suture works as a "shock absorber" in the skull and provides a niche for SSCs/MSCs of craniofacial bones [47,47a,47b].

PROCESS OF BONE REPAIR

Because of their primary functions, bones experience various degrees of damage, ranging in severity from microfractures to fractures that completely disrupt tissue continuity. Most small and mechanically stable fractures heal by intramembranous bone formation, whereas large and unstable fractures also involve endochondral bone formation in which fibrocartilages and soft callus are newly generated near the fracture site to bridge bone fragments. Bone repair requires mobilization, proliferation, and differentiation of mesenchymal cells to allow deposition of mineralized matrix at the injury site. The periosteum and the bone marrow are two major sources for these cells, although other sources are also possible. Periosteal and bone marrow stromal cells differentially contribute to bone repair by providing chondrocytes and osteoblasts. In complete bone fracture, periosteal cells robustly become chondrocytes, whereas bone marrow stromal cells can also become chondrocytes but to a much lesser extent [48]. In addition, in a special circumstance of neonatal angulated fractures, a bidirectional growth plate similar to the synchondroses of the cranial base forms at the concave side of the fracture to realign the fragments [49]. Therefore, periosteal cells retain the capability to regenerate chondrocytes and growth plate—like structures after fracture, particularly in early life. These periosteal cells are located in the cambium layer immediately outside the osteoblasts lining the bone surface. Similar to chondrocytes in the growth plate, chondrocytes in the soft callus are also likely to become osteoblasts in subsequent ossification [50]. Therefore, it is possible that osteoblasts repairing the gap within or between bones are generated through two distinct mechanisms of chondrocyte-mediated and non—chondrocyte-mediated pathways.

Skeletal/Mesenchymal Stem Cells for Bone Repair

Bone repair requires the mobilization of stem cells to allow deposition of mineralized matrix at the injury site. Two major sources for these stem cells are found in the periosteum and the bone marrow and differentially contribute to bone repair. Lineage-tracing experiments using a unicortical transplantation model and subsequent fracture experiments indicate that periosteal stem cells are the major source of chondrocytes in callus, whereas bone marrow stromal cells also contribute to callus chondrocytes but to a much lesser extent [48]. A lineage-tracing study using *Prx1-creERt* reveals that these periosteal cells give rise to some of the chondrocytes and osteoblasts in the fracture callus [51]. Another effective marker for periosteal cells is αSMA. *αSMA-creERt*-marked periosteal cells expand in response to fracture and start to differentiate into chondrocytes and osteoblasts [45]. These lineage-tracing studies support the idea that a subset of periosteal cells behaves as stem cells and becomes the principal source of chondrocytes and osteoblasts during fracture repair. The contribution of bone marrow stromal cells to fracture repair has also been investigated. Bone marrow stromal cells derived from perinatal *Osx-creERt*[+] cells or *LepR-cre*[+] cells contribute to chondrocytes in the fracture callus [44,52]. However, because descendants of both of these *cre*-marked populations also include periosteal cells, these studies cannot conclusively exclude periosteal involvement. As in bone growth, chondrocytes in the soft callus may well become a source of osteoblasts in subsequent ossification. A study shows that *Acan-creERt* has been used to mark chondrocytes in the fracture callus, and they subsequently become *Col1(2.3 kb)*-GFP[+] osteoblasts in the fracture site [50]. Despite these recent significant advances, further detailed analyses need to be undertaken to delineate the characteristics of stem cells for bone repair. Use of additional transgenic markers for molecular characterization of the cells and their heterogeneity will be essential to understand the contribution of each stem cell population to fracture repair. The fundamental properties of periosteal stem cells and how they differ from bone marrow stromal cells need clarification.

STEM CELLS AND CELL LINEAGES FOR MUSCULOSKELETAL REGENERATION

A critical question for tissue engineers is what types of "stem cells" they should use to initiate efficient regeneration of musculoskeletal tissues of their interest. A currently prevailing notion is that omnipotent bone marrow−derived "SSCs" or "MSCs" can be readily applied for musculoskeletal regeneration. However, it is possible that this model is too simplistic, and "stem cells" need to be custom-designed depending on the target tissue that the tissue engineers want to regenerate. For example, human bone marrow−derived SSCs/MSCs do not differentiate into chondrocytes after transplantation in immunodeficient mice in vivo. The differentiation potential of SSCs/MSCs is generally limited to cell types found in their native tissue. It is therefore important for tissue engineers to understand how putative endogenous stem cell populations of each tissue differentiate into unique types of differentiated skeletal cells during development.

Colony-Forming Unit Fibroblasts and Skeletal/Mesenchymal Stem Cells

The bulk of existing knowledge on stem cells of the skeletal lineage in adult life has been built on experiments using human and rodent bone marrow cells. Traditionally, in vitro bone marrow cell culture and subsequent heterotopic transplantation into immunodeficient mice have been used as the gold standard to identify these stem cells [53]. It has been proposed that these putative stem cells are found among adventitial reticular cells on bone marrow sinusoids (perisinusoidal cells) or pericytes on arterioles (periarteriolar cells); these cells are preferentially located around the bone marrow vasculature. The first discovery that bone marrow may include stem cells capable of making bones was made when bulk human bone marrow cells were subcutaneously transplanted into immunodeficient mice [54]. They formed hybrid ossicles that contained osteoblasts and stromal cells of the donor origin and blood cells of the recipient origin. Bone marrow cells from rodents show similar properties. CFU-Fs, which are defined as cells capable of adhering to a plastic culture dish and establishing colonies, were later identified as cells responsible for heterotopic ossicle formation [55]. Therefore, stem cells residing in human and rodent adult bone marrow are capable of reconstituting bone marrow in a new environment. However, in this traditional approach, these stem cells can be identified only after expansion in cell culture, and their location and properties cannot be clarified in a native setting. More sophisticated approaches using cell surface markers and cell sorting technologies identified that the CD146$^+$ fraction of human bone marrow cells contain all CFU-Fs [56]. CD146$^+$ cells meet the criteria of "SSCs," as they can be serially transplanted and generate CD146$^+$ cells on secondary transplantation. These CD146$^+$ cells correspond to adventitial reticular cells lining bone marrow sinusoids in vivo. Furthermore, CD51(αV integrin)$^+$PDGFRα^+ cells represent a small subset of CD146$^+$ cells with even more enriched colony-forming activity [57]. Therefore, these findings have established the idea that SSCs/MSCs, which can generate bone and stroma on heterotopic transplantation, are identified among CFU-Fs and are typically found among CD146$^+$ perivascular stromal cells.

In Vivo Lineage-Tracing Experiments in Mice and Insights Into Adult Skeletal/Mesenchymal Stem Cells: Developmental Tools for Tissue Engineers

Much of the work on SSCs/MSCs has been driven by the ideas of regenerative medicine, the desire to find cells capable of restoring function to human bones.

Not only a distinct but also an interesting series of studies of SSCs have been driven by questions about the normal physiological functions of putative stem cells. However, the tools that allow isolation of putative stem cells for transplant do not easily allow the answer to questions about normal physiological roles.

The gold standard for investigating such stem cell functions in vivo is lineage-tracing experiments using transgenic mice. This approach typically relies on the *cre-loxP* technology to permanently mark cells of interest using a bigenic system. *Cre* recombinase is expressed in a promoter-specified manner in the first transgenic line and acts on the reporter locus of the second transgenic line. In the reporter locus, multiple sequences directing the addition of polyA sequences, as well as translation termination codons in all three reading frames ("STOP" sequences), which halt continued transcription and translation of reporter genes, are flanked by *loxP* sites. The reporter gene becomes expressed under the direction of a ubiquitously active promoter when *cre* recombinase removes the "STOP" sequences. In a modified version, the *cre* recombinase is covalently bound to the ligand-binding domain of the estrogen receptor (*creERt*) that has been mutated so that tamoxifen, but not estradiol, can bind and change its tertiary structure. Translocation of the *creERt* complex to the nucleus is dependent on the presence of 4-hydroxy-tamoxifen (4-OHT), which an active form of tamoxifen produced after being metabolized in the liver. Therefore, in the *creERt* system, tamoxifen administration can temporarily activate *cre-loxP* recombination only for 24—48 h until 4-OHT is cleared away from cells (Fig. 1.4). Recombination in the reporter locus is irreversible;

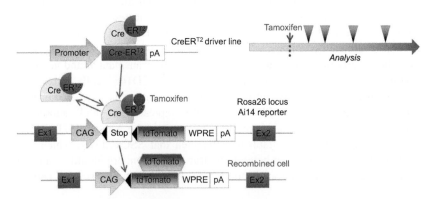

FIGURE 1.4 A lineage-tracing model based on tamoxifen-inducible creER^{T2} system. A lineage-tracing model based a tamoxifen-inducible creER^{T2} system for lineage tracing. creER^{T2} recombinase is expressed by a named promoter. It excises the stop codons in the Rosa26 reporter locus only in the presence of tamoxifen. Once the stop sequences are removed, the targeted cells permanently express tdTomato in a ubiquitously active CAG promoter-dependent manner. Tamoxifen stays active around 48 h after injection. *Modified from Ono W, et al. Nat Commun 2016;7:11277.*

therefore, the reporter gene is continually expressed in the targeted cells and their descendants as long as they survive, even after the promoter that drove expression of *cre* recombinase is no longer active.

Several different versions of the modified *Rosa26* reporter locus are available, including *R26R*-LacZ (encoding β-galactosidase), *R26R*-YFP (yellow fluorescent protein), *R26R*-tdTomato (encoding tandem dimer of red fluorescent protein, DsRed), and *R26R*-Confetti. The last locus encodes four different fluorescent proteins (nuclear GFP, YFP, tdTomato and CFP (cyan fluorescent protein)), in which one of them becomes stochastically expressed on *cre-loxP* recombination. It is noteworthy to mention here that each version has its own advantages and disadvantages. For example, although *R26R*-Confetti locus provides four distinct colors that facilitate in vivo clonal analysis, this version is relatively insensitive to *cre-loxP* recombination because of the complexity of the transgene design.

This lineage-tracing technology has been extensively applied to establish progenitor−descendant relationships in an unperturbed native environment. To achieve this purpose, it is essential to identify promoters whose activities are in a narrow array of desirable cell types, and particularly, promoters without any activity in putative descendant cells. An increasing number of studies have used this approach to reveal new aspects of SSCs/MSCs in vivo. The transgenic tools used in these studies include *Gremlin1-creERt* [58], *Col2a1-creERt* and *Sox9-creERt* [59], *Nestin-creERt* [42], *neural/glial antigen 2 (NG2)-creERt* [60], *αSMA-creERt* [45], and interferon-inducible *Mx1-cre* [61]. Of these lines, interferon-inducible *Mx1-cre* and *Gremlin1-creERt* mark precursors of osteoblasts and chondrocytes, but not adipocytes. By contrast, *Col2a1-creERt* and *Sox9-creERt* mark cells that populate all three lineages. These differences demonstrate that these cells are self-renewing but variable in the number of lineages they can populate. A challenge for the future will be to study the relationships between the stem cells marked by these different promoters.

These genetic models have uncovered important aspects of skeletal/mesenchymal stem cell biology. However, heterogeneity of cell populations of interest marked by a promoter-based approach complicate overall interpretation of presented data. This is particularly true in studies using a "constitutively active" version of *cre* recombinase. The fundamental difference between "constitutively active" *cre* and "inducible" *creERt* requires close attention; unlike the latter, the former induces recombination whenever the promoter becomes active (no temporal factor controls *cre* activities). If that promoter becomes active in other cell types at a late phase during lineage development, then the possible relationships between the different cell types marked by a reporter cannot be delineated. Particularly, we will need to develop more sophisticated inducible genetic tools that can specifically mark putative SSCs/MSCs residing in adult bone marrow to better define the characteristics of these cells.

Skeletal/Mesenchymal Stem Cells for Development and Growth

Growing data suggest that distinct types of SSCs/MSCs exist to support the explosive growth of bones in early life. These "growth-associated" stem cells appear to be different from well-studied adult bone marrow SSCs/MSCs mentioned earlier. Insights into these "growth-associated" SSCs/MSCs have been generated in mice based on in vivo lineage-tracing experiments. The transcription factor Prx1 is expressed by cells in the lateral plate mesoderm during early limb development [62]. The notion that the lateral plate mesoderm provides precursors for all limb mesenchymal cells is supported by the finding that *Prx1-cre* marks essentially all mesenchymal cells in limb bones, including osteoblasts, chondrocytes, and stromal cells, but not cells in skeletal muscles. Interestingly, Prx1 expression becomes confined to the periosteum in the cambium layer during the postnatal period, immediately outside the layer of osteoblasts lining the bone surface. A lineage-tracing study using *Prx1-creERt* reveals that these periosteal cells give rise to some of the chondrocytes and osteoblasts in fracture callus [51], suggesting the possibility that Prx1-expressing cells in the periosteum include stem cells.

Other MSCs are marked by genes expressed in mesenchymal condensation. *Sox9-cre*, for example, first marks cells of mesenchymal condensations and then marks essentially all chondrocytes, perichondrial cells, and osteoblasts [63]. These fate-mapping studies indicate that SSCs/MSCs for development arise locally within the bone primordia. Type II collagen (Col2a1) starts to be expressed in mesenchymal condensations, but at a slightly later stage and in a slightly smaller domain than Sox9, and continues to be expressed robustly within the cartilage template especially in chondrocytes. *Col2a1-cre* also marks essentially all chondrocytes, perichondrial cells, and osteoblasts and, in that way, resembles *Sox9-cre*, therefore marks virtually all osteochondroprogenitors. Moreover, similar osteochondroprogenitors at a later stage of development or even during the postnatal period can be marked by the *Col2a1* promoter/enhancer as skeletal cells marked by *Col2a1-creERt* at various stages of development continue to provide chondrocytes, osteoblasts, bone marrow stromal cells, and adipocytes for a long period [59]. Therefore, growth-associated SSCs/MSCs may be found among cells closely associated with chondrocytes. The idea that hypertrophic chondrocytes can be precursors of cells that can become osteoblasts has a growing body of evidence using lineage-tracing strategy. Cells marked by *type X collagen (ColX)-creERt* or *ColX-creERt* can differentiate into osteoblasts [34,50]. It is intriguing to speculate that a differentiation pathway through hypertrophic chondrocytes might be one of the primary mechanisms that "growth-associated" SSCs/MSCs give rise to osteoblasts. How these growth plate—derived stem cells differ from those found in the periosteum is an important agenda for the future.

The perichondrium is also considered to be the site where SSCs/MSCs can be identified. Cells in the osteogenic perichondrium adjacent to hypertrophic chondrocytes express Osx. These perichondral cells marked by *Osx-creERt* invade into the cartilage template together with blood vessels and subsequently become osteoblasts and stromal cells inside the ossification center [33]. Interestingly, while $Osx\text{-}creERt^+$ fetal perichondrial cells that move to bone marrow stroma can stay there only transiently and are replaced by other cells, early postnatal $Osx\text{-}creERt^+$ cells move to and stay in bone marrow stroma persistently for a long time [52]. Adult $Osx\text{-}creERt^+$ cells do not have such capability. These findings indicate that progenitor cells that can be marked by *Osx-creERt* at a specific time of development are recruited to a perivascular location in bone marrow and become stromal cells, some of which may later behave as mesenchymal progenitors. The underlying reason for differential expression of Osx in a specific subtype of progenitors at a specific time is unknown and requires further studies. The varying behavior of cells marked with Osx expression at various times is a reminder of the likely heterogeneity of stem cells marked simply by expression of one particular gene.

Additional insights have been obtained from extensive cell sorting experiments in mice. "Mouse skeletal stem cells," which can generate bone, cartilage, and stroma on heterotopic transplantation, have been identified by a combination of cell surface markers, including αV integrin (CD51), Thy (CD90), endoglin (CD105), and OX2 (CD200) and found in proximity to the growth plate at an early postnatal period [64].

Therefore, these recent studies support the idea that chondrocytes or cells closely related to them (particularly perichondrial cells) in the growth plate may include a population of "growth-associated" SSCs/MSCs that may be the primary source of osteoblasts and stromal cells during endochondral bone ossification. How these and other growth-associated stem cells relate to CFU-Fs or PαS cells found in adult bone marrow is an important agenda for the future.

CONCLUSIONS

In this chapter, we have provided a brief overview on the process of skeletal development, and the recent progress on stem cells and cell lineages for musculoskeletal regeneration. The emerging concept that multiple distinct types of stem cells support bone growth, maintenance and repair might be particularly relevant for tissue engineers to improve their modalities to achieve better regeneration. More detailed mechanisms regarding how these stem cells are regulated have not been understood. It will be important to identify the molecular mechanism of cell fate specification and differentiation so that tissue engineers can robustly take advantage of the variety of SSCs.

REFERENCES

[1] Brent AE, Tabin CJ. Developmental regulation of somite derivatives: muscle, cartilage and tendon. Curr Opin Genet Dev October 2002;12(5):548–57. PMID:12200160. eng.

[2] Fleming A, Kishida MG, Kimmel CB, Keynes RJ. Building the backbone: the development and evolution of vertebral patterning. Development May 2015;142(10):1733–44. PMID:25968309. eng.

[3] Cooper KL, Hu JK, Ten Berge D, Fernandez-Teran M, Ros MA, Tabin CJ. Initiation of proximal-distal patterning in the vertebrate limb by signals and growth. Science May 2011;332(6033):1083–6. PMID:21617075. PMCID:PMC3258580. eng.

[4] Roselló-Díez A, Ros MA, Torres M. Diffusible signals, not autonomous mechanisms, determine the main proximodistal limb subdivision. Science May 2011;332(6033):1086–8. PMID:21617076. eng.

[5] Towers M, Wolpert L, Tickle C. Gradients of signalling in the developing limb. Curr Opin Cell Biol April 2012;24(2):181–7. PMID:22169676. Epub 2011/12/13. eng.

[6] Zeller R, López-Ríos J, Zuniga A. Vertebrate limb bud development: moving towards integrative analysis of organogenesis. Nat Rev Genet December 2009;10(12):845–58. PMID:19920852. eng.

[7] Bhatt S, Diaz R, Trainor PA. Signals and switches in Mammalian neural crest cell differentiation. Cold Spring Harb Perspect Biol February 2013;5(2). PMID:23378583. PMCID:PMC3552505. Epub 2013/02/01. eng.

[8] Frisdal A, Trainor PA. Development and evolution of the pharyngeal apparatus. Wiley Interdiscip Rev Dev Biol 2014 ;3(6):403–18. PMID:25176500. PMCID:PMC4199908. Epub 2014/08/29. eng.

[9] Graham A. Development of the pharyngeal arches. Am J Med Genet A June 2003;119A(3):251–6. PMID:12784288. eng.

[10] Ray P, Chapman SC. Cytoskeletal reorganization drives mesenchymal condensation and regulates downstream molecular signaling. PLoS One 2015;10(8):e0134702. PMID:26237312. PMCID:PMC4523177. Epub 2015/08/03. eng.

[11] Kronenberg HM. Developmental regulation of the growth plate. Nature May 2003;423(6937):332–6. PMID:12748651. eng.

[12] Akiyama H, Chaboissier MC, Martin JF, Schedl A, de Crombrugghe B. The transcription factor Sox9 has essential roles in successive steps of the chondrocyte differentiation pathway and is required for expression of Sox5 and Sox6. Genes Dev November 2002;16(21):2813–28. PMID:12414734. PMCID:PMC187468. eng.

[13] Bi W, Deng JM, Zhang Z, Behringer RR, de Crombrugghe B. Sox9 is required for cartilage formation. Nat Genet May 1999;22(1):85–9. PMID:10319868. eng.

[14] Kan A, Tabin CJ. c-Jun is required for the specification of joint cell fates. Genes Dev March 2013;27(5):514–24. PMID:23475960. PMCID:PMC3605465. eng.

[15] Kozhemyakina E, Lassar AB, Zelzer E. A pathway to bone: signaling molecules and transcription factors involved in chondrocyte development and maturation. Development March 2015;142(5):817–31. PMID:25715393. PMCID:PMC4352987. eng.

[16] Gao B, Yang Y. Planar cell polarity in vertebrate limb morphogenesis. Curr Opin Genet Dev August 2013;23(4):438–44. PMID:23747034. PMCID:PMC3759593. Epub 2013/06/05. eng.

[17] Li Y, Dudley AT. Noncanonical frizzled signaling regulates cell polarity of growth plate chondrocytes. Development April 2009;136(7):1083–92. PMID:19224985. PMCID:PMC 2685929. Epub 2009/02/18. eng.

[18] Cooper KL, Oh S, Sung Y, Dasari RR, Kirschner MW, Tabin CJ. Multiple phases of chondrocyte enlargement underlie differences in skeletal proportions. Nature March 2013;495(7441):375−8. PMID:23485973. PMCID:PMC3606657. Epub 2013/03/13. eng.

[19] Kobayashi T, Soegiarto DW, Yang Y, Lanske B, Schipani E, McMahon AP, et al. Indian hedgehog stimulates periarticular chondrocyte differentiation to regulate growth plate length independently of PTHrP. J Clin Invest July 2005;115(7):1734−42. PMID:15951842. PMCID: PMC1143590. Epub 2005/06/09. eng.

[20] Mak KK, Kronenberg HM, Chuang PT, Mackem S, Yang Y. Indian hedgehog signals independently of PTHrP to promote chondrocyte hypertrophy. Development June 2008;135(11):1947−56. PMID:18434416. Epub 2008/04/23. eng.

[21] St-Jacques B, Hammerschmidt M, McMahon AP. Indian hedgehog signaling regulates proliferation and differentiation of chondrocytes and is essential for bone formation. Genes Dev August 1999;13(16):2072−86. PMID:10465785. PMCID:PMC316949. eng.

[22] Schipani E, Lanske B, Hunzelman J, Luz A, Kovacs CS, Lee K, et al. Targeted expression of constitutively active receptors for parathyroid hormone and parathyroid hormone-related peptide delays endochondral bone formation and rescues mice that lack parathyroid hormone-related peptide. Proc Natl Acad Sci USA December 1997;94(25):13689−94. PMID:9391087. PMCID:PMC28367. eng.

[23] Lanske B, Karaplis AC, Lee K, Luz A, Vortkamp A, Pirro A, et al. PTH/PTHrP receptor in early development and Indian hedgehog-regulated bone growth. Science August 1996;273(5275):663−6. PMID:8662561. eng.

[24] Vortkamp A, Lee K, Lanske B, Segre GV, Kronenberg HM, Tabin CJ. Regulation of rate of cartilage differentiation by Indian hedgehog and PTH-related protein. Science August 1996;273(5275):613−22. PMID:8662546. eng.

[25] Ornitz DM, Marie PJ. Fibroblast growth factor signaling in skeletal development and disease. Genes Dev July 2015;29(14):1463−86. PMID:26220993. PMCID:PMC4526732. eng.

[26] Kronenberg HM. The role of the perichondrium in fetal bone development. Ann NY Acad Sci November 2007;1116:59−64. PMID:18083921. eng.

[27] Blitz E, Sharir A, Akiyama H, Zelzer E. Tendon-bone attachment unit is formed modularly by a distinct pool of Scx- and Sox9-positive progenitors. Development July 2013;140(13):2680−90. PMID:23720048. Epub 2013/05/29. eng.

[28] Chen X, Macica CM, Dreyer BE, Hammond VE, Hens JR, Philbrick WM, et al. Initial characterization of PTH-related protein gene-driven lacZ expression in the mouse. J Bone Miner Res January 2006;21(1):113−23. PMID:16355280. Epub 2005/10/10. eng.

[29] Shapiro F, Holtrop ME, Glimcher MJ. Organization and cellular biology of the perichondrial ossification groove of ranvier: a morphological study in rabbits. J Bone Joint Surg Am September 1977;59(6):703−23. PMID:71299. eng.

[30] Bianco P, Cancedda FD, Riminucci M, Cancedda R. Bone formation via cartilage models: the "borderline" chondrocyte. Matrix Biol July 1998;17(3):185−92. PMID:9707341. eng.

[31] Nakashima K, Zhou X, Kunkel G, Zhang Z, Deng JM, Behringer RR, et al. The novel zinc finger-containing transcription factor osterix is required for osteoblast differentiation and bone formation. Cell January 2002;108(1):17−29. PMID:11792318. eng.

[32] Maes C. Signaling pathways effecting crosstalk between cartilage and adjacent tissues: seminars in cell and developmental biology: the biology and pathology of cartilage. Semin Cell Dev Biol February 2017;62:16−33. PMID:27180955. Epub 2016/05/12. eng.

[33] Maes C, Kobayashi T, Selig MK, Torrekens S, Roth SI, Mackem S, et al. Osteoblast precursors, but not mature osteoblasts, move into developing and fractured bones along with invading blood vessels. Dev Cell August 2010;19(2):329−44. PMID:20708594. PMCID:PMC3540406. eng.

[34] Yang L, Tsang KY, Tang HC, Chan D, Cheah KS. Hypertrophic chondrocytes can become osteoblasts and osteocytes in endochondral bone formation. Proc Natl Acad Sci USA August 2014;111(33):12097−102. PMID:25092332. PMCID:PMC4143064. Epub 2014/08/04. eng.

[35] Kusumbe AP, Ramasamy SK, Adams RH. Coupling of angiogenesis and osteogenesis by a specific vessel subtype in bone. Nature March 2014;507(7492):323−8. PMID:24646994. PMCID:PMC4943525. Epub 2014/03/12. eng.

[36] Sugiyama T, Kohara H, Noda M, Nagasawa T. Maintenance of the hematopoietic stem cell pool by CXCL12-CXCR4 chemokine signaling in bone marrow stromal cell niches. Immunity December 2006;25(6):977−88. PMID:17174120. eng.

[37] Keller JR, Ortiz M, Ruscetti FW. Steel factor (c-kit ligand) promotes the survival of hematopoietic stem/progenitor cells in the absence of cell division. Blood September 1995;86(5):1757−64. PMID:7544641. eng.

[38] Kozhemyakina E, Zhang M, Ionescu A, Ayturk UM, Ono N, Kobayashi A, et al. Identification of a prg4-expressing articular cartilage progenitor cell population in mice. Arthritis Rheumatol May 2015;67(5):1261−73. PMID:25603997. PMCID:PMC4414823. eng.

[39] Abad V, Meyers JL, Weise M, Gafni RI, Barnes KM, Nilsson O, et al. The role of the resting zone in growth plate chondrogenesis. Endocrinology May 2002;143(5):1851−7. PMID:11956168. eng.

[40] Stern T, Aviram R, Rot C, Galili T, Sharir A, Kalish Achrai N, et al. Isometric scaling in developing long bones is achieved by an optimal epiphyseal growth balance. PLoS Biol August 2015;13(8):e1002212. PMID:26241802. PMCID:PMC4524611. Epub 2015/08/04. eng.

[41] Omatsu Y, Sugiyama T, Kohara H, Kondoh G, Fujii N, Kohno K, et al. The essential functions of adipo-osteogenic progenitors as the hematopoietic stem and progenitor cell niche. Immunity September 2010;33(3):387−99. PMID:20850355. eng.

[42] Méndez-Ferrer S, Michurina TV, Ferraro F, Mazloom AR, Macarthur BD, Lira SA, et al. Mesenchymal and haematopoietic stem cells form a unique bone marrow niche. Nature August 2010;466(7308):829−34. PMID:20703299. PMCID:PMC3146551. eng.

[43] Morikawa S, Mabuchi Y, Kubota Y, Nagai Y, Niibe K, Hiratsu E, et al. Prospective identification, isolation, and systemic transplantation of multipotent mesenchymal stem cells in murine bone marrow. J Exp Med October 2009;206(11):2483−96. PMID:19841085. PMCID:PMC2768869. Epub 2009/10/19. eng.

[44] Zhou BO, Yue R, Murphy MM, Peyer JG, Morrison SJ. Leptin-receptor-expressing mesenchymal stromal cells represent the main source of bone formed by adult bone marrow. Cell Stem Cell August 2014;15(2):154−68. PMID:24953181. PMCID:PMC4127103. Epub 2014/06/19. eng.

[45] Grcevic D, Pejda S, Matthews BG, Repic D, Wang L, Li H, et al. In vivo fate mapping identifies mesenchymal progenitor cells. Stem Cell February 2012;30(2):187−96. PMID:22083974. PMCID:PMC3560295. eng.

[46] Yoshida T, Vivatbutsiri P, Morriss-Kay G, Saga Y, Iseki S. Cell lineage in mammalian craniofacial mesenchyme. Mech Dev September−October 2008;125(9−10):797−808. PMID:18617001. Epub 2008/06/20. eng.

[47] Zhao H, Feng J, Ho TV, Grimes W, Urata M, Chai Y. The suture provides a niche for mesenchymal stem cells of craniofacial bones. Nat Cell Biol April 2015;17(4):386−96. PMID:25799059. PMCID:PMC4380556. Epub 2015/03/23. eng.

[47a] Wilk et al., Stem Cell Rep April 2017;8:933−46.

[47b] Maruyama et al., Nat Commun 7:10526. https://doi.org/10.1038/ncomms10526.

[48] Colnot C. Skeletal cell fate decisions within periosteum and bone marrow during bone regeneration. J Bone Miner Res February 2009;24(2):274−82. PMID:18847330. PMCID: PMC3276357. eng.

[49] Rot C, Stern T, Blecher R, Friesem B, Zelzer E. A mechanical Jack-like Mechanism drives spontaneous fracture healing in neonatal mice. Dev Cell October 2014;31(2):159−70. PMID:25373776. eng.

[50] Zhou X, von der Mark K, Henry S, Norton W, Adams H, de Crombrugghe B. Chondrocytes transdifferentiate into osteoblasts in endochondral bone during development, postnatal growth and fracture healing in mice. PLoS Genet December 2014;10(12):e1004820. PMID:25474590. PMCID:PMC4256265. Epub 2014/12/04. eng.

[51] Kawanami A, Matsushita T, Chan YY, Murakami S. Mice expressing GFP and CreER in osteochondro progenitor cells in the periosteum. Biochem Biophys Res Commun August 2009;386(3):477−82. PMID:19538944. PMCID:PMC2742350. Epub 2009/06/16. eng.

[52] Mizoguchi T, Pinho S, Ahmed J, Kunisaki Y, Hanoun M, Mendelson A, et al. Osterix marks distinct waves of primitive and definitive stromal progenitors during bone marrow development. Dev Cell May 2014;29(3):340−9. PMID:24823377. PMCID:PMC4051418. eng.

[53] Bianco P. "Mesenchymal" stem cells. Annu Rev Cell Dev Biol 2014;30:677−704. PMID:25150008. Epub 2014/08/18. eng.

[54] Friedenstein AJ, Piatetzky-Shapiro II , Petrakova KV. Osteogenesis in transplants of bone marrow cells. J Embryol Exp Morphol December 1966;16(3):381−90. PMID:5336210. eng.

[55] Castro-Malaspina H, Gay RE, Resnick G, Kapoor N, Meyers P, Chiarieri D, et al. Characterization of human bone marrow fibroblast colony-forming cells (CFU-F) and their progeny. Blood August 1980;56(2):289−301. PMID:6994839. eng.

[56] Sacchetti B, Funari A, Michienzi S, Di Cesare S, Piersanti S, Saggio I, et al. Self-renewing osteoprogenitors in bone marrow sinusoids can organize a hematopoietic microenvironment. Cell October 2007;131(2):324−36. PMID:17956733. eng.

[57] Pinho S, Lacombe J, Hanoun M, Mizoguchi T, Bruns I, Kunisaki Y, et al. PDGFRα and CD51 mark human nestin+ sphere-forming mesenchymal stem cells capable of hematopoietic progenitor cell expansion. J Exp Med July 2013;210(7):1351−67. PMID:23776077. PMCID:PMC3698522. Epub 2013/06/17. eng.

[58] Worthley DL, Churchill M, Compton JT, Tailor Y, Rao M, Si Y, et al. Gremlin 1 identifies a skeletal stem cell with bone, cartilage, and reticular stromal potential. Cell January 2015;160(1−2):269−84. PMID:25594183. PMCID:PMC4436082. eng.

[59] Ono N, Ono W, Nagasawa T, Kronenberg HM. A subset of chondrogenic cells provides early mesenchymal progenitors in growing bones. Nat Cell Biol December 2014;16(12):1157−67. PMID:25419849. PMCID:PMC4250334. eng.

[60] Kunisaki Y, Bruns I, Scheiermann C, Ahmed J, Pinho S, Zhang D, et al. Arteriolar niches maintain haematopoietic stem cell quiescence. Nature October 2013;502(7473):637−43. PMID:24107994. PMCID:PMC3821873. Epub 2013/10/09. eng.

[61] Park D, Spencer JA, Koh BI, Kobayashi T, Fujisaki J, Clemens TL, et al. Endogenous bone marrow MSCs are dynamic, fate-restricted participants in bone maintenance and regeneration. Cell Stem Cell March 2012;10(3):259−72. PMID:22385654. PMCID:PMC3652251. eng.

[62] Logan M, Martin JF, Nagy A, Lobe C, Olson EN, Tabin CJ. Expression of Cre Recombinase in the developing mouse limb bud driven by a Prxl enhancer. Genesis June 2002;33(2):77−80. PMID:12112875. eng.

[63] Akiyama H, Kim JE, Nakashima K, Balmes G, Iwai N, Deng JM, et al. Osteo-chondroprogenitor cells are derived from Sox9 expressing precursors. Proc Natl Acad Sci USA October 2005;102(41):14665−70. PMID:16203988. PMCID:PMC1239942. Epub 2005/10/03. eng.

[64] Chan CK, Seo EY, Chen JY, Lo D, McArdle A, Sinha R, et al. Identification and specification of the mouse skeletal stem cell. Cell January 2015;160(1−2):285−98. PMID:25594184. PMCID:PMC4297645. eng.

Chapter 2

The Mechanics of Skeletal Development

Astrid Novicky[a], Soraia P. Caetano-Silva[a], Behzad Javaheri[a],
Andrew A. Pitsillides[a]
The Royal Veterinary College, London, United Kingdom

INTRODUCTION

The mechanics of skeletal development have been viewed from two major perspectives. One focused on early shaping of emergent musculoskeletal structures to furnish potential for movement. The second centered on the role exerted by embryo muscle contraction and movement mechanics on skeletal morphogenetic form. These viewpoints are intimately linked, yet their reinforcing frameworks may be considered distinct. The first is underpinned by the influence of cell-derived forces on their extracellular environment, culminating in a tissue-level impact; as in the mechanics generated by growth differentials in rapidly expanding tissues [1]. The latter is instead dominated by mechanical forces generated by muscle contraction. When regarded in this way, it is vital whether the exogenously derived influence exerted by muscular activity is considered an "additional" contributor, acting on a developmental "blueprint," or as an evolutionarily conserved component essential for the emergence of functionally proficient musculoskeletal system [2]. Herein we will focus on how early specification of skeletal form is related to later acquisition of skeletal function and how movement mechanics influence specific constituents of the embryo limb skeletal system.

Embryo movement emerges surprisingly early and has been studied for over a century [3]. These movements are initially sporadic and random and only later develop coordination. The idea that mechanical stimuli engendered by such movement can influence *postnatal* skeletal architecture is not new; it was documented by Galileo Galilei when he suggested that the shape of bones is related to loading [4]. Later, Roux (1881) proposed "the theory of functional adaptation" highlighting the primary importance of a load-bearing stimulus in

---

[a]These authors contributed equally to this work.

Developmental Biology and Musculoskeletal Tissue Engineering. https://doi.org/10.1016/B978-0-12-811467-4.00002-4

producing an appropriate bone architecture. Julius Wolff later speculated that a functionally adaptive mechanism was responsible for matching bone mass and architecture to the mechanical loads to which it was subjected [5]. He proposed that "every change in the form and function of bones, or of their function alone, is followed by certain definite changes in their internal architecture and equally definite secondary alteration in their external conformation, in accordance with mechanical laws."

Building on these ideas, Harold Frost suggested that a given externally applied mechanical loading stimulus leads to deformation, which initiates adaptive changes in bone architecture and mass [6]. He speculated that strain was the controlling stimulus for mechanoadaptation of bone and that this mechanism was governed by a strain-controlled feedback loop; which he compared with a thermostat, coining the term "mechanostat" that served a physiological function [6]. These ideas formed the basis for major advances in our appreciation of the role of mechanics in bone homeostasis during postnatal growth, maturation, and aging [7]. They have not, however, yet been addressed to define how embryonic elements of the skeletal system first attain functional competence during development. The notion that adaptive mechanical inputs critically influence and coordinate embryo skeletal development within a framework of an expanding range of movements is now receiving much attention. It is tempting, to view these mechanical influences as epigenetic, allowing for selective gene activation in response to environmental signals to produce phenotypic complexity in morphogenesis.

In this chapter, we will (1) focus on early patterning and later morphogenetic events particularly in chondro-osseous skeletal development, (2) consider the role of mechanics in normal development and how it might inform our understanding of osteoarthritis (OA), and (3) examine the evidence that there are distinct and critical phases when the emerging skeletal system acquires receptiveness to movement's impact and whether there might be interplay between signaling for growth and mechanics [8]. We will introduce the hypothesis that specific genes responsible for skeletal "mechanosensitivity" can adjust skeletal architecture under mechanical regulation to introduce phenotypic plasticity [9] and that these modifications in the embryo may have evolved as an additional source of species-specific phenotypic skeletal diversity.

SKELETAL DEVELOPMENT

Earliest shaping of the embryo germ layers results from cell division and progressive structural remodeling, beginning with egg fertilization. Mechanical forces mold this process from the start, as they can influence egg activation, early asymmetric cell divisions, and the establishment of initial embryonic polarity. It has been proposed that even the very opening intracellular signaling event may be considered to be activated mechanically, as the sperm penetrates the physical barrier of the egg using spring forces [1].

Musculoskeletal development is initiated by the formation of cell clusters within mesenchymal cell regions, derived from the lateral plate mesoderm of the early limb bud. These cells become chondrocytes and secrete a matrix, rich in type II collagen and the proteoglycan, aggrecan. Chondrocyte proliferation and matrix production increase the size of this cartilage anlage ("mold"). Later, chondrocytes in the center stop proliferating and begin hypertrophy to generate a leading figure for bone formation; the hypertrophic chondrocytes in the center of the anlagen direct mineralization of the surrounding matrix, attract both blood vessels and chondroclasts, and also direct cells in the surrounding perichondrium to become osteoblasts, within the primary center of ossification. It has been the general consensus that once hypertrophic chondrocytes have finalized the construction of the calcified cartilage template for bone formation, they undergo apoptotic cell death. Recent studies indicate, however, that instead some hypertrophic chondrocytes undergo transdifferentiation to osteoblasts and ultimately osteocytes (see below) [10]. The calcified hyaline cartilage template is regardless used as a scaffold for bone formation by newly developed osteoblasts to create the primary bone spongiosa [11]. Simultaneously, neighboring chondrocytes continue to undergo proliferation to enlarge the precursor cartilage mold for bone formation, increasing bone length. While the bone grows further, secondary centers of ossification are formed toward the ends of the emerging bone element; these occur mostly in the postnatal period and are similarly preceded by chondrocyte hypertrophy, blood vessel invasion, and cartilage calcification. The tissue sandwiched in-between primary and secondary ossification centers forms a distinct plate of cells called the growth plate [11]. It is critical that our appreciation of these initial events remains cognizant of the massive expansion of the limb that occurs during these early developmental stages.

The development of muscles starts with the intense proliferation of myoblasts in the dermomyotome, which terminally differentiate into myocytes. These myocytes later begin to express actin and myosin and together form myofibers; the functional contractile units of the muscle. Muscles are built and developed both pre- and postnatally: during both the primary myogenic phase, which occurs in the embryo, and the secondary fetal myogenic phase, when postnatal satellite cells acquire this precursor role, for example, in response to exercise and injury [12]. Skeletal, smooth, and cardiac muscles are formed during embryonal development. In contrast to smooth and cardiac muscles, which are being formed together with walls of blood vessels and viscera and the heart, skeletal muscles develop together with bone.

Tendons are formed by the construction of parallel bundles of collagen fibers, arranged in a gelatinous extracellular matrix (ECM) initially rich in mucopolysaccharides. Tensile forces generated at the tendon ends straighten the collagen fibers, and additional load causes stretching of the straightened strands and, therefore, the tendons get stiffer as they lengthen. Forces exerted by contracting muscles are tensile, as they effectively stretch the tendon. In long tendons, muscle is attached to cartilage skeleton at E12.5, and elongation of tendon from E13.5 onward occurs in response to physical tensional signals from muscle and skeleton. The molecular signals that mediate tenocyte responses to physical stimuli are incompletely defined, yet Scx, TGF-β, and FGF seem to represent strong candidate mediators [13].

Early events in musculoskeletal development clearly involve fundamental changes in cell—cell and cell—matrix relationships within and between the emerging tissues. These need to coordinate perfectly within the context of considerable cell division, expansion, and migration of distinct cell pools during progressive structural remodeling. Perfect control over organizing cell apoptosis, polarity, clustering, and hypertrophy is also required. Changes in the quantity, quality, and distribution of the surrounding ECM need to facilitate both these events and the emergence of distinct tissue characteristics. Events such as cellular invasion for blood vessel formation, myocyte merger to form myofiber contractile units, formation of parallel ECM collagen fiber bundles required for the generation of tensile forces at tendon insertions and the elaboration cell-free joint cavities for friction-free articulation are all crucial in the overall emergence of the musculoskeletal system. The earliest of these changes appear to be governed by cell-derived force generation and cell shaping.

CELL-DERIVED FORCES

Mechanical stress force at the integrin—ECM interfaces recruits and phosphorylates proteins that link actin to myosin proteins [14]. These forces have been measured in 3D hydrogel scaffolds, using confocal microscopy-based monitoring of cell-mediated reorganization of fluorescent beads, to create strain maps of traction-mediated matrix and cell contractility. They reveal that the greatest stresses (0.1-5Kpa) occur at the tips of long thin pseudopodia; intracellular 3D stresses were first measured by Legant et al. [15]. Early studies in zebra fish gastrulation found that modulation of these forces with inhibitors of cytoskeletal function led to diminished mass cell movement (epiboly) and thickening of the epiblast cell layer, indicating that this process is fundamentally reliant on mechanical cell deformation [16]. Intriguingly, ECM stiffness and cell contractility has also been shown to modulate mesenchymal stem cells (MSC) commitment between neurogenic, adipogenic, or chondrogenic lineages [17]. One mechanism by which such mechanical forces may be "sensed" is via the primary cilium; a structure composed of an axonemal "scaffold" and basal body proteins (BBPs), which has been shown to respond to fluid flow, osmotic pressures, and tensile strains [18]. BBP mutations are found in human congenital ciliopathies, including autosomal-dominant polycystic kidney disease, in which primary cilia mechanosensation function was originally established [19].

Cells use force in different ways, to compress, elongate, twist, or tension their ECM or each other at the piconewton force scale, which is sufficient to guide essential cell migration or ECM adhesion [20]. Such cell forces can significantly impact development by influencing morphogenesis, migration, ECM adhesion, proliferation, differentiation, homeostasis, and even inflammatory cascades. It is notable that nonmuscle cells also generate such contractile forces even in an emerging musculoskeletal system [21] to generate intracellular forces that act on the ECM for structural organization. Thus, integrin adhesion at cell−ECM interfaces anchors and shapes actin cytoskeletal architecture [22] even during migration to generate mechanical forces via the actin−myosin network. Cadherin adhesion proteins transmit intracellularly generated traction forces to reinforce this mechanical integrity across neighboring cells and contribute to planar cell polarity, morphogenesis, and cell migration via Wnt and receptor tyrosine kinase pathways; with known roles in osteoblast- and chondrogenesis [23].

The cytoskeleton is a dynamic hierarchical intracellular filamentous actin, intermediate filament, and microtubulular scaffold. At one level, tubulins polymerize into microtubules that support movement of motor proteins such as kinesins and dyneins across cell compartments [24]. A triad of vimentin, keratin, and lamin protein monomer intermediate filaments connect the nucleus with the endoplasmic reticulum, mitochondria, and Golgi apparatus, providing another level of cellular structural integrity. Actin monomers form filamentous complexes with myosin (myosin II) to give form to the cytoskeletal contractile apparatus [25], to anchor into protein clusters, including focal adhesions, at the cell membrane thus linking the cytoskeleton via transmembrane integrin receptors to the ECM. Structural deformations and rearrangements of ECM and intracellular structure result from force application to the cell−ECM unit, and microtubule polymerization/depolymerization can lead to push and pull forces. This emphasizes the crucial juxtapositioning of these structures at the very interface of cell force generation and responsiveness to externally applied forces.

Integrins serve as central regulators in both these events; they orient with head domains connecting to ECM and cytoplasmic tails binding to focal adhesion kinase (FAK) and paxillin, within the cell membrane, assembled on a talin, vinculin, zyxin, vasodilator-stimulated phosphoprotein, and α-actinin-based stratum [26]. ECM structures change very rapidly in early development to restructure mechanical properties to, for example, achieve regulation of stem cell differentiation. Cell−ECM transduction pathways transmit force to focal adhesion complexes (FACs), where talin strengthens actin-integrin and FAC linkage and where talin/vinculin clustering increases FAK in response to substrate tension [21,27,28]. Rho family GTPases also serve a role in generating cytoskeletal forces, cell polarity, migration, and cell shape [29]. Cell−ECM integrin adhesions thus remodel to form large FACs that function as biochemical/biomechanical signaling centers, containing FAK, ERK, JNK, Src, Ras, and Raf with roles in cell migration, proliferation, and differentiation [30]. This places ECM mechanics centrally and leads to the question: what comes first to guide these early embryonic events in the musculoskeletal system, the specific ECM encountered, or the regulated ECM production behavior of cells? The answer to this question is unlikely simple.

Later skeletogenic events appear to involve in continued participation of growth-generated cellular strains and now also responses to muscle contraction−generated loading [21]. Thus, the fibroblastic layer that emerges between skeletal primordia and surrounding mesenchyme, to form perichondrium, is a region that at the same time facilitates rapid expansion (in girth) that conforms to a growth-restraint model and also contributes to directing skeletal growth predominantly along the longitudinal axis of the skeletal element [31].

THE EMERGENCE OF A CHONDRO-OSSEOUS CELLULAR CONTINUUM

Can hypertrophic chondrocytes become osteoblasts and contribute to the osteogenic lineage? This question, which has been asked more than a century [32], has recently been readdressed to yield new answers. The process of endochondral ossification is predominantly understood in the context of its function in long bone development. After their initial condensation, limb bud mesenchymal cells differentiated to form a cartilage model by undergoing chondrogenesis; a multistep differentiation process during which chondrocytes within the cartilage model form distinct morphological and functional groups, which correspond to maturation status; adopting resting, proliferating, mature, and hypertrophic states. These states correspond to phases in which the precursor pool is maintained, an expansion in cell number, a halt on division and entry into differentiation and matrix production, and finally, a dramatic increase in size occur [33]. The hypertrophic chondrocytes are characterized by the expression of matrix metalloprotease 13 (MMP13) , type X collagen, and vascular endothelial growth factor (VEGF) that are therefore hallmarks of their terminal maturation status. It has been assumed until recently that these markers may be considered hallmarks that precede their apoptotic cell death [34]. This heralds the replacement of cartilage by bone, which is initiated by osteoblasts and accompanied by vascular invasion of the ECM [10]. It also precedes the degradation of the cartilage matrix and its subsequent replacement by a marrow cavity, the promotion of vascularization, and recruitment of osteoblast precursors from surrounding tissues followed by osteoid bone matrix deposition and mineralization [35].

Recent studies using a range of cell-tracking strategies have now shown, however, that hypertrophic chondrocytes can survive all of these phases to become osteoblasts and ultimately undergo osteocytogenesis, as they are

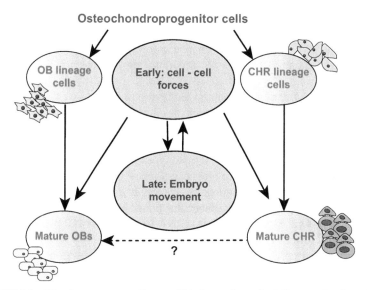

FIGURE 2.1 Chondro-osseous continuum. This figure shows the influence of early and late mechanical forces in regard to osteoblast (OB) and chondrocyte (CHR) development. New models explain that hypertrophic, mature chondrocytes might not undergo apoptosis but transdifferentiate into osteoblasts.

embedded within the bone matrix, thus contributing and populating trabecular and, at least to some extent, mature cortical bone. This survival of hypertrophic chondrocytes during normal endochondral bone formation was also found to extend to pathological states, as it has also become evident that postnatal hypertrophic chondrocytes can also contribute directly to bone repair (see Fig. 2.1). These data are consistent with studies showing that grafted cartilage supports bone regeneration in vivo [35].

This notion that hypertrophic chondrocytes transform into osteoblasts was further supported by an exploration of stem cell behaviors in the endochondral cartilage found in bone healing, which revealed an activation of the pluripotent transcription factor, Oct4A—an established stem cell marker thought to be responsible for cellular reprogramming—in vascularizing tissues and hypertrophic chondrocytes [35]. Despite a lack of clarity regarding the function of Oct4A, it appears that chondrocytes may dedifferentiate, in a manner akin to that described for induced pluripotent cells, to regain progenitor capabilities that is consistent with the mechanism described by Song and Tuan [36]. Elucidation of the cellular reprogramming required to direct transition from cartilage to bone is a key target; however, the substantial overlap that exists between markers of hypertrophic chondrocytes and osteoblasts represent a significant hurdle in achieving this aim [32]. These difficulties in discerning the cellular reprogramming mechanisms are not altogether surprising, however, since chondrocytes and osteoblasts are known to share common osteochondroprogenitor origins during limb development. These discoveries nonetheless provide a conceptual change with regard to the origins of osteoblasts of endochondral bone, with important implications for bone biology [10].

Indeed, little attention has been given before to the origin of cells used for characterization of osteoblast behavior in vitro. To begin to address the possibility that osteoblasts from different bone types may exhibit divergence in their in vitro behavior, we compared osteoblasts from structurally distinct subchondral, trabecular, and cortical bone types from singular skeletal vicinity from within the same patients to find specific bone type—related differences in growth, differentiation, and angiogenic potential of cultured osteoblasts [37]. We found that osteoblasts from trabecular bone showed slower proliferation, but higher RUNX2, SP7, and BSP-II mRNA levels, TNAP mRNA and protein activity, and lower TNFRSF11B:TNFSF11 mRNA ratios compared with subchondral and cortical bone. In contrast, subchondral osteoblasts showed higher levels of VEGF-A mRNA and protein release, implying more intimate vascular relationships for this bone type [37]. These findings are consistent with data showing that the response of trabecular, cortical, and subchondral bone to in vivo challenges is not always identical [38], with distinct transcriptional signatures in osteoblasts from long bone and calvaria sites [39] and with osteoblasts from within a single bone structure displaying differing gene transcription depending on anatomical location [40]. These data point to inherently distinct osteoblast behaviors in different bone types. The possibility that these behaviors reflect progenitor origins related to the crossing of a chondro-osseous continuum or that this might underpin mechanical differences in bone behavior has not yet been explored.

MECHANICS OF NORMAL DEVELOPMENT—DISTINCT AND CRITICAL PHASES

Securing efficient and appropriate animal locomotion is the fundamental purpose of the musculoskeletal system. It is not surprising therefore that mechanical loading is the ultimate adaptive driving force behind the assembly of multiple tissues into a harmonious functional system that meets this purpose. It seems somewhat circular, therefore, that this mechanical driving force must itself arise from embryo movement or one of its downstream consequences (e.g., environmental reaction force); either way embryonic muscle contraction is responsible for engendering mechanical stress on the embryo's skeleton. Many publications have shown the importance of mechanical stimuli in normal skeletal development [41]. The importance of mechanical signals in musculoskeletal development have mostly been highlighted by studies linking restricted fetal movement to developmental abnormalities in a range of animal models and from observations in human embryos [42].

Limb movements are recordable in embryonic chick from day 4.5, in mouse embryos at day 12.5, and in human embryos at 7.5 weeks of gestation [41]. The initiation of movement at these stages correlates with establishment of physical contacts between motor axons and presumptive muscle cells; it is relatively uncoordinated initially, gaining greater coordination at later developmental stages [43,44]. The mechanoadaptive role exerted by movement is known to be critical for both joint formation and in the regulation of endochondral ossification. Although it is essential to pinpoint that movement's effects on these processes serve as evidence that mechanical signals can be transformed into growth signals [45], we will concentrate on these specific processes and use them as examples through which we can make more speculative, generalized conclusions regarding the role of mechanics in musculoskeletal development. A recent review provides an excellent evaluation of these relationships across a broader range of emerging musculoskeletal tissues [46].

It is evident from our discussions (above) that the range of forces experienced in vivo include those that are generated intrinsically by neighboring cells, the cytoskeleton, the cell surface—ECM interface, and also those applied extrinsically by muscle contraction. Although it is possible to model the strains generated by these emerging contractions of the developing musculature, it currently remains impossible to measure them directly in the embryonic skeleton. Strategies do exist, however, that allow us to observe the impact of removing or overstimulating movement during embryo development; including paralysis with decamethonium bromide (DMB) and pancuronium bromide, generation of hyperactivity with 4-aminopyridine (4-AP) treatment, the use of mutant mouse models, and most recently the indirect regulation of movement by variation of incubation temperature in poikilothermic species.

Murray, Fell, and Canti [47—49] were the pioneers of research in the first of these areas, dating back to the 1920s and 1930s. They used a range of experimental models, including explanted limbs from chick embryos. Their cultivation of the explanted chick limbs ex vivo revealed that the early limb patterning of skeletal elements took place without modification, independently of an external mechanical input in the absence of the limb's normally contiguous structures including skeletal muscle; this has been nicely validated by in vitro mechanical stimulation of similar explants to demonstrate that both under- and overstimulation is detrimental [50]. Although not explicitly monitored, the studies by Fell and Canti also indicated that initial specification of the presumptive joint's position is independent of movement. In stark contrast these intriguing studies, and others since, have reported that later joint cavitation events are reliant on muscle contraction from embryo movement. Interestingly, Drachman and Sokoloff and others reported [51,52] that pharmacologically imposed in ovo paralysis with DMB or type A botulinum toxin in chick embryos resulted in a failure of cavitation and led to cartilaginous fusion of limb elements across the knee, toe, and ankle joint areas. More recent studies by us and others in chicks and in "muscleless" mutant mice show very similar outcomes, although in the latter some joints appear to achieve this terminal cavitation event even in the absence of contracting limb musculature [52,53]. Finite element analyses have demonstrated that forces originating external to the mammalian embryo may account for joint cavitation when muscle development is abnormal [54]. Nonetheless, the differential effect of such immobility in certain joints suggests that the response of skeletal progenitors to locomotor mechanics involves an interaction between mechanical forces and location-specific regulatory factors [55]. Morphometry allied to local assays of proliferation and mRNA expression suggests that locomotor mechanics likely generate the patterns that guide developing joint shape morphogenesis [56]. Thus, joint development is clearly mechano-dependent and clearly exemplifies the separation of early intrinsically regulated events, controlled by cell-derived forces, and later extrinsically governed events, regulated by movement mechanics.

This appears to be consistent with the long-held view that joint development can be separated into two steps: interzone specification and later cavitation [57]. During joint interzone specification, chondrogenic markers such as SOX9 and COL2a1 are thought to decline as expression of GDF5, WNT4, and WNT9a are activated. The signaling mechanisms underlying the role of movement in joint development has been the subject of many studies. Kahn et al. suggested that the absence of embryo movement in mutant mice, with either aberrant muscle development or functionless muscles, leads to failure of interzone cells to express interzone-specific markers [58]. Kahn et al. extended these observation by showing that the cartilaginous fusion of limb elements in response to limb paralysis occurs because of expression of factors associated with chondrogenesis, including Sox9 and Col2a1, leading to differentiation of progenitor cells at the interzone into chondrocytes [58]. In addition, other studies have reported that the immobilization of embryonic chick limbs leads to selective changes in growth factor expression, and importantly, a reduction in the local cellular synthesis of hyaluronan (HA), which in normally active limbs takes place to facilitate the separation of skeletal elements either side of the forming joint cavity [59,60,61]. This regulation by locomotor mechanics appears to involve a mechanism that leads movement-induced activation of the MEK-ERK-1/2 pathway at the presumptive joint line. These data indicate that the role of mechanics at the developing joint line is to promote local production and retention of HA by a mechanism reliant on the mechanical induction of MEK-ERK-1/2 pathway activity [59]. Additional studies showed that embryonic immobilization restricts HA production because of significant reduction in uridine diphosphoglucose dehydrogenase, an enzyme essential in HA synthesis [62]; activity and cell surface HA-receptor (CD44) expression, and HA binding [57,62]. Intriguingly, the constitutive activation of the MEK-ERK signaling cascade, as seen in articular surface cells in the cavitating joint of a moving limb, is considered a hallmark of pathological, inflammatory cell signaling.

Further exploration of the mechanisms by which mechanics contribute to joint cavitation has revealed even more intimate links with pathways more traditionally linked to inflammatory signaling. Our data have demonstrated similar patterns of activation for p38^{MAPK} [63] and that COX-2 (cyclooxygenase 2, essential in prostaglandin production) is also highly expressed in joint line articular surface cells, in which it serves as a key regulator of HA synthesis and binding. These observations are somewhat difficult to reconcile as they imply that normal movement−dependent processes involved in the development of the joint exploit pathways thought "classically" to be involved in lymphocyte activation and the promotion of proinflammatory cytokine production.

Together with earlier ground-breaking work, these studies indicate that local mechanics may play little role in the specification of joint location, but instead is essential in promoting the separation of skeletal elements in the later joint cavitation process that facilitates articulation, suggesting that mechanics is vital at late but not at early stages of musculoskeletal development. They emphasize, however, that it is not yet known how chondrocytes sense the local mechanical signals generated by contracting muscles leading to activation of these and other Wnt-related pathway activation evident at the forming joint. They do not explain, either, how the major anatomical features of the adult hip are present before cavitation of the joint [64]. Our unpublished data and other studies indicate that the joint interzone specifying gene profiles are retained in early-stage paralyzed limbs but that later joint cavitation stages are critically dependent on muscle contraction for normal development. In line with this, chondrogenic differentiation of early limb bud mesenchyme in high-density micromass culture rapidly progress to hypertrophy, suggesting that they have lost the local signals required to direct formation of stable cartilage [65].

More recent studies have reinforced this general assertion, providing further evidence that embryonic paralysis (with associated failure in development of limb musculature) during early developmental stages does not significantly affect endochondral ossification or consequently bone length [41]. Along these lines, we recently explored if the critical role for mechanoadaptation is also confined to the later phase of long-bone elongation during development and, hence, whether endochondral ossification during only the later stages of skeletal development are affected by the absence of mechanical cues. This was investigated using two distinct forms of pharmacologically induced immobilization in the chick model, which evoke either rigid of flaccid paralysis. These studies have shown that the conversion of cartilage into bone is slowed in specific skeletal elements by immobilization of embryonic chickens and, hence, that the rate of calcification is increased in response to mechanical loading only after, and not before, day 14 of development (gestation 21 days) [52]. This influence of movement on endochondral ossification correlated both with the targeting of selective growth zones along the limb and with a regulatory effect of mechanical loading on growth plate chondrocyte proliferation in vivo. This suggests that mechanical stimuli are necessary for recruitment and proliferation of immature chondrocytes in the growth plate. In embryonic growth cartilage, reduced proliferation has previously been reported in chicks immobilized with DMB and mouse models of paralysis [66].

It is therefore notable that in both immobilized zebra fish, which show a significant reduction in the size of all pharyngeal cartilage elements, and in muscle-deficient mouse embryos, the total number of chondrocytes does not appear to be altered by this absence of muscle contraction [67]. Chondrocyte intercalation into columns in the proliferative zone of the growth plate is, however, abnormal in these models, indicating that cell polarity is likely also dependent on mechanical stimuli. ECM production by growth plate chondrocytes is also influenced by mechanical loading, with implications for promotion of chondrocyte differentiation and structural and mechanical integrity of developing skeletal elements. It is tempting to speculate, based on these data concerning joint formation and long-bone growth in the embryo, that early events governing limb formation and patterning occur independently of externally derived mechanical forces generated by muscle contraction and that embryonic skeletal development acquires its dependence on mechanical input resulting from movement only at the later developmental stages.

Other structures that show a similar switch in mechanical dependence include the bone eminences, protrusions that are present in long bones. These protrusions integrate tendons and the skeleton and act to distribute the load to minimize stresses at the enthesis. These structures start to form after cartilage differentiation and bone formation have taken place by endochondral ossification [68]. If loading is not present during later stages of development, chondrocyte proliferation at these locations is decreased and eminences are lost. Thus, muscle contractions are needed for proper skeletal morphogenesis [67].

Another interesting structure that shows mechanical regulation is the furcula, equivalent to a fusion of the paired clavicles in many mammals and the single interclavicle in most reptiles. Postnatally, this bone serves mechanical stabilizing roles in the shoulder joint and acts as a mechanical spring during flight. We investigated how individual clavicular and interclavicular aspects of furcula growth respond to both rigid and flaccid paralysis in chick embryos and to hypermotility (by changing incubation temperature) in crocodilian embryos. We found that growth rates in individual aspects of a single bone differ in their mechanical regulation; the interclavicle required both static and dynamic locomotor mechanical components for normal growth, while static loading preserved most aspects of clavicle growth. This suggests that locomotor mechanics can shape these structures during development to suit postnatal roles [69].

It is therefore vital to stress that this influence of movement mechanics is not hard-wired into the later stages of development across all elements of the entire developing skeleton. For example, our recent study in chick embryos showed that longitudinal expansion of the femur by endochondral ossification is much more sensitive than the tibia to embryonic hypoloading. Thus, late-stage immobilization exerts differential effects on limb proportions in response to altered embryo movement [52], suggesting that there is evolutionary targeting of some growth plates by mechanoregulatory mechanisms to influence the emergence of limb anatomy. Our recent study in West African dwarf crocodiles reported that thermal regulation of embryo movement also selectively controls the endochondral ossification in only some skeletal elements by modulating chondrocyte proliferation in specific growth plates, to achieve similar manipulation of limb proportions during embryo growth [52].

We have exploited this targeting of specific skeletal elements in gene array analyses. Our studies in the chick compared elements exhibiting a requirement for movement mechanics to attain optimal growth and those that do not, across developmental stages during which this sensitivity was acquired and also compared their responses with the imposition of immobilization (at late stages). These studies revealed a transcriptional signature heavily centered on the downregulation of the mechanistic target of rapamycin (mTOR) pathway as a crucial component in the acquisition of sensitivity to movement mechanics specifically in the chick femur [52]. The mTOR complex has diverse roles in regulating cell proliferation and growth (see below) and appears to be the primary pathway altered during the onset of sensitivity to movement mechanics.

In addition to showing that the mTOR-associated genes were not, however, significantly altered in their expression by immobilization itself, we found that a number of traditionally "myogenic" genes associated with MEF2c were downregulated in response to immobilization in all the elements we examined. These genes are associated with actin-mediated cell contraction and have previously been implicated in the mechanotransduction process during embryonic limb growth [52]. Deeper data evaluation showed that although the skeletal muscle genes that were downregulated in response to immobilization were very similar between limb elements, a number of genes associated with the regulation of chondrogenesis were only altered in expression in the mechanosensitive femur. Upstream analysis to link this combined pattern of up/downregulated set of RNAs with preexisting signatures identified that these changes are consistent with modified Wnt3a activity in the femur in response to immobilization [52]. Transcriptome profiling of the developing humerus in muscleless mutant embryos also identified downregulation of genes linked to cytoskeletal architecture and cell signaling via the Wnt pathway (and its target gene, Cd44), to highlight signaling via the Wnt signaling pathway as a potential point of integration of mechanical signals [70]. Together, these data suggest that genes associated with actin-mediated cell contraction are regulated in chondrocytes by embryo movement and may contribute to sensing and transducing mechanical signals. Our data also suggest that Wnt3a is involved in coordinating chondrocyte responses to embryo movement in growth plate cartilage and that these responses are heavily linked with prior downregulation of mTOR.

NF-κB, mTOR, AND WNT SIGNALING

NF-κB and mTOR

It is evident that inflammation and mechanics play roles both in physiological and pathological processes. Both inflammation and mechanics have, for example, been prescribed central roles in the progression of OA. For a long time, researchers and clinicians indeed tried to define the disease as either inflammatory or mechanical in its etiology, whereas more recent studies have shown that these contributions are perhaps more likely synergistic. Although still not completely resolved, inflammation can be evoked in response to mechanical stress, as an attempt to compensate for and, repair mechanical insult/injury. Many molecular factors including MEK-ERK, P38^{MAPK}, and COX-2 are part of these inflammatory cascades, wherein NF-κB (nuclear factor kappa-light-chain-enhancer of activated B cells) serves an evolutionary-conserved and pivotal function, responsible for regulating both immune and inflammatory responses; it is also linked to metabolic disorders, such as obesity and diabetes. The NF-κB family contains five members (NF-κB1 [p50], NF-κB2 [p52], RelA [p65], RelB, and c-Rel), which are essential for regulation of cell survival, apoptosis, invasion, migration and is, therefore, an emerging target for drug discovery in cancer, inflammation, and autoimmune diseases. NF-κB is ubiquitously expressed in the cytoplasm and when inactive is associated with its inhibitor, IkBα. Activation of the transcription factor is mediated by the IκB kinase (IKK) complex, which is formed of the catalytic IKKα and IKKβ subunits and the regulatory subunit NF-kappa-B essential modulator (NEMO). On activation, the IKK complex is phosphorylated to release NF-κB from its inhibitory IkBα association. Activated NF-κB (RelA, RelB, or c-Rel) forms heterodimers with p50 and translocates to the nucleus where it regulates gene transcription.

Of particular interest herein, NF-κB exhibits an interaction with mTOR. Indeed, it has been shown that NF-κB is located downstream of Akt and that inhibition of mTORC1 reduces NF-κB activity. On the other hand, IKK interacts with the mTOR complex, both activated mTORC1 and mTORC2, and hence NF-κB appears to be novel regulator for mTOR activation [71,72]. Cell proliferation can be inhibited, and apoptosis promoted, by the administration of CmpdA, the novel IKKβ inhibitor [73]. Additionally, the mTORC2 complex and IKKα phosphorylate and activate Akt [74]. Akt/PKB is a serine/threonine protein kinase, which comprises a central regulator of cell growth, survival, and proliferation.

It is therefore of importance that the Akt1 isoform plays a major role in embryonic development, fetal growth, and survival. Akt/PKB is activated by receptor tyrosine kinases, such as PDGF-R (platelet-derived growth factor receptor), insulin, and insulin growth factor 1. Activation of Akt via these growth factors, in turn signals via mTOR to regulate autophagy and imposes growth restriction in circumstances when only low amino acid levels are available. The major targets of Akt are thus regulators of cell survival/death such as IKK-NF-κB and p53, regulators of cell cycle such as cyclin D1 and of protein synthesis and angiogenesis such as mTOR. Akt inhibits the mTOR (mTORC1) pathway; mTOR is a serine/threonine PI3K-like kinase family member responsible for coordinating cell growth and division in response to nutritional and growth factor status. The mTOR complexes, mTORC1 and mTORC2, activate lipogenesis and suppress autophagy, proliferation, and cytoskeletal organization [75]. The mTORC1 complex is sensitive to rapamycin, is a key mediator of growth factors and insulin, and comprises Raptor (regulatory-associated protein of mTOR), mTOR catalytic and Gβ-like protein subunits; its best characterized substrates (S6K1/S6K2) control ribosome biogenesis, mRNA translation, cell growth, and autophagy. In contrast, the mTORC2 complex is rapamycin-resistant. This complex is composed of Rictor (rapamycin-insensitive companion of mTOR), mTOR and kinase-interacting protein (mSin1) subunits. Interestingly, phosphorylation of mTORC2 activates Akt, even in serum-starved media which increases NF-κB activation. As inhibition of mTORC1 decreases NF-κB activity, these data together suggest an overall stimulation by mTORC1/2 of NF-κB activity (see Fig. 2.2). Rapamycin strongly inhibits mTORC1 but only mildly affects mTORC2 even after prolonged exposure and not in all cell/tissue types [75].

In the context of our findings concerning the control of endochondral ossification in the growing elements of the developing limb, we find that a downregulation of mTOR-related pathway signaling components occurs on their sensitization to movement mechanics [70]. These data suggest that differential intrinsic mTOR activity in individual growth plates determine if their growth can exhibit responsiveness to external mechanical cues, derived from movement mechanics, or is instead regulated only by intrinsic factors (such as amino acid availability and growth factors). mTOR's regulation of protein synthesis in the cell cycle shows its potential to impact events in the proliferative zone. Our exploration of immobilization-induced changes in the proliferative zone indeed suggests that chondrocytes are arrested in the G2 phase, which requires rapid protein synthesis in preparation for mitosis. Our observation that proliferative zone chondrocytes in the femur reach S phase at a normal rate (they incorporate BrdU during DNA replication) yet either fail or take longer time to undergo mitosis strengthens this connection. These

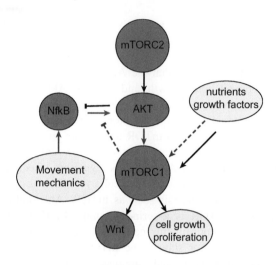

FIGURE 2.2   Interaction of mTOR and NfkB in development and movement. Movement mechanics influence cell growth and proliferation by acting via the NfkB—Akt—mTORC1 pathway (*blue arrows*). Nutrients and growth factors also enhance mTORC1, which leads to a negative feedback control, inhibiting NfkB and therefore not allowing mechanical regulation to take place (*dashed, red arrows*).

observations lead us to hypothesize that downregulation of intrinsic mTOR activity reduces growth in the absence of movement and appears to allow the capacity for mechanical stimuli to, instead, drive cell cycle progression through the proliferative zone [70]. Previous studies indicate that selection for intrinsic growth regulation can indeed occur at the expense of adaptability to mechanical loading [76]. It is currently not clear how mTOR signaling would interact with Wnt3a mediation of the growth response to regulate mechanical signaling. Selection for different levels of intrinsic mTOR activity may offer an explanation for why different elements are more susceptible to growth regulation by mechanical stimuli in different species.

Wnt Family

Wnt proteins are secreted, lipid-modified glycoproteins that allow for communication between cells. They regulate cell growth, function, differentiation, and cell death. Wnt proteins play a central role in bone development, modeling, and remodeling. Interestingly, mutations in LRP5 (low-density lipoprotein receptor-related protein 5), a coreceptor for Wnt, leads to alteration in bone mass in human, whereas LRP5 loss-of-function mutations are associated with skeletal fragility and low bone mineral density. The Wnt/β-catenin pathway increases bone mass through an anabolic induction of osteoblastogenesis and renewal of stem cells, which made it an interesting target for drug development in skeletal diseases. Interestingly, Wnt/β-catenin signaling inhibits adipogenesis and might therefore show clinical importance in bone fracture healing. GSK3 (glycogen synthase kinase 3 located downstream of Wnt) gives further evidence for the bone mass−increasing effects of Wnt signaling. Pharmacological inhibition of GSK3 with LiCl in C5BL/6 mice increased bone formation and osteoblast numbers.

Wnt signaling also plays a crucial role in the development of mesoderm, axial patterning, and the segmentation into somites. It has also been shown that Wnt9a and Wnt4 are responsible for suppressing the formation of ectopic cartilage during joint formation. A study has shown that in mice lacking Wnt9, normal joint formation takes place; however, subsequent chondrogenesis is inhibited and mineralization of the joints occurs, which leads to the development of OA [77]. Further studies had established that the locomotor mechanics required for joint morphogenesis involves mechanically induced Wnt activation to control local cell proliferation and migration in developing zebrafish larvae [78]. These studies also defined regions of high strain as the locations at which Wnt signalling is most active during the joint formation process [78]. The idea that Wnt signaling can be activated by movement mechanics has now been extended to show its regulatory function in tissue remodeling, cell migration, wound closure, and tissue destruction by metalloproteases in the lung [79].

ROLE OF MECHANICS IN OSTEOARTHRITIS (TRADITIONAL)

OA, a progressive syndrome affecting the entire joint, is characterized structurally by sclerosis of subchondral bone, articular cartilage degradation, and bony lesions formation (osteophytes). Late-stage OA is associated with inflammation, pain, progressive loss of function, and life-altering disability. On onset and during OA progression, all of the joint's components exhibit significant changes in cellular activity, functional properties, and ultimately structural alterations. Although OA affects multiple joint components, it is conceivable that its initiation targets different single tissues, propagating to others due to close interaction and cross talk, to culminate in the consistent hallmark pattern of characteristics. It is vital to point out that this occurs against a backdrop of continued locomotor mechanical load-bearing.

Physiological loading is a primary determinant of homeostasis across the joint's osteochondral unit. Joints withstand millions of physiological load cycles that do not lead to any adverse osteochondral consequences. In fact, normal physiological load acts to coordinate function across the unified joint unit to maintain homeostatic catabolic and anabolic remodeling. Load transmission coordinately engages all of its musculoskeletal components including muscle, bone, ligaments/tendons, and articular cartilage, as well as nerves and vasculature. Deviations from this coordinated balance arise because, for example, of aging, trauma, and obesity to perturb joint homeostasis and to provide further irreversible mechanical challenge to cartilage [80]. On the surface, this appears little different from the central role ascribed to locomotor mechanics in the emergence of a functionally competent skeletal system at later developmental stages in the embryo; it makes an elucidation of the underpinning mechanisms even more salient.

Although physiological activity levels maintain normal cartilage integrity, a deviation to nonphysiological load levels leads to OA cartilage degradation and bone changes. Defining levels of activity as "physiological" is not an easy task, however, as the load engendered has multiple parameters (e.g., duration, magnitude, and frequency) and is further complicated in any in vivo studies by systemic variables (e.g., gender, age, hormonal status, and genetics). Identifying the molecular basis of this threshold between "adaptive" and pathological locomotor mechanics is likely to yield new chondroprotective targets.

In response to loading, resident chondrocytes experience a diverse range of biophysical signals that include stress, strain, fluid flow and pressure, changes in the tissue fixed charge density, electrokinetic effects, and changes in shape and volume [81]. Although the mechanisms of cartilage mechanoadaptation are not fully elucidated, earlier work had shown that mechanical signals inhibit multiple aspects in the IL-β-induced signaling cascade, downstream of IκB kinase activation. These regulate IκB-α/-β degradation/synthesis and promote IκB-α shuttling to export nuclear NF-κB, which terminates NF-κB's transcriptional inflammatory signaling [82], implying an antiinflammatory signaling role for locomotor mechanics.

Transactivation of CITED2 (transcriptional factor) is also induced by dynamic load in vivo [83] and by fluid shear and intermittent hydrostatic pressure in vitro [83,84] to suppress matrix-metalloproteinase (MMP) expression and matrix degradation [83]. Other studies have found that controlled loading exerts protective effect via reduction in the activity of ADAMTS (a disintegrin and metalloproteinase with thrombospondin motifs) and suppression of inflammatory cytokines (IL-6, TNF-α) and mediators (COX-2, prostaglandin E2, and nitric oxide). Furthermore, mechanical loading enhances antiinflammatory signaling via IL-4 and IL-10 [85]. Roles for calcium, integrin-ECM, and primary cilia signaling in mechanical antiinflammatory chondrocyte signaling are also indicated [86].

On the other hand, many clinical and animal studies indicate that abnormal, nonphysiological loading (both high and low) produces deleterious joint alterations [87]. The most obvious example is in traumatic loading during which impact forces likely cause direct damage to cartilage ECM [88]. Animal OA models using ligament transection, meniscectomy, and loading produce structural changes comparable with those observed in human OA [89]. These traditional views of the role of mechanical and inflammatory signaling pathways in OA have failed to identify any specific new treatment strategies, other than to reinforce the relationships with physiotherapeutic mechanical manipulation approaches and the continued use of classic antiinflammatory pain relief therapies.

More recently, TRPV4 (transient receptor potential vanilloid 4, a Ca2+-preferred membrane ion channel) was found to act as a transducer of load-induced chondrocyte matrix biosynthesis and that its inhibition prevented the proanabolic/anticatabolic effects of loading; indeed, chemical TRPV4 activation even in the absence of load mimics the anabolic/anticatabolic effects of moderate loading [90]. Nevertheless, the precise relationships between locomotor mechanics and OA etiology and pathology remain incompletely understood.

OSTEOARTHRITIS DISEASE PROCESSES IN THE CONTEXT OF MECHANICAL FUNCTION (NOVEL)

Another relatively recent advance concerns the idea that one of the pivotal changes in chondrocytes is their switching in OA from a stable, articular cartilage phenotype to the more transient phenotype normally associated with chondrocytes resident in the growth plate [2,91]. Thus, the acquisition of the more dynamic, growth plate—like behaviors, as opposed to those much more long-lived and unchanging normally seen in articular cartilage chondrocytes, destabilizes articular cartilage through the rapid proliferation, hypertrophy, and apoptosis that this switch entails. The chondrocyte processes associated with these events in OA thus include a change in type II collagen, aggrecan, and SOX9 expression and increases in expression and activity of MMP13. These chondrocyte OA changes are also linked, akin to growth plate, with elevated recruitment of osteoblasts and osteoclasts with increases in bone remodeling indices and with elevated neovascularization and mineralization levels.

This switching to growth cartilage behaviors is perhaps best exemplified by their hypertrophy both in OA cartilage and by the endochondral ossification, which drives the formation of osteophyte bony lesions and subchondral sclerosis. It is plausible that these events result in cartilage thinning and the joint space narrowing, which characterizes the X-ray radiographic appearance of OA. Furthermore, these alterations in ECM and articular chondrocyte phenotype are also particularly evident in the detection of type X collagen, a distinctive hypertrophic chondrocyte marker, and raised Indian hedgehog (IHH), osteocalcin, CD36, and alkaline phosphatase levels in chondrocytes, particularly those near the osteochondral junction in OA joints.

Stable to transient chondrocyte phenotype switching in OA results in the reinduction of several key pathways of endochondral ossification, and (as discussed above) these are clearly subject to control in the embryonic skeleton at least by locomotor mechanics. Given that these OA changes, characterized by growth plate chondrocyte-like behaviors, occur against a backdrop of continued locomotor mechanical load-bearing in the adult, it becomes intriguing that the same molecular signatures recapitulated in OA are also driven by mechanics in the emerging embryonic skeleton for the attainment of optimal rates of endochondral ossification and longitudinal bone growth. How this interchange between the apparently discordant "transient" and inherently "stable" articular cartilage chondrocyte phenotypes contributes to OA is not yet fully established, but it is tempting to speculate that it driven in OA by locomotor mechanics.

This intriguing possibility is complicated further by data indicating that articular cartilage and growth plate chondrocytes do not share a common origin [92], suggesting that this chondrocyte switching can occur across a range of cellular contexts. For example, we have recently reported that the molecular articular cartilage changes in a model of spontaneous OA in STR/Ort mice are similarly characterized by aberrant redeployment of these endochondral ossification processes [93]. In this model, this link to aberrant endochondral ossification is further strengthened by our observations showing significantly modified longitudinal growth kinetics in this strain, which suggest that such abnormalities in endochondral ossification may predict mechanisms underlying OA pathology. These findings reinforce the idea that the etiology of these growth cartilage-related pathological OA characteristics—which are shared with those driven in the growth plate by movement of embryonic limbs—might indeed also rely on locomotor mechanics. These considerations may indeed be vital to many of the observations that have been made in OA joints, which are serving their "normal" mechanical load-bearing function.

Indeed, OA is often characterized by frequent new rounds of proliferation in individual chondrocyte lacunae, and by newly identified roles for mTOR in OA processes. Raised ECM production rates are also evident in early OA and hastening of cartilage-to-bone transition seen at the cartilage—subchondral bone interface. It is therefore tempting to speculate that activation of all of these transient growth/hypertrophic chondrocyte behaviors in OA joints are connected to those driven in developing embryo growth plates by movement and that they exploit common mechanoregulatory mechanisms.

Disrupted cartilage homeostasis in OA induces profound phenotypic chondrocyte alterations, which promote the synthesis of a subset of factors that damage cartilage and target other joint tissues. For example, chondrocyte-derived inflammatory cytokines, chemokines, alarmins, nitric oxide, adipokines and prostanoids, and their cell surface receptors, contribute to at least the later OA stages to achieve complex modulation of catabolic and anabolic pathways [86]. Intriguingly, while these inflammatory markers are linked to OA, so is expression postnatally of some factors that rely on locomotor mechanics for their expression during joint formation. For instance, expression of CD44 by chondrocytes allows their ECM interaction and contributes to their transduction of homeostatic mechanical signaling to maintain cartilage integrity. Increased CD44 expression in articular cartilage chondrocytes has previously been linked with severity of OA in human joints [94]. Chondrocyte

MEK-ERK-1/2 pathway activation is also intimately linked with elevated levels of MMP and ADAMTS expression that are strongly associated with OA development [2,95]. Perhaps the reliance on locomotor mechanics for raised levels of CD44 and MEK-ERK pathway activation in developing articular cartilage surface cells in the embryonic limb is connected by common mechanoregulatory mechanisms to their similar modification in OA joints. Studies are required to address this because proof of such a connection would significantly alter our strategies for preventing, treating, and even reversing OA pathobiology.

Constitutive activation of MEK-ERK, p38^{MAPK}, and COX-2 pathways exploits signaling centered on NF-κB. Regulation of these pathways— conventionally linked with inflammation—by movement in normal joint development and in OA chondrocytes, suggests that the cellular features linked to OA onset and progression might be the product of local mechanical challenge. Recent findings indicate that the natural OA that arises spontaneously in STR/Ort mice [96] is also intimately linked to an "inflammatory" articular chondrocyte gene signature. Our gene array analyses of articular cartilage chondrocytes from mice at various stages of natural OA progression (pre/at/post OA onset) in these mice indicate that a transcriptional signature dominated by NF-κB pathway signaling is activated even before OA onset and also governs chondrocyte behavior at stages of overt OA pathology [97]. These data together suggest that the NF-κB activation, which predisposes cartilage to OA, is ignited by locomotor mechanics.

NF-κB family members are indeed known to orchestrate mechanical, inflammatory, and stress-activated processes. Two pivotal kinases, IKKα and IKKβ, activate NF-κB dimerization to regulate expression of specific target genes involved in ECM remodeling and vitally, terminal chondrocyte differentiation. Indeed, it has been established that IKKα functions in vivo to control hypertrophic differentiation and MMP activity and thus represents a novel target in OA therapy [98]. Other pathways that may shift the phenotype of normal articular chondrocytes, disrupting homeostasis and causing aberrant proinflammatory and catabolic gene expression, include DDR2, discoidin domain receptor 2 and syndecan-4, as well as the transcription factors NF-κB, C/EBPβ, ETS, Runx2, and HIF-2α. Low-grade "inflammation" detected in the proteome/transcriptome of human OA synovial fluid/membrane consolidates this view by identifying the membrane attack complex-mediated arm of complement, and its colocalization with MMP13 and activated ERK in chondrocytes, as crucial in OA pathogenesis [99]. Discriminating the contributions of these pathways to the initiation and progression of OA is therefore likely pivotal [100]. Indeed, this link we propose between OA chondrocyte changes and those contributing to embryonic movement-dependent joint formation processes is likely best explored by comparing loaded with unloaded cartilage regions, where OA vulnerability likely diverges on the basis of local mechanics.

CONCLUSION

This idea that skeletal development may be "phasic," definable by virtue of reliance on movement, is now supported by many lines of investigation. Herein (as in an elegant recent review) we elaborate on the division of many aspects of musculoskeletal system development into at least two distinct phases [8]. Broadly, the first involves progenitor cell specification, organization, and patterning and is independent of locomotor mechanics and a second that involves differentiation, maturation, and maintenance, which is highly sensitive to movement. Similar relationships seem to be conserved in endochondral ossification and skeletal configuration, joint, muscle, and tendon development and at emerging tendon—bone and muscle—tendon interfaces. It appears, from our examples in the joint and in endochondral ossification that the signaling pathways responsible for the emergence of this mechanical sensitivity, interplay significantly with growth and appear reliant on its metabolic suppression to furnish a potential for mechanics to exert an influence. Finally, we show that at least two of these pathways are recapitulated in the osteoarthritic joint suggesting that their appearance in the disease context is similarly reliant on mechanical loading.

REFERENCES

[1]   Mammoto T, Ingber DE. Mechanical control of tissue and organ development. Development 2010;137(9):1407—20.

[2]   Pitsillides AA, Beier F. Cartilage biology in osteoarthritis—lessons from developmental biology. Nat Rev Rheumatol 2011;7(11):654—63.

[3]   Hamburger V, Balaban M. Observations and experiments on spontaneous rhythmical behavior in the chick embryo. Dev Biol 1963;6:533—45.

[4]   Galilei G. Two new sciences (translated by Drake, S. 1974). Madison, WI: University of Wisconsin Press; 1638.

[5]   Wolff J. The law of bone remodelling (Das Gesetz der Transformation der Knochen, Kirchwald). Berlin: Springer-Verlag; 1892.

[6]   Frost HM. The mechanostat: a proposed pathogenic mechanism of osteoporoses and the bone mass effects of mechanical and nonmechanical agents. Bone Miner 1987;2(2):73—85.

[7]   Lanyon LE. Using functional loading to influence bone mass and architecture: objectives, mechanisms, and relationship with estrogen of the mechanically adaptive process in bone. Bone 1996;18(Suppl. 1):37S—43S.

[8]   Pitsillides AA. Early effects of embryonic movement: 'a shot out of the dark'. J Anat 2006;208(4):417—31.

[9]   Pollard AS, McGonnell IM, Pitsillides AA. Mechanoadaptation of developing limbs: shaking a leg. J Anat 2014;224(6):615—23.

[10] Yang L, Tsang KY, Tang HC, Chan D, Cheah KS. Hypertrophic chondrocytes can become osteoblasts and osteocytes in endochondral bone formation. Proc Natl Acad Sci U S A 2014;111(33):12097—102.

[11] Kronenberg HM. Developmental regulation of the growth plate. Nature 2003;423(6937):332—6.

[12] Chen JC, Goldhamer DJ. Skeletal muscle stem cells. Reprod Biol Endocrinol 2003;1:101.

[13] Yang G, Rothrauff BB, Tuan RS. Tendon and ligament regeneration and repair: clinical relevance and developmental paradigm. Birth Defects Res C Embryo Today 2013;99(3):203—22.

[14] Carey SP, Charest JM, Reinhart-King CA. Forces during cell adhesion and spreading: implications for cellular homeostasis, vol. 4; 2010. p. 29—69.

[15] Legant WR, Miller JS, Blakely BL, Cohen DM, Genin GM, Chen CS. Measurement of mechanical tractions exerted by cells in three-dimensional matrices. Nat Methods 2010;7(12):969—71.

[16] Piccolo S. Developmental biology: mechanics in the embryo. Nature 2013;504(7479):223—5.

[17] Yang Y, Wang K, Gu X, Leong KW. Biophysical regulation of cell behavior—cross talk between substrate stiffness and nanotopography. Engineering 2017;3(1):36—54.

[18] Hoey DA, Chen JC, Jacobs CR. The primary cilium as a novel extracellular sensor in bone. Front Endocrinol (Lausanne) 2012;3:75.

[19] Kathem SH, Mohieldin AM, Nauli SM. The roles of primary cilia in polycystic kidney disease. AIMS Mol Sci 2014;1(1):27—46.

[20] Eyckmans J, Boudou T, Yu X, Chen CS. A hitchhiker's guide to mechanobiology. Dev Cell 2011;21(1):35—47.

[21] Arvind V, Huang AH. Mechanobiology of limb musculoskeletal development. Ann N Y Acad Sci 2017;1409:18—32.

[22] Shawky JH, Davidson LA. Tissue mechanics and adhesion during embryo development. Dev Biol 2015;401(1):152—64.

[23] Marie PJ, Hay E, Modrowski D, Revollo L, Mbalaviele G, Civitelli R. Cadherin-mediated cell-cell adhesion and signaling in the skeleton. Calcif Tissue Int 2014;94(1):46—54.

[24] Sheetz MP. Microtubule motor complexes moving membranous organelles. Cell Struct Funct 1996;21:369—73.

[25] Hatami-Marbini H, Mofrad MRK. Cytoskeletal mechanics and cellular mechanotransduction: a molecular perspective. 2010.

[26] Kanchanawong P, Shtengel G, Pasapera AM, Ramko EB, Davidson MW, Hess HF, et al. Nanoscale architecture of integrin-based cell adhesions. Nature 2010;468(7323):580—4.

[27] Critchley R. Cytoskeletal proteins talin and vinculin in integrin-mediated adhesion. Biochem Soc Trans 2004;32:831—6.

[28] Humphries JD, Wang P, Streuli C, Geiger B, Humphries MJ, Ballestrem C. Vinculin controls focal adhesion formation by direct interactions with talin and actin. J Cell Biol 2007;179(5):1043—57.

[29] Sit ST, Manser E. Rho GTPases and their role in organizing the actin cytoskeleton. J Cell Sci 2011;124(Pt 5):679—83.

[30] Sun Z, Guo SS, Fassler R. Integrin-mediated mechanotransduction. J Cell Biol 2016;215(4):445—56.

[31] Wolpert L. Limb patterning: reports of model's death exaggerated. Curr Biol 2002;12:R628—30.

[32] Javaheri B, Caetano-Silva S, Kanakis I, Bou Gharios G, Pitsillides A. The chondro-osseous continuum: is it possible to unlock the potential assigned within? Front Bioeng Biotechnol 2018;6(28).

[33] Tsang KY, Chan D, Cheah KS. Fate of growth plate hypertrophic chondrocytes: death or lineage extension? Dev Growth Differ 2015;57(2):179—92.

[34] Little DG, Ramachandran M, Schindeler A. The anabolic and catabolic responses in bone repair. J Bone Joint Surg 2007;89(B):425—33.

[35] Bahney CS, Hu DP, Taylor AJ, Ferro F, Britz HM, Hallgrimsson B, et al. Stem cell-derived endochondral cartilage stimulates bone healing by tissue transformation. J Bone Miner Res 2014;29(5):1269−82.

[36] Song L, Tuan RS. Transdifferentiation potential of human mesenchymal stem cells derived from bone marrow. FASEB J 2004;18(9):980−2.

[37] Shah M, Gburcik V, Reilly P, Sankey RA, Emery RJ, Clarkin CE, et al. Local origins impart conserved bone type-related differences in human osteoblast behaviour. Eur Cell Mater 2015;29:155−75. discussion 75−76.

[38] Lavigne P, Benderdour M, Lajeunesse D, Reboul P, Shi Q, Pelletier JP, et al. Subchondral and trabecular bone metabolism regulation in canine experimental knee osteoarthritis. Osteoarthritis Cartilage 2005;13(4):310−7.

[39] Rawlinson SC, McKay IJ, Ghuman M, Wellmann C, Ryan P, Prajaneh S, et al. Adult rat bones maintain distinct regionalized expression of markers associated with their development. PLoS ONE 2009;4(12):e8358.

[40] Candeliere GA, Liu F, Aubin JE. Individual osteoblasts in the developing calvaria express different gene repertoires. Bone 2001;28(4):351−61.

[41] Pollard AS, Pitsillides AA. Mechanobiology of embryonic skeletal development. Mechanobiology: Exploitation for Medical Benefit 2017:101−14.

[42] Verbruggen SW, Kainz B, Shelmerdine SC, Hajnal JV, Rutherford MA, Arthurs OJ, et al. Stresses and strains on the human fetal skeleton during development. J Royal Soc Interface 2018;15(138).

[43] Martin P. Tissue patterning in the developing mouse limb. Int J Dev Biol 1990;34(3):323.

[44] Suzue T. Movements of mouse fetuses in early stages of neural development studied in vitro. Neurosci Lett 1996;218(2):131−4.

[45] Tidball JG. Mechanical signal transduction in skeletal muscle growth and adaptation. J appl physiol 2005;98(5):1900−8.

[46] Felsenthal N, Zelzer E. Mechanical regulation of musculoskeletal system development. Development 2017;144(23):4271−83.

[47] Murray PDF. An experimental study of the development of the limbs of the chick. Proc Linn Soc N S W 1926;51:179−263.

[48] Murray PDF, Selby D. Intrinsic and extrinsic factors in the primary development of the skeleton. Dev Gene Evol 1930;122(3):629−62.

[49] Fell HB, Canti RG. Experiments on the development in vitro of the avian knee-joint. Proceedings of the royal society of London series B. Biol Sci 1934;116(799):316−51.

[50] Chandaria VV, McGinty J, Nowlan NC. Characterising the effects of in vitro mechanical stimulation on morphogenesis of developing limb explants. J Biomech 2016;49(15):3635−42.

[51] Drachman DBSL. The role of movement in embryonic joint development. Dev Biol 1966;14:401−20.

[52] Pollard AS, Charlton BG, Hutchinson JR, Gustafsson T, McGonnell IM, Timmons JA, et al. Limb proportions show developmental plasticity in response to embryo movement. Sci Rep 2017;7:41926.

[53] Rolfe RA, Kenny EM, Cormican P, Murphy P. Transcriptome analysis of the mouse E14.5 (TS23) developing humerus and differential expression in muscle-less mutant embryos lacking mechanical stimulation. Genom Data 2014;2:32−6.

[54] Nowlan NC, Dumas G, Tajbakhsh S, Prendergast PJ, Murphy P. Biophysical stimuli induced by passive movements compensate for lack of skeletal muscle during embryonic skeletogenesis. Biomechanics Model Mechanobiol 2012;11(1−2):207−19.

[55] Nowlan NC, Bourdon C, Dumas G, Tajbakhsh S, Prendergast PJ, Murphy P. Developing bones are differentially affected by compromised skeletal muscle formation. Bone 2010;46(5):1275–85.

[56] Roddy KA, Prendergast PJ, Murphy P. Mechanical influences on morphogenesis of the knee joint revealed through morphological, molecular and computational analysis of immobilised embryos. PLoS ONE 2011;6(2):e17526.

[57] Pitsillides AA, Ashhurst DE. A critical evaluation of specific aspects of joint development. Dev Dyn 2008;237(9):2284–94.

[58] Kahn J, Shwartz Y, Blitz E, Krief S, Sharir A, Breitel DA, et al. Muscle contraction is necessary to maintain joint progenitor cell fate. Dev Cell 2009;16(5):734–43.

[59] Bastow ER, Lamb KJ, Lewthwaite JC, Osborne AC, Kavanagh E, Wheeler-Jones CP, et al. Selective activation of the MEK-ERK pathway is regulated by mechanical stimuli in forming joints and promotes pericellular matrix formation. J Biol Chem 2005;280(12):11749–58.

[60] Kavanagh E, Church VL, Osborne AC, Lamb KJ, Archer CW, Francis-West PH, et al. Differential regulation of GDF-5 and FGF-2/4 by immobilisation in ovo exposes distinct roles in joint formation. Dev Dynam 2006;235(3):826–34.

[61] Pitsillides AA, Archer CW, Prehm P, Bayliss MT, Edwards JC. Alterations in hyaluronan synthesis during developing joint cavitation. J Histochem Cytochem 1995;43(3):263–73.

[62] Edwards JC, Wilkinson LS, Jones HM, Soothill P, Henderson KJ, Worrall JG, et al. The formation of human synovial joint cavities: a possible role for hyaluronan and CD44 in altered interzone cohesion. J Anat 1994;185(Pt 2):355–67.

[63] Lewthwaite JC, Bastow ER, Lamb KJ, Blenis J, Wheeler-Jones CP, Pitsillides AA. A specific mechanomodulatory role for p38 MAPK in embryonic joint articular surface cell MEK-ERK pathway regulation. J Biol Chem 2006;281(16):11011–8.

[64] Nowlan NC, Sharpe J. Joint shape morphogenesis precedes cavitation of the developing hip joint. J Anat 2014;224(4):482–9.

[65] Saha A, Rolfe R, Carroll S, Kelly DJ, Murphy P. Chondrogenesis of embryonic limb bud cells in micromass culture progresses rapidly to hypertrophy and is modulated by hydrostatic pressure. Cell Tissue Res 2017;368(1):47–59.

[66] Roddy KA, Kelly GM, van Es MH, Murphy P, Prendergast PJ. Dynamic patterns of mechanical stimulation co-localise with growth and cell proliferation during morphogenesis in the avian embryonic knee joint. J Biomech 2011;44(1):143–9.

[67] Shwartz Y, Farkas Z, Stern T, Aszodi A, Zelzer E. Muscle contraction controls skeletal morphogenesis through regulation of chondrocyte convergent extension. Dev Biol 2012;370(1):154–63.

[68] Blitz E, Viukov S, Sharir A, Shwartz Y, Galloway JL, Pryce BA, et al. Bone ridge patterning during musculoskeletal assembly is mediated through SCX regulation of Bmp4 at the tendon-skeleton junction. Dev Cell 2009;17(6):861–73.

[69] Pollard A, Boyd S, McGonnell I, Pitsillides A. The role of embryo movement in the development of the furcula. J Anat 2017;230(3):435–43.

[70] Rolfe RA, Nowlan NC, Kenny EM, Cormican P, Morris DW, Prendergast PJ, et al. Identification of mechanosensitive genes during skeletal development: alteration of genes associated with cytoskeletal rearrangement and cell signalling pathways. BMC Genomics 2014;15(1):48.

[71] Gao Y, Gartenhaus RB, Lapidus RG, Hussain A, Zhang Y, Wang X, et al. Differential IKK/NF-kappaB activity is mediated by TSC2 through mTORC1 in PTEN-Null prostate cancer and tuberous sclerosis complex tumor cells. Mol Cancer Res 2015;13(12):1602–14.

[72] Dan HC, Antonia RJ, Baldwin AS. PI3K/Akt promotes feedforward mTORC2 activation through IKKalpha. Oncotarget 2016;7(16):21064−75.

[73] Li Z, Yang Z, Passaniti A, Lapidus RG, Liu X, Cullen KJ, et al. A positive feedback loop involving EGFR/Akt/mTORC1 and IKK/NF-kB regulates head and neck squamous cell carcinoma proliferation. Oncotarget 2016;7(22):31892−906.

[74] Jones RG, Pearce EJ. MenTORing immunity: mTOR signaling in the development and function of tissue-resident immune cells. Immunity 2017;46(5):730−42.

[75] Schreiber KH, Ortiz D, Academia EC, Anies AC, Liao CY, Kennedy BK. Rapamycin-mediated mTORC2 inhibition is determined by the relative expression of FK506-binding proteins. Aging Cell 2015;14(2):265−73.

[76] Rawlinson SC, Murray DH, Mosley JR, Wright CD, Bredl JC, Saxon LK, et al. Genetic selection for fast growth generates bone architecture characterised by enhanced periosteal expansion and limited consolidation of the cortices but a diminution in the early responses to mechanical loading. Bone 2009;45(2):357−66.

[77] Fuerer C, Nusse R, Ten Berge D. Wnt signalling in development and disease. Max Delbruck Center for molecular medicine meeting on Wnt signaling in development and disease. EMBO Rep 2008;9(2):134−8.

[78] Brunt LH, Begg K, Kague E, Cross S, Hammond CL. Wnt signalling controls the response to mechanical loading during zebrafish joint development. Development 2017;144(15):2798−809.

[79] Villar J, Cabrera NE, Valladares F, Casula M, Flores C, Blanch L, et al. Activation of the Wnt/beta-catenin signaling pathway by mechanical ventilation is associated with ventilator-induced pulmonary fibrosis in healthy lungs. PLoS ONE 2011;6(9):e23914.

[80] Buckwalter J, Mankin H. Articular cartilage: degeneration and osteoarthritis, repair, regeneration, and transplantation. Instr Course Lect 1998;47:487−504.

[81] Sanchez-Adams J, Leddy HA, McNulty AL, O'Conor CJ, Guilak F. The mechanobiology of articular cartilage: bearing the burden of osteoarthritis. Curr Rheumatol Rep 2014;16(10):451.

[82] Nam J, Aguda BD, Rath B, Agarwal S. Biomechanical thresholds regulate inflammation through the NF-κB pathway: experiments and modeling. PLoS ONE 2009;4(4):e5262.

[83] Leong DJ, Li YH, Gu XI, Sun L, Zhou Z, Nasser P, et al. Physiological loading of joints prevents cartilage degradation through CITED2. FASEB J 2011;25(1):182−91.

[84] Yokota H, Goldring MB, Sun HB. CITED2-mediated regulation of MMP-1 and MMP-13 in human chondrocytes under flow shear. J Biol Chem 2003;278(47):47275−80.

[85] Sun HB. Mechanical loading, cartilage degradation, and arthritis. Ann N Y Acad Sci 2010;1211(1):37−50.

[86] Houard X, Goldring MB, Berenbaum F. Homeostatic mechanisms in articular cartilage and role of inflammation in osteoarthritis. Curr Rheumatol Rep 2013;15(11):375.

[87] Guilak F. Biomechanical factors in osteoarthritis. Best Pract Res Clin Rheumatol 2011;25(6):815−23.

[88] Radin EL, Martin RB, Burr DB, Caterson B, Boyd RD, Goodwin C. Effects of mechanical loading on the tissues of the rabbit knee. J Orthop Res 1984;2(3):221−34.

[89] Kuyinu EL, Narayanan G, Nair LS, Laurencin CT. Animal models of osteoarthritis: classification, update, and measurement of outcomes. J Orthop Surg Res 2016;11(1):19.

[90] O'Conor CJ, Leddy HA, Benefield HC, Liedtke WB, Guilak F. TRPV4-mediated mechanotransduction regulates the metabolic response of chondrocytes to dynamic loading. Proc Natl Acad Sci U S A 2014;111(4):1316−21.

[91]   Staines KA, Pollard AS, McGonnell IM, Farquharson C, Pitsillides AA. Cartilage to bone transitions in health and disease. J Endocrinol 2013;219(1):R1−12.

[92]   Ito MM, Kida MY. Morphological and biochemical re-evaluation of the process of cavitation in the rat knee joint: cellular and cell strata alterations in the interzone. J Anat 2000;197(4):659−79.

[93]   Staines KA, Madi K, Mirczuk SM, Parker S, Burleigh A, Poulet B, et al. Endochondral growth defect and deployment of transient chondrocyte behaviors underlie osteoarthritis onset in a natural murine model. Arthritis Rheumatol 2016;68(4):880−91.

[94]   Zhang F-J, Luo W, Gao S-G, Su D-Z, Li Y-S, Zeng C, et al. Expression of CD44 in articular cartilage is associated with disease severity in knee osteoarthritis. Mod Rheumatol 2013;23(6):1186−91.

[95]   Appleton CTG, Usmani SE, Mort JS, Beier F. Rho/ROCK and MEK/ERK activation by transforming growth factor-α induces articular cartilage degradation. Lab Invest 2010;90(1):20−30.

[96]   Staines KA, Poulet B, Wentworth DN, Pitsillides AA. The STR/ort mouse model of spontaneous osteoarthritis—an update. Osteoarthritis Cartilage 2017;25(6):802−8.

[97]   Poulet B, Ulici V, Stone TC, Pead M, Gburcik V, Constantinou E, et al. Time-series transcriptional profiling yields new perspectives on susceptibility to murine osteoarthritis. Arthritis Rheum 2012;64(10):3256−66.

[98]   Olivotto E, Otero M, Marcu KB, Goldring MB. Pathophysiology of osteoarthritis: canonical NF-kappaB/IKKbeta-dependent and kinase-independent effects of IKKalpha in cartilage degradation and chondrocyte differentiation. RMD Open 2015;1(Suppl. 1):e000061.

[99]   Wang M, Shen J, Jin H, Im HJ, Sandy J, Chen D. Recent progress in understanding molecular mechanisms of cartilage degeneration during osteoarthritis. Ann N Y Acad Sci 2011;1240:61−9.

[100]   Goldring MB, Otero M. Inflammation in osteoarthritis. Curr Opin Rheumatol 2011;23(5):471−8.

Chapter 3

Development, Tissue Engineering, and Orthopedic Diseases

Linda J. Sandell[1,2]

[1]*Washington University School of Medicine, St. Louis, MO, United States;* [2]*Washington University School of Engineering and Applied Science, St. Louis, MO, United States*

This chapter will summarize the field as seen from the viewpoint of a biologist and cite some examples of challenges in the field and a perspective on where the field is moving. Secondly, it will provide some examples of how developmental principles have been used to understand and potentially treat orthopedic diseases in a broader view. An excellent review from both the tissue engineering and biology sides was published in 2017 [1]. Classically, the goal of tissue engineering was to treat orthopedic disease by making a tissue replacement as good as or better than the biologic one, often using principles derived from developmental biology. The tissue could be manufactured from components such as plastics or ceramics or made biologically by cells such as stem cells differentiated into tissues or mature cells 3D-printed into the shape of an organ. Interestingly, few of these tissue-engineered products are available for clinical use. There is certainly no lack of ingenuity and creativity in the tissue engineering field; however, making tissues as good as the original, keeping them in place and functioning over time, and overcoming regulatory hurdles has proven a higher bar than expected. However, there are some successes to point to. In 2011, scientists at the Karolinska Institute were able to implant an artificial trachea made from a synthetic scaffold seeded with patient's stem cells [2]. This was based on an established proof of concept by Langer where they grew an outer ear from bovine chondrocytes and an ear-shaped scaffold on the back of a mouse [3]. Cancedda and colleagues [4,5] used bone marrow—derived stem cells in an osteoconductive scaffold to repair a large segmental bone defect. The reconstruction of a human mandible was reported using a bone-muscle flap in vivo where the patient served as his/her own bioreactor [6]. These early trials relied on the differentiated cells (i.e., chondrocytes making more chondrocytes), porous scaffolds that provided

53

a substrate for cell differentiation from adult stem cells (osteoconductive or osteoinductive scaffolds), or induction of unknown tissue differentiation events. Most tissue engineering was impeded by lack of knowledge of the native developmental processes. Now much more attention is paid to understanding the developmental processes involved in differentiation of cells and tissues, renewal or repair of mature tissues, stem cell generation, stem cell reservoirs, and stem cell differentiation to specific tissues. The focus on developmental processes has also had an impact on potential treatment of disease. For example, one theory of the pathogenesis of osteoarthritis (OA) is that the cells in articular cartilage continue to differentiate into hypertrophic cells, a characteristic of growth plate hypertrophic chondrocytes. This theory has directed studies geared toward inhibition of this growth plate hypertrophic phenotype, maintaining the cells in a mature chondrocyte phenotype [7].

Another hurdle that must be considered is that particularly in a disease situation, the tissue milieu requiring replacement may be very different from the developmental milieu. In development, you have many tissues interacting to help form the new tissue. In the injury situation, there may be detrimental forces that act against developing normal healthy tissues: such detrimental forces could be local inflammatory components, altered biomechanics, or systemic factors.

The development of almost all tissue requires specific sets of genes from patterning genes and gene families (e.g., HOX, fibroblast growth factor [FGF], Wnt) to differentiating genes (e.g. TGF-β, BMPs, Hedgehogs) to specific transcription factors that in combination with cofactors control expression of the differentiated cell phenotype (e.g., SOX9, MYOD, P-PARPgamma) and repress unwanted phenotypes. I will touch on a few paradigms that are relevant to the musculoskeletal field.

DEVELOPMENTAL PARADIGMS

Insights From Amphibian Limb Regeneration

The Mexican Axolotl is one of the few tetrapod species that is capable of regenerating complete skeletal elements in injured adult limbs. The regeneration of a limb requires signaling feedback loops between mesenchymal and epithelial tissue layers resulting in the induction of organ fields with temporal and spatial restriction. The mesenchymal component is responsible for imposing regional (positional) specificity: for example if foot mesenchyme is juxtaposed with wing epithelium in chicken embryos, scales and claw structures form [8]. If mature dermis is grafted from the tail to the forelimb of an adult salamander and a regenerative response is induced, the result is formation of tail-like structures [9]. In regeneration of the adult amphibian mesenchymal cells that provide positional cures originate from the mature connective tissues. Interestingly, the amputation of boneless amphibian forelimbs results in the regeneration of the skeletal elements distal to the amputation plane revealing that nonskeletal cells can undergo metaplasia to regenerate skeletal tissue in the limb [10]. If differentiated skeletal elements are present, they do not contribute to the regenerated skeletal elements [11], possibly due to the dense connective tissue surrounding them unless freed by native enzymes.

Tanaka and colleagues as well as many other laboratories [12] have worked out the developmental cell and molecular biology that controls regeneration in these species (Fig. 3.1). Surprisingly, the cells of the cut limb including muscle, bone, nerve, tendon, cartilage, etc., "convert" to a mass of limb-bud—like mesenchymal blastema cells that has the competence to produce the correct diversity of tissue types in the right place at the right time. In contrast to what was expected, these cells undergo a reprogramming process that involves losing the mature phenotype, proliferating, and differentiating anew. Tanaka's work has shown that limb amputation initiates a wound-related epidermis activation that secretes factors that induce the first cell cycle. Recently, the Tanaka group has also demonstrated that blood-clotting proteases cleave and activate blood-derived BMPs to promote BMP signaling—dependent cell cycle reentry for myofiber dedifferentiation [13]. Other tissues such as injured nerve and macrophages provide additional factors such as FGFs and BMPs that sustain proliferation. During this blastemal induction process, connective tissue cells become proliferative and also start to express position-specific factors such as MEIS in the upper arm and FGF8 or SHH in other regions. This sets up a cross-inductive positive feedback loop sustaining growth and differentiation. Understanding and harnessing these regenerative processes will provide sophisticated approaches to repairing or engineering tissues and provide potential treatments for orthopedic diseases.

Wnts and Skeletal Development

How can knowledge of developmental processes generating tissue be applied to tissue engineering and treatment of orthopedic diseases? The answer lies in at least two areas of research. First, to replace orthopedic structures, tissue engineers could try to recapitulate development in a bioreactor to make the tissue from undifferentiated or partially differentiated cells. This would require developing an artificial milieu where the "stem cells" could develop into mature tissue, say bone, cartilage, or tendon. This has proven to be very difficult but has been greatly aided by the development of biocompatible scaffolds to provide a template and substrate for the cells, as we saw above in cartilage [14]. The second way that developmental processes could be used is to harness the regenerative potential of the tissue itself, recruiting new cells to produce the tissue in situ. Interestingly, our knowledge of developmental processes and recruitment of stem cells, although extensive, is not complete. However, we can cite a few systems where potentially useful information exists.

The first paradigm is the role of Wnt signaling pathway in skeletal development [15,16]. The Wnt signaling cascades have essential roles in

(A) Canonical signalling via β-catenin (inhibition of chondrogenesis)

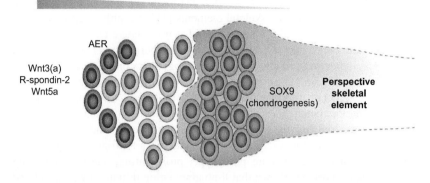

Noncanonical signalling via NFAT (longitudinal growth)

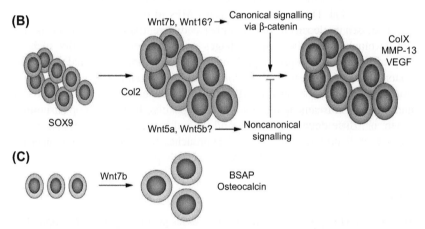

FIGURE 3.1 The varied roles of Wnts in skeletal development. (A) At the early phase of skeletogenesis, expression of Wnts, such as Wnt3 and Wnt5a, at the AER keeps the progenitor cells (red cells) in a proliferative state and prevents differentiation into chondrocytes (yellow cells). Absence of sufficient Wnt signaling in cells further along the concentration gradient allows them to express the chondrogenic transcription factor SOX9 (green cells). Condensations of these chondrocytes occur distant from the Wnt source and form the cartilage template for the prospective skeletal elements. Canonical, β-catenin-dependent Wnt signaling by Wnt3a, with coactivation by R-spondin-2, has been linked to inhibition of chondrogenesis, whereas noncanonical signaling, induced by Wnt5a and involving NFAT, has been associated with proximal–distal growth. (B) During endochondral bone formation, Wnt signaling is critical in the final differentiation steps toward chondrocytes hypertrophy when cells start expression ColX, MMP13, and VEGF (range cells); canonical Wnt signaling stimulates this process. Chondrocyte differentiation might be triggered by Wnt7b and Wnt16 that are expressed in developing chondrocytes, which are positive for Col2 (large green cells). By contrast, noncanonical cascades triggered by Wnt5a or Wnt5b might inhibit chondrocyte hypertrophy. (C) During direct bone formation, Wnts stimulate the differentiation from osteoblast precursors toward mature osteoblasts, which produce BSAP and osteocalcin. Wnt7b has been specifically associated with this process. *AER*, apical ectodermal ridge; *BSAP*, bone-specific alkaline phosphatase; *Col2*, collagen type II; *ColX*, collagen type X; *MMP13*, matrix metalloproteinase 13; *NFAT*, nuclear factor of activated cells; *VEGF*, vascular endothelial growth factor.

development, growth, and homeostasis of joints and the skeleton. Wnts are extracellular signaling molecules that signal the cell to proliferate or differentiate. There are 19 Wnt molecules and a large variety of receptors, inhibitors, and activators. They signal through at least three pathways; two are canonical signaling via β-catenin, and one is noncanonical, by definition, through a calcium signaling intermediate. In addition, like all morphogens in developmental biology, there is a concentration gradient in the developing tissue that offers spatial and temporal control of cellular differentiation patterns. Wnts are posttranslationally modified during synthesis in the endoplasmic reticulum by the addition of a palmitoyl lipid molecule and traffic through the Golgi apparatus guided by another membrane protein Wntless. The Wnt system of signals, receptors, and inhibitors has a very wide variety of effects on many different cell differentiation systems from the skeletal system to the cardiovascular system. Even in the skeletal system, certain Wnts can induce bone and cartilage hypertrophy (Fig. 3.2), and some come to the rescue upon injury [17]. This article from the Dell'Accio laboratory demonstrates that the processes that occur in response to an injury are not necessarily the same as the developmental process. They have shown that a new Wnt—Wnt16—is responsive to injury in cartilage and acts through the Wnt pathway to stimulate restoration of the injured superficial zone cells of articular cartilage. This is an important paradigm for the use of developmental biology in tissue engineering and for disease treatment: different signal pathways will be involved in every aspect of tissue development, homeostasis, and injury—so there is no "one-size-fits-all" approach.

Hedgehog and Tendon—Bone Development

A different developmental paradigm is being used in a more complex tissue engineering approach, forming tissues that connect one tissue to another, such as myotendinous junctions that connect muscle to tendon and the enthesis that connects tendon to bone. This tissue is often injured likely because of stress amplification at the enthesis due to the stiffness mismatch between relatively compliant tendon and stiff bone [18]. In addition, matrix synthesis of mineralized fibrocartilage is extremely low in the enthesis, making it more difficult for the enthesis to heal effectively in the adult. Building on initial developmental biology studies of Galatz and colleagues [19], it was recently shown that hedgehog signaling is crucial for proper enthesis cell differentiation and maturation [20]. Hedgehog gain-of-function in tendon cells leads to the expression of fibrocartilaginous matrix proteins within the tendon midsubstance, whereas loss-of-function leads to reductions of these proteins within the enthesis and reduced mineralization. Although Indian hedgehog (IHH) expression has been shown at neonatal time points, recent evidence indicates that IHH is involved in the transition from unmineralized to mineralized fibrochondrocytes [21]. Rowe and colleagues used various knockout mice

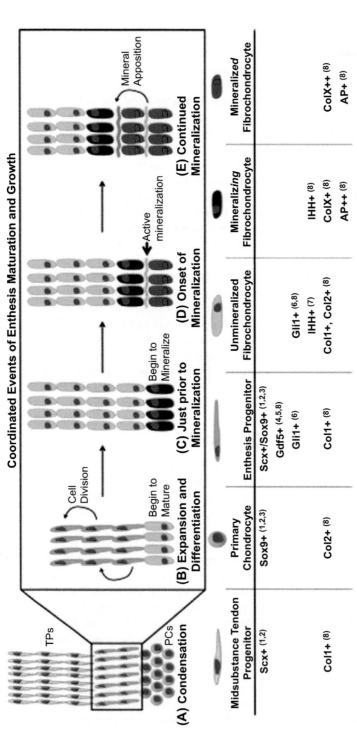

FIGURE 3.2 Model for cellular expansion and maturation during enthesis growth and development. Based on current work and the work of others, the authors suggest five stages of maturation within the enthesis during growth and development, with cell markers for each stage. (A) At condensation, there are three distinct progenitor populations at the region where tendons attach to cartilage: midsubstance tendon progenitors(TPs), enthesis progenitors, and primary chondrocytes (PCs) of the epiphysis. The enthesis progenitors give rise to unmineralized fibrochondrocytes, mineralizing fibrochondrocytes, and mineralized fibrochondrocytes of the more mature enthesis. (B) After condensation, these progenitors clonally expand and cells at the base begin to mature into unmineralized fibrochondrocytes. (C) Before the onset of mineralization, the mineralizing fibrochondrocytes are the base express AP and ColX as they begin to mineralize. (D) This leads to the active mineralization stage where mineralization can be detected. (E) Mineral apposition continues and cells that are embedded within the mineralized fibrocartilage express ColX. These sequential events lead to a maturation gradient with the most mature cells at the bottom and immature cells at the top (closer to mid-substances). *Taken from Lories et al., Fig. 3, Nat Rev Rheumatol 2013.*

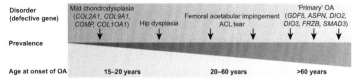

| Disorder (defective gene) | Mild chondrodysplasia (COL2A1, COL9A1, COMP, COL10A1) | Hip dysplasia | Femoral acetabular impingement ACL tear | | 'Primary' OA (GDF5, ASPN, DIO2, DIO3, FRZB, SMAD3) |

FIGURE 3.3 OA can be viewed as a continuum in terms of prevalence, age of onset, and types of genes that are involved in the etiology. Obvious changes in extracellular matrix function owing to mutations in matrix proteins can result in chondrodysplasias with associated OA. Early-onset OA can also result from suboptimal hip alignment, such as dysplasia or impingement. Even more subtle alterations conferred on the joint by changes in growth factor signaling required for morphogenesis during development might underlie "primary" OA. Linking early-onset OA with late-onset disease provides a new conceptual framework to understand the pathogenesis of OA. Although posttraumatic OA is not considered to have a genetic basis, many properties of the joint structure, such as suboptimal tissue composition or architecture, could increase susceptibility to OA after trauma, or even susceptibility to injury. *ACL*, anterior cruciate ligament; *OA*, osteoarthritis. *Fig. 2 from Dyment et al.*

to develop a map of the coordinated sequence of events in enthesis growth and development, including expansion of the progenitor pool, spatiotemporal changes in matrix synthesis, cell maturation from unmineralized to mineralized fibrochondrocytes, and signaling pathways that regulated these processes (Fig. 3.3). These studies indicate that successful repair may require either tissue-resident or implanted progenitors to anchor collagen to bone, synthesis of a fibrocartilaginous extracellular matrix (e.g., collagens and proteoglycans) and finally mineralization of the fibrocartilage to produce a solid anchor. Therefore the principles described below using multistep tissue engineering processes with different cells, different stimulating factors, and different bioreactor conditions will be implemented.

Stem Cells in Tissue Engineering

The use of stem cells either endogenous isolated from various adult tissues, i.e., bone marrow, synovium, adipose, or the use of mature cells that have been dedifferentiated into iPS (induced pluripotent stem—see below) cells has been very attractive in the pursuit of differentiated tissues. It is now thought that all stem cells are derived from blood vessel pericytes cells that are particularly rich in bone marrow and adipose [22]. The development of a tissue requires the classical mechanism of cell differentiation, whereby a committed progenitor cell proceeds down a well-ordered pathway of an immature cell type to a more mature cell type. The mature cell carries out the differentiated function for a specific time: in the case of blood cells, the time is quite short (days), in the case of cartilage or nerve cells, the time is very long (sometimes a lifetime). Tissues with short-lived cells that line blood or bone have elaborate mechanisms of replacing themselves. In the first iteration of tissue engineering, endogenous progenitors or progenitors expanded ex vivo were thought to be able to repair damaged or impaired tissues or organs. Fairly, simple media were used to expand and differentiate stem cells into chondrocytes (chondrogenic media), bone (osteogenic media), and other tissues. These initial attempts generally used a proliferation factor such as FGF2, and induction factor, TGF-β or BMPs, and other conditions that allowed for high levels of protein synthesis. In the cartilage field, most of these experiments, although they could demonstrate nice chondrogenic-like phenotype in cell culture, when implanted into animals, eventually became fibrous or proceeded to form bone, recapitulating endochondral ossification. Developmental biology is certainly now suggesting more sophisticated approaches where different factors are delivered at different times in multilayered 3D tissue differentiation bioreactors (for an extensive review, see Ref. [23]).

A second method has gained more prominence recently and that is to harness the process of epithelial-to-mesenchymal transition (EMT) or endothelial-to-mesenchymal transition (EndMT) to provide progenitors for repair and regeneration [24]. EMT occurs in biological processes such as gastrulation, cardiogenesis, and fibrosis. EndMT is involved in development and tissue fibrosis and recently has been implicated in musculoskeletal biology and pathology. For tissue engineering and regenerative medicine, these processes may aid in finding new methods for reducing fibrosis and identifying novel plastic progenitor populations for tissue repair. We have mentioned reprogramming earlier in relation to regeneration in newts or salamanders. However, to maintain tissue mass and function, differentiated mature cells will constantly be resupplied from a committed progenitor cell pool, thereby generating natural tissue regeneration. For example, the iPS cells mentioned earlier, the committed progenitor is asked to participate in processes to which they would not normally contribute, that is, cellular reprogramming, where an adult cell is induced to take on a pluripotent state. Reprogramming involves resetting many epigenetic markers, including microRNAs (miRNAs) and histone modifications. This occurs in iPS cell generation with the transduction of transfection of Yamanaka transcription factors, such as OCT3/4, SOX2, KLF4, and mYC [25]. After reprogramming, the cells can be stimulated to undergo differentiation into various types of somatic cell.

Two examples of EndMT in musculoskeletal tissues are the capacity of endothelial cells to contribute to muscle regeneration, and heterotopic bone formation. The first is the activation of muscle quiescent mesenchymal mononuclear progenitors called satellite cells. After injury, these cells are activated, differentiate, and fuse together to repair muscle damage. As these cells are difficult to culture and become depleted with age, for tissue engineering, another cell type with endothelial qualities, the mesoangioblast, may be a good choice. These cells are associated with the walls of large vessels and can differentiate into muscle cells [26]. The second example involves the role of endogenous endothelial cells in bone formation and was discovered in context of the genetic disease, fibrodysplasia ossificans progressiva. In this disease, muscle becomes ossified at the slightest stress [27] and is caused by constitutive activation of ALK2, a member of the TGF-β superfamily. This group has shown that endothelial cells of the Tie2 lineage and not myogenic or mesenchymal cells of the MyoD lineage contribute to heterotopic bone formation via EndMT. This also can occur in cell lines where the cells can become multipotent on activation [28]. Although this finding is controversial, the possibility of reprogramming by EndMT is intriguing.

DEVELOPMENT AND ORTHOPEDIC DISEASES: HOW ARE DEVELOPMENTAL PARADIGMS USED TO UNDERSTAND AND POTENTIALLY TREAT ORTHOPEDIC DISEASES?

Osteoarthritis and Osteoporosis

The most prominent orthopedic diseases are OA and osteoporosis (OP). In both of these broad diseases, developmental paradigms have been used to explain the pathobiology of the disease process and even used to treat the disease. However, the targets of treatment in each disease are very different. In OA, the aim is to slow or inhibit cartilage degeneration with the potential of restoring cartilage and joint integrity. In OP where bones become fragile and subject to fracture, the goal is to decrease bone loss and increase bone deposition.

In 2012, I published a review on development and OA relating the developmental aspects of OA as revealed by genetic studies to synovial joint development [29]. Mutations in genes associated with rare skeletal malformation syndromes predispose their carriers to the early development of OA, such as COL2A1, COL9A1, COMP, and COL10A1 [30]. Polymorphisms in other susceptibility genes are associated with OA without causing overt skeletal abnormalities [29]. These associations include single nucleotide polymorphisms in genes known for their roles in skeletal development, such as members of the TGF-β, asporin, bone morphogenetic protein (BMP, SMAD3, GDF5), and Wnt (FRZB, DOT1L) signaling pathways (Fig. 4 from Sandell, 2012). These genes could also be associated with the ability of cartilage to repair, a process that recapitulates many aspects of the developmental process. Among these genes, the genetic association with GDF5 seems to be the most consistent and reproducible. GDF5 is involved in synovial joint development, maintenance, and repair where GDF5-expressing progenitor cells populate the interzone domain during synovial joint development. Recent evidence indicates that GDF5 may play an instructive role in many joint tissues depending on when during development they are recruited into the joint [31]. In OA, there are specific polymorphisms in the GDF5 gene causing an imbalance in the type of allele expressed. However, it has been discovered that the expression of the allelic imbalance of GDF5 is regulated at the level of methylation of the gene, and methylation is an epigenetic target, which at this time cannot be specifically targeted to the GDF5 gene by known agents [32]. Interestingly, a new method of targeting genes, using CRISPR-cas9, may allow for the regulation of methylation at specific sites.

Wnts play a variety of important roles in both bone and joint development and maintenance and repair. Wnt 3A via canonical signaling (through β-catenin) stimulates osteogenesis and inhibits chondrogenesis [15]. Wnt 5A stimulates the transition from growth plate—resting chondrocytes to hypertrophic chondrocytes. The effect on OA appears to be confusing: generalized inhibition of Wnt increases experimental OA [33,34], whereas overexpression of Wnt3A also increases experimental OA [35].

Wnt signaling may be a good target for treatment of OA as first suggested by van den Berg and colleagues [36], where they demonstrated that overexpression of the Wnt antagonist WISP1 induced expression of matrix metalloproteinases and aggrecanase, hallmarks of the OA cartilage-destructive phenotype. All will agree that maintenance of a balanced and appropriate expression of Wnts and Wnt regulators is critical. When that balance is disturbed as in OA, cartilage destruction and bone changes can result.

Insights into both joint disease, as seen in OA, and bone fragility, as seen in OP, may be obtained from the molecular signaling pathways that together coordinate the development of the entire synovial joint. To form the joint, progenitor cells of different tissue types express genes that belong to the same principal signaling pathways. For example, GDFs, Wnts, and BMP inhibitors are all expressed in the presumptive joint compartments. Second, the regulatory relationships between the principal signaling pathways are reiteratively used during development of the joint. A signaling axis of Wnt-Fgf from within the joint-forming compartment opposes chondrogenic signals from Ihh-Bmp to promote specification and differentiation of specialized tissues at the joint—bone interface. In synovial joints, the balance between Wnt and Ihh signaling promotes formation of intermediate tissues, such as hyaline cartilage and fibrous connective tissues by limiting chondrogenic potential. Third, the same signal pathway is used for the same function in distinct joint types. In the synovial joint, progenitor cells are sourced from outside and move into the joint space. In the long term, it is possible that these developmental paradigms may be harnessed to improve bone while they also inhibit cartilage degeneration.

Sclerostin (SOST) is another member of the Wnt pathway. It is produced by both osteoblasts and chondrocytes and is an endogenous inhibitor of Wnt signaling, binding to the outside of the cell along with the LRP5/6 and Frizzled coreceptors on the cell surface and inhibiting canonical signaling. This signal inhibits osteoblast differentiation, proliferation, and activity, resulting in reduced osteoblastic bone formation [37]. The inhibition of SOST is a potential therapeutic target, first for OP, targeted by circulating antibodies. Romosozumab is the first Phase I in-human study of inhibition of SOST where serum levels of SOST were reduced in healthy individuals [38]. A Phase III study is currently underway (ClinicalTrials.gov identifier: NCT01575834). Blosozumab is currently being tested for changes in lumbar spine BMD increase and is in Phase I and Phase II trials. Please see ClinicalTrials.gov for updates on SOST inhibition for OP.

In summary, we are at the early stages of harnessing development process for understanding and treating orthopedic diseases. Further translational studies will have to address the relationship between development and repair in orthopedic tissues and the decision of exactly which molecules to target and the specificity of delivering these drugs.

REFERENCES

[1] Marcucio RS, Qin L, Alsberg E, Boerckel JD. Reverse engineering development: crosstalk opportunities between developmental biology and tissue engineering. J Orthop Res 2017;35(11):2356−68.

[2] Jungebluth P, Macchiarini P. Stem cell-based therapy and regenerative approaches to diseases of the respiratory system. Br Med Bull 2011;99:169−87.

[3] Cao Y, Vacanti JP, Paige KT, Upton J, Vacanti CA. Transplantation of chondrocytes utilizing a polymer-cell construct to produce tissue-engineered cartilage in the shape of a human ear. Plast Reconstr Surg 1997;100(2):297−302. discussion 3−4.

[4] Quarto R, Mastrogiacomo M, Cancedda R, Kutepov SM, Mukhachev V, Lavroukov A, et al. Repair of large bone defects with the use of autologous bone marrow stromal cells. N Engl J Med 2001;344(5):385−6.

[5] Cancedda R, Giannoni P, Mastrogiacomo M. A tissue engineering approach to bone repair in large animal models and in clinical practice. Biomaterials 2007;28(29):4240−50.

[6] Warnke PH, Wiltfang J, Springer I, Acil Y, Bolte H, Kosmahl M, et al. Man as living bioreactor: fate of an exogenously prepared customized tissue-engineered mandible. Biomaterials 2006;27(17):3163−7.

[7] Nishimura R, Hata K, Takahata Y, Murakami T, Nakamura E, Yagi H. Regulation of cartilage development and diseases by transcription factors. J Bone Metab 2017;24(3):147−53.

[8] Cairns J, Saunders JW. The influence of embryonic mesoderm on the regional specification of epidermal derivatives in the chick. J Exp Zool 1954;127:221−48.

[9] Glade RW. Effects of tall skin, epidermis, and dermis on limb regeneration in Triturus viridescens and Siredon mexicanum. J Exp Zool 1963;152:169−93.

[10] Wigmore P. Regeneration from half lower arms in the axolotl. J Embryol Exp Morphol 1986;95:247−60.

[11] McCusker CD, Diaz-Castillo C, Sosnik J, Anne QP, Gardiner DM. Cartilage and bone cells do not participate in skeletal regeneration in *Ambystoma mexicanum* limbs. Dev Biol 2016;416(1):26−33.

[12] Tanaka EM. The molecular and cellular choreography of appendage regeneration. Cell 2016;165(7):1598−608.

[13] Wagner I, Wang H, Weissert PM, Straube WL, Shevchenko A, Gentzel M, et al. Serum proteases potentiate BMP-induced cell cycle Re-entry of dedifferentiating muscle cells during newt limb regeneration. Dev Cell 2017;40(6):608−17. e6.

[14] Muhonen V, Salonius E, Haaparanta AM, Jarvinen E, Paatela T, Meller A, et al. Articular cartilage repair with recombinant human type II collagen/polylactide scaffold in a preliminary porcine study. J Orthop Res 2016;34(5):745−53.

[15] Lories RJ, Corr M, Lane NE. To Wnt or not to Wnt: the bone and joint health dilemma. Nat Rev Rheumatol 2013;9(6):328—39.

[16] Nalesso G, Sherwood J, Bertrand J, Pap T, Ramachandran M, De Bari C, et al. WNT-3A modulates articular chondrocyte phenotype by activating both canonical and noncanonical pathways. J Cell Biol 2011;193(3):551—64.

[17] Nalesso G, Thomas BL, Sherwood JC, Yu J, Addimanda O, Eldridge SE, et al. WNT16 antagonises excessive canonical WNT activation and protects cartilage in osteoarthritis. Ann Rheum Dis 2017;76(1):218—26.

[18] Liu Y, Schwartz AG, Birman V, Thomopoulos S, Genin GM. Stress amplification during development of the tendon-to-bone attachment. Biomech Model Mechanobiol 2014;13(5):973—83.

[19] Galatz L, Rothermich S, VanderPloeg K, Petersen B, Sandell L, Thomopoulos S. Development of the supraspinatus tendon-to-bone insertion: localized expression of extracellular matrix and growth factor genes. J Orthop Res 2007;25(12):1621—8.

[20] Schwartz AG, Long F, Thomopoulos S. Enthesis fibrocartilage cells originate from a population of Hedgehog-responsive cells modulated by the loading environment. Development 2015;142(1):196—206.

[21] Dyment NA, Breidenbach AP, Schwartz AG, Russell RP, Aschbacher-Smith L, Liu H, et al. Gdf5 progenitors give rise to fibrocartilage cells that mineralize via hedgehog signaling to form the zonal enthesis. Dev Biol 2015;405(1):96—107.

[22] Caplan AI. New MSC: MSCs as pericytes are sentinels and gatekeepers. J Orthop Res 2017;35(6):1151—9.

[23] Lenas P, Luyten FP. An emerging paradigm in tissue engineering: from chemical engineering to developmental engineering for bioartifical tissue formation through a series of unit operations that simulate the in vivo successive developmental stages. Ind Eng Chem Res 2011;50:482—522.

[24] Schindeler A, Kolind M, Little DG. Cellular transitions and tissue engineering. Cell Reprogram 2013;15(2):101—6.

[25] Takahashi K, Yamanaka S. Induction of pluripotent stem cells from mouse embryonic and adult fibroblast cultures by defined factors. Cell 2006;126(4):663—76.

[26] Cossu G, Bianco P. Mesoangioblasts—vascular progenitors for extravascular mesodermal tissues. Curr Opin Genet Dev 2003;13(5):537—42.

[27] Shore EM, Kaplan FS. Insights from a rare genetic disorder of extra-skeletal bone formation, fibrodysplasia ossificans progressiva (FOP). Bone 2008;43(3):427—33.

[28] Medici D, Shore EM, Lounev VY, Kaplan FS, Kalluri R, Olsen BR. Conversion of vascular endothelial cells into multipotent stem-like cells. Nat Med 2010;16(12):1400—6.

[29] Sandell LJ. Etiology of osteoarthritis: genetics and synovial joint development. Nat Rev Rheumatol 2012;8(2):77—89.

[30] Xu L, Peng H, Glasson S, Lee PL, Hu K, Ijiri K, et al. Increased expression of the collagen receptor discoidin domain receptor 2 in articular cartilage as a key event in the pathogenesis of osteoarthritis. Arthritis Rheum 2007;56(8):2663—73.

[31] Shwartz Y, Viukov S, Krief S, Zelzer E. Joint development involves a continuous influx of Gdf5-positive cells. Cell Rep 2016;15(12):2577—87.

[32] Reynard LN, Bui C, Canty-Laird EG, Young DA, Loughlin J. Expression of the osteoarthritis-associated gene GDF5 is modulated epigenetically by DNA methylation. Hum Mol Genet 2011;20(17):3450—60.

[33]   Zhu M, Chen M, Zuscik M, Wu Q, Wang YJ, Rosier RN, et al. Inhibition of beta-catenin signaling in articular chondrocytes results in articular cartilage destruction. Arthritis Rheum 2008;58(7):2053−64.

[34]   Wu Q, Kim KO, Sampson ER, Chen D, Awad H, O'Brien T, et al. Induction of an osteoarthritis-like phenotype and degradation of phosphorylated Smad3 by Smurf2 in transgenic mice. Arthritis Rheum 2008;58(10):3132−44.

[35]   Zhu M, Tang D, Wu Q, Hao S, Chen M, Xie C, et al. Activation of beta-catenin signaling in articular chondrocytes leads to osteoarthritis-like phenotype in adult beta-catenin conditional activation mice. J Bone Miner Res 2009;24(1):12−21.

[36]   van der Kraan PM, Blaney Davidson EN, Blom A, van den Berg WB. TGF-beta signaling in chondrocyte terminal differentiation and osteoarthritis: modulation and integration of signaling pathways through receptor-Smads. Osteoarthritis Cartilage 2009;17(12):1539−45.

[37]   Billings PC, Fiori JL, Bentwood JL, O'Connell MP, Jiao X, Nussbaum B, et al. Dysregulated BMP signaling and enhanced osteogenic differentiation of connective tissue progenitor cells from patients with fibrodysplasia ossificans progressiva (FOP). J Bone Miner Res 2008;23(3):305−13.

[38]   Padhi D, Jang G, Stouch B, Fang L, Posvar E. Single-dose, placebo-controlled, randomized study of AMG 785, a sclerostin monoclonal antibody. J Bone Miner Res 2011;26(1):19−26.

Chapter 4

Limb Synovial Joint Development From the Hips Down: Implications for Articular Cartilage Repair and Regeneration

Maurizio Pacifici, Rebekah S. Decker, Eiki Koyama
The Children's Hospital of Philadelphia, Philadelphia, PA, United States

INTRODUCTION

The synovial joints of the limbs play obvious, essential, and nonredundant roles in skeletal movement, body locomotion, and quality of life. Each joint is anatomically diverse and exquisitely fitted to the location it occupies and the specific movements it permits and sustains [1−3]. As an example, the shape, organization, and size of the knee are not only strikingly different from those of the shoulder, but the contours, orientation, and location of skeletal elements forming the knee and shoulder joints are diverse and unique as well, thus allowing the knee to sustain larger biomechanical loads and forward motion while enabling the shoulder with the broadest flexibility, range of motion, and reach among all joints [4,5]. The diversity of movement and organization are also reflected in the diversity of specific anatomical components present in different joints, as exemplified by the meniscus and patella in the knee and the acromioclavicular ligament and clavicular artic-ulation in the shoulder [4,5]. Little is known about how each joint is formed at the appropriate skeletal location and how each joint attains its unique shape, organization, and composition. With regard to anatomical location, studies originally suggested that joints form at the boundaries and intersec-tion points among Hox gene expression patterns [6,7], but this has not been confirmed yet. However, we did show that Hox genes influence joint shape character and composition as indicated by the fact that compound ablation of

Developmental Biology and Musculoskeletal Tissue Engineering. https://doi.org/10.1016/B978-0-12-811467-4.00004-8

all Hox11 paralogous genes turned the elbow joint into a knee-like structure displaying a patella [8,9].

Despite their arresting and intriguing diversities, the limb joints do share a number of tissues and general structural features. The joints all contain articular cartilage that provides resilience and frictionless motion, a capsule that insulates the joint from the surrounding tissues and environment and delimits the synovial fluid-filled cavity, and intrajoint ligaments that connect the opposing articulating surfaces and stabilize and orientate motion [10]. Articular cartilage is able to provide resilience owing to its very abundant extracellular matrix composed of collagen fibrils with tensile strength and proteoglycan aggregates with reversible water- and ion-binding capacity and mechanical elasticity and may also contain progenitor cells [11,12]. The capsule does insulate the joint from the body environment, but its inner synovial surface and lining cells produce essential joint lubricants including phospholipids and Prg4/lubricin and may also represent a reservoir of cell progenitors [13−15]. The intrajoint ligaments are not only important for stabilizing movement and kinematics but are also rich in blood vessels and nerves and thus contribute to nutrition as well as space and motion proprioception [16,17].

Although the basis for the morphological and architectural diversity among limb joints remains largely obscure [2,18], there have been significant recent advances in understanding the following: the nature of embryonic joint progenitor cells; the function of several signaling proteins and cues involved in joint initiation and progression; the possible ways in which tissues such as articular cartilage acquire their phenotype; and the nature and roles of macromolecules involved in joint cavitation and lubrication. A number of authoritative and comprehensive reviews of these and related aspects of joint developmental biology have appeared recently and provide important and critical information in this rich, complex, and broadly relevant area of biomedical and clinical research [19−21]. In the present review, we will concentrate on the most recent studies in limb joint development, paying particular attention to the origin of joint progenitor cells and their diversification, the roles of certain key signaling and growth factor proteins including GDF5, and the possible mechanisms by which articular cartilage acquires its functional multizone organization over postnatal life. Curiously, most of these and previous studies have been carried out on medial and distal limb joints including knee, elbow, and digits, but far less is known about the developmental biology of the shoulder and hip joints. Of course, this is of surprising, given that both joints are clinically relevant and affected by numerous pathologies [4,22]. Thus, we include a series of previously unpublished data from our group on the development of the hip joint in mouse. The data reveal that this joint follows a developmental program distinct from that of its more distal counterparts that may be because of being at the embryological and anatomical boundary and interface between the axial and appendicular skeleton. We conclude this review by identifying and discussing implications from the developmental studies above that may aid in the design of more effective strategies to repair and regenerate joint tissues and supplant the notoriously poor native healing capacity of joint tissues [23,24].

ORIGIN OF JOINT PROGENITORS

The skeletal elements of the limbs all derive from mesenchymal progenitors that originate in the lateral mesoderm in the embryonic flank and enter, invade, and populate the limb buds as they elongate distally under the direction of the apical ectodermal ridge and the zone of polarizing activity [25−27]. Those progenitors give rise to several additional limb tissues including perichondrium, ligaments, tendons, blood vessels, and other connective tissues, whereas the limb skeletal muscles derive from distinct progenitors originating in the somites. The earliest visible sign of impending skeletogenesis is the formation of mesenchymal cell condensations within the core region of the limb buds that preconfigure and delineate the future skeletal elements [26]. As the name implies, the cell condensation phase involves the following: the formation of tight cell−cell associations and interactions via several mechanisms including cell adhesion molecules [28,29]; the minimization of pericellular and extracellular matrix; and the complete elimination of blood vessels and capillaries previously occupying the condensation sites [30,31], the latter step being required for skeletogenesis to proceed [32]. The vessel-less condensations then undergo chondrogenesis and lay down the cartilaginous blueprint of the limb skeleton, in line with the fact that low-oxygen tension favors chondrogenic cell differentiation [33,34]. The resulting skeletal framework is composed of uninterrupted cartilaginous rods in which there is no obvious sign of proximal (elbow and knee) and distal (carpal/tarsal/digit) joints, a striking phenomenon recognized by embryologists long ago [26,35]. The onset of joint formation becomes apparent soon afterward with the emergence of mesenchymal cells specifically located at each prospective joint site at midgestation (about E12.5 in mouse embryo elbow/knee sites), leading those embryologists to name the resulting tissue "the interzone" [36,37]. Holder was among the first to ask what the function of the interzone may actually be, namely whether it is a simple signpost demarcating the joint site or whether its cells have subsequent roles in joint formation [36]. When he microsurgically removed the interzone tissue mass from the prospective elbow joints in chick embryos in ovo, the joints failed to form and cavitate, whereas the opposing cartilaginous long bones remained fused to each other, providing the first clear evidence that the interzone cells are needed for joint formation. It is interesting to note that the sequela of developmental events starting from uninterrupted cartilaginous rods and proceeding to mesenchymal interzones and cavitated synovial joints may reflect evolutionary processes [38,39] by which limb joints may have evolved stepwise from symphysis-like structures to fully cavitated and synovial fluid—rich joints. Indeed, even in extant salamanders, the knee and wrist joints are not cavitated, and the articulating site is filled with fibroblastic, interzone-like mesenchymal cells connecting and bridging the flanking bones [40,41]. However, what are the specific roles of the interzone in synovial joint development?

Some answers to such a fundamental question have come from mammalian genetic cell lineage tracing-and-tracking studies. The first one was conducted by Rountree et al. [42] based on their previous studies showing that growth and differentiation factor 5 (*Gdf5*) (a member of the transforming growth factor β [TGF-β] superfamily) is one of the earliest genes specifically expressed in emerging limb joint interzones [43]. Using *Gdf5-Cre* mice, they created and mated with *RosaR26R-LacZ* reporter mice, these authors found that the *Gdf5+* (*LacZ*-positive) cells remained preferentially associated with knee joint tissues and only those tissues over developmental time, implying for the first time that *Gdf5+* interzone cells represent a special cohort of progenitors with the exclusive capacity to persist at joint sites and generate joint tissues. We established collaborative efforts with that group and extended those studies to show that *Gdf5+* cells gave rise to articular cartilage, intrajoint ligaments, synovial lining, and inner half of the capsule in joints as diverse as knee, elbow, carpal/tarsal elements, and digits, and similar cells constituted the articulating cartilage in wrist elements [44]. The data led us to conclude that *Gdf5+* cells and their progeny have the ability to produce a number of joint tissues within diverse limb joints, lending further support to the notion that the cells are determined and committed progenitors endowed with joint tissue formation capacity.

Interzone cells initially develop at sites previously occupied by chondrocytes and may thus be their direct descendants via dedifferentiation [45,46]. This notion has been supported by work with transgenic mice expressing tamoxifen-inducible Cre recombinase (*CreER*) under the direction of the skeletal progenitor cell master gene *Sox9*. Temporal analysis of compound $Sox9\text{-}CreER^{T2};R26R$ mouse embryos receiving tamoxifen once at E11.5 indicated that *Sox9+* (*LacZ*-positive) cells constituted both the interzone and adjacent cartilaginous elements in prospective knees by E13.5 and thereafter [47]. Similar conclusions were reached by lineage and fate map analyses of limb mesenchymal cells demarcated by the expression of *doublecortin* (Dcx) [48]. The TGF-β type II receptor (*Tgfbr2*) is needed for joint development [49], and a new twist to the understanding of joint development emerged by monitoring the behavior of *Tgfbr2*-expressing cells. Using transgenic *Tgfbr2-βGal-GFP-BAC* reporter mice, Li et al. found that *Tgfbr2+* cells were first restricted to the dorsal and ventral regions of E13.5 limb joints but were undetectable in the central region of the interzone [50].

With increasing developmental time, *Tgfbr2+* cells became apparent in synovial lining, meniscal surface, ligaments, and groove of Ranvier, leading the authors to conclude that *Tgfbr2+* cells emerge at specific times and sites and act as joint progenitor subsets involved in the development of certain tissues within given joint niches [20,51].

A very recent study by Shwartz et al. [52] represents a significant step ahead in this research area. The authors created knockin *Gdf5-CreER^T2* mice, mated them with reporter mice, and found that induction by a single tamoxifen injection at E10.5 failed to elicit *Gdf5+* (*tdTomato*-positive) cells over the entire interzone by E11.5 and that only the most epiphyseal cells were labeled by E18.5 in knee, elbow, and metacarpalphalangeal joints. However, when tamoxifen was given three consecutive times at E11.5, E13.5, and E15.5, *Gdf5+* cells populated most joint tissues by E18.5. Additional experiments using a single tamoxifen injection and comparison of reporter expression at subsequent times versus real-time endogenous *Gdf5* expression pointed to a highly dynamic behavior of joint-forming cells, with sets of cells turning *Gdf5* expression on and others turning it off while maintaining *dtTomato* expression and contributing differentially to joint tissue formation. Mitotic activity monitored by BrdU incorporation suggested that *Gdf5+* cell proliferation decreased significantly between E13.5 and E18.5. The authors concluded that not all joint-forming cells are direct descendants of the dedifferentiated *Sox9+* chondrocytes constituting the incipient interzone. Rather, additional *Sox9+* progenitors would be recruited from the joint's immediate surroundings over time and would generate *Gdf5+* cohorts contributing to distinct tissue joint development. Local tissue genesis and involvement of surrounding local progenitors in joint development were also suggested by data in several previous studies [53–55].

The important data and insights in the Shwartz et al. study [52] correlate well with data in a very recent study from our group [56] in which we carried out additional analyses of joint progenitors, using *Gdf5-Cre* mice and our new BAC-based *Gdf5-CreER^T2* line mated with *Confetti* reporter mice in which green (GFP), yellow (YFP), red (RFP), or cyan (CFP) fluorescent protein can be alternatively activated on *Cre* action [57]. Using *Gdf5-Cre* mice, we observed that *Gdf5+* cells expressing different reporters were present in joint tissues only and were scattered throughout the several cell layers of articular cartilage (*ac*) and joint tissues such as meniscus (*me*) at postnatal day 0 (P0) (not shown) and P14–P21 knees (Fig. 4.1A and B). As there was no dominance of large single-color clones, the proliferation rate of each progenitor cell was likely to be low at those stages. This trend was confirmed using unbiased *RosaCreER;Confetti* mice injected postnatally with tamoxifen once and harvested over time. For example, injection at P7 and harvest at P14 showed the presence of mostly single articular chondrocyte expressing one reporter (Fig. 4.1C, bird-eye view of P14 tibia plateau). In addition, we mated *Gdf5-CreER^T2* mice with single-color

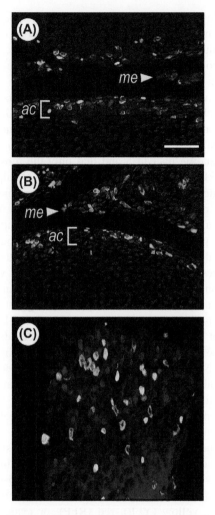

FIGURE 4.1 Distribution and characteristics of fluorescent reporter-expressing cells in knee joint tissues in juvenile mice. (A and B): Longitudinal sections through peripheral (A) and central (B) portions of knee joints from P14 *Gdf5-Cre* mice mated with *Confetti* reporter mice. In these reporter mice, Cre-mediated recombination leads to alternative expression of green (GFP), yellow (YFP), red (RFP), or cyan (CFP) fluorescent proteins. Note that the reporter-expressing cells were exclusively present in joint tissues including femoral and tibial articular cartilage (*ac*) and meniscus (*me*), verifying that *Gdf5+* lineage cells represent a specialized cohort of progenitors producing joint tissues only. (C) Bird-eye view of whole-mount tibial plateau articular cartilage in unbiased *RosaCreER;Confetti* mice injected postnatally with tamoxifen at P7 and harvested at P14. Note that most of the reporter-expressing cells were single and did not produce large groups of same reporter cells during the 7 day period since tamoxifen injection, implying that the proliferation rate of the entire cell population was rather low. Bar in (A) for (A–C): 50 μm.

reporter mice, injected them with tamoxifen once at E13.5, E15.5, or E17.5 and analyzed the compound mice at P0 or P28. We found that *Gdf5+* cells generated at each time point remained largely confined to site of origin and had not produced large migrating progenies over time, indicating that they became committed to specific sites and produced local tissues. Together, the above studies support the important notion that although the overall *Gdf5+* cell population is responsible for joint formation [44], the origin, behavior, and dynamics of the cells are very complex and their roles diverge over developmental time and location, resulting in generation of distinct joint tissues. What remains unclear is how such diversification occurs and, in particular, whether it reflects intrinsic mechanisms and distinct progenitor subcohorts within the overall *Gdf5+* population or may depend on temporal, positional, and other local cues operating on a developmentally malleable and pluripotent cell population.

SIGNALING PROTEINS AND THEIR ROLES IN JOINT FORMATION

The mechanisms by which chondrocytes located at each prospective joint site undergo dedifferentiation and contribute to mesenchymal interzone formation have attracted much interest but remain unclear [18]. Some of those cells will give rise to articular cartilage and thus will have to reundergo chondrogenesis and cartilage formation along with additional *Gdf5+* progenitors generated locally and recruited from the surroundings. This striking process (occurring in strict temporal and spatial manners) requires fine-tuned, diverse, and temporally set mechanisms. Many signaling proteins and transcription factors with distinct and often opposing properties (such as anti- and prochondrogenic) are in fact expressed in the interzone and during progression of joint development where they have essential and distinct roles [2,58]. Just to cite major examples, global loss or conditional ablation of *Noggin*, *β-catenin*, *Wnt4/Wnt9a*, *Tgfbr2*, or *TAK1* disrupt limb joint development, including partial or total lack of cavitation. These defects likely reflect substandard interzone determination and formation, precocious or excess chondrogenic differentiation within the prospective joint site, and/or substandard joint tissue differentiation, leading to partial or complete skeletal fusion [44,49,55,59−61]. Joint failure and fusion were also observed in mouse embryo mutants lacking the zing finger transcription factors *Osr1* and *Osr2* in which expression of such interzone-associated genes as *Gdf5*, *Wnt4*, and *Wnt9a* was initiated but not sustained [62]. *Wnt9a* was found to be regulated by *c-Jun*, which when ablated, caused a derangement of both Wnt signaling and limb joint development [63]. It is of interest to point out that gross abnormalities of joint formation and even fusion were seen in mouse embryos lacking Indian hedgehog (*Ihh*) or *Sox5* and *Sox6* [64,65].

Defects were also seen after conditional activation of constitutive hedgehog signaling within joints starting in prenatal mice, indicating that both excess and substandard hedgehog signaling are detrimental [66]. *Ihh* is first expressed in the incipient diaphysis of long bone cartilaginous anlagen as early as E11.5 in mouse embryos, and then expression continues in the prehypertrophic zone of growth plate [67]. *Sox5* and *Sox6* are master regulators of chondrogenic cell differentiation [68]. Thus, these important studies have revealed a link and mutual interdependence between joint initiation and progression to functioning and activities of growth plate chondrocytes in developing long bone shafts, although details about underlying regulatory mechanisms and the nature of cross talk and interactions remain to be worked out.

By being among the earliest phenotypic traits of interzone cells, Gdf5 has deservedly attracted a great deal of research interest and activity. It was originally shown that *Gdf5* ablation in mice deranged joint formation particularly in the digits, whereas ablation of both *Gdf5* and *Gdf6* markedly affected elbow and knee joint development also [69]. The greater sensitivity of autopod joint development to *Gdf5* action (and lack of redundancy by *Gdf6*) is also reflected in the significant number of human pathologies in hands and feet caused by *GDF5* mutations, including brachydactyly-type C2 or A2 caused by loss-of-function *GDF5* mutations, and symphalangism (SYM1) and multiple synostosis syndromes (SYNS) due to gain-of-function mutations [70,71]. Interestingly, Kingsley and coworkers recently analyzed the regulatory elements controlling *Gdf5* expression in different joints in the appendicular and axial skeleton of mice [72]. They identified multiple enhancers distributed over the entire *Gdf5* locus—including regions upstream and downstream of the coding exons—that specify and differentially regulate *Gdf5* expression in the following: axial joints; proximal and distal limb joints; and even within portions or areas of a given joint. The large flanking regions were required to rescue normal joint development in *Gdf5* mutants, and orthologs of these enhancers were previously associated with common osteoarthritis (OA) risk in people [73]. The study is a major contribution to the field. Its data and insights uncover the sheer complexity of regulatory mechanisms operating in different joints and also suggest that *Gdf5* is part of key morphogenetic mechanisms by which each joint acquires its shape, contours, and organization. Hence, when Gdf5 biological function is deranged by mutations, certain joints are particularly affected as is the case of the autopod syndromes cited earlier. Based on differential *Gdf5* expression in developing chick embryo limb joints, we previously suggested that Gdf5 may exert morphogenetic roles [2].

There is, however, a lingering, major, and perplexing issue regarding Gdf5 biological properties and roles. As summarized above, *Gdf5* is first expressed by incipient mesenchymal interzone cells concomitant with local chondrocyte dedifferentiation, but tests of *Gdf5* activity in several in vitro systems have indicated that the protein is prochondrogenic [74]. Indeed, gain-of-function *GDF5* mutants from patients with a pronounced form of multiple synostosis syndrome 1 (SYNS1) were found to be massively prochondrogenic when tested in mouse embryo limb mesenchymal micromass cultures or after implantation in chick embryos and to be resistant to Noggin inhibition [75]. Thus, it is perplexing that a seemingly prochondrogenic factor is normally expressed in conjunction with the emergence of mesenchymal interzone cells and dedifferentiated chondrocytes. Insights into this long-standing conundrum have come from a very interesting and comprehensive recent study in which the biological properties of recombinant human GDF5 were compared with those of its family member bone morphogenetic protein 2 (BMP2) by in vitro and in vivo experimentation [76]. The premise of the work was that both GDF5 and BMP2 are important for joint and skeletal development, growth, and function, but the mechanisms by which each exerts selective action remain unclear. BMP2 and GDF5 bind with similar affinity to such cell surface type II receptors as BMPR-II, ActR-II and ActR-IIB, and BMP2 also interacts with similar affinity to the type I receptors BMPR-IA and BMPR-IB [77]. However, GDF5 exhibits an over 10-fold greater affinity for BMPR-IB than BMPR-IA, and this specificity is due to a single amino acid—Arg57—present in its prehelix loop [70,78]. The authors carried out extensive high-resolution X-ray analyses of ligand-receptor ectodomain crystals to clarify the basis of interaction specificity. A telling finding was that when interacting with GDF5, the important $\beta1\beta2$ loop in BMPR-IB was in an open conformation providing sufficient space for the large side chain of Arg57 in GDF5, whereas the loop of BMPR-IA appeared to adopt a closed conformation after GDF5 interaction, providing important hints to ligand-receptor binding selectivity. In biological tests, the authors found that whereas both GDF5 and BMP2 stimulated alkaline phosphatase (APase) activity in the chondrogenic cell line ATDC-5, only BMP2 did so in C2C12cells and indeed, GDF5 was able to antagonize BMP2-induced APase during cotreatment of C2C12 cultures. Importantly, BMP2, but not GDF5, induced heterotopic ossification (an endochondral process) when implanted in rat muscles, and GDF5 antagonized the action of BMP2 when coimplanted. These and other very important data led the authors to conclude that GDF5 can act as a BMP2 antagonist and does so in a context-dependent manner. Given that several BMPs (including BMP2) are expressed in limb joint interzone [67,69], GDF5 could normally inhibit their skeletogenic potentials, thus fine-tuning the effects of local pro- and anti-chondrogenic factors, protecting the initial mesenchymal character of the interzone and allowing joint formation and cell differentiation to unfold in physiologic spatiotemporal manners. GDF6 (also known as BMP13) could have similar properties [79]. Interestingly, as the authors pointed out, Arg57 is mutated to leucine in some patients with proximal symphalangism (SYM1), and GDF5[R57L] shows enhanced binding to BMPR-IA and lower ability to discriminate between the two type I receptors, thus behaving more like BMP2 [70,78]. In doing so, GDF5^{R57L} could thus tilt the delicate phenotypic balance within the mesenchymal interzone, accelerate chondrogenesis, and cause phalangeal joint fusion in SYM1 patients. Tellingly, ablation of *Bmpr-1b* causes defects in mouse joint development [80]. Given that only certain joints are affected by SYM1 or related syndromes, the data point again to the existence of subtle but critical joint-specific mechanisms, and the intriguing challenge for the future will be to understand exactly what these GDF5-dependent mechanisms are and how they operate selectively and specifically within each joint [72].

ARTICULAR CARTILAGE GENESIS AND MORPHOGENESIS

Articular cartilage is the spring and shock absorber of the joints, but it is far from being a simple tissue. In large adult joints such as the knee, the tissue displays a characteristic stratified organization consisting of: a thin superficial zone abutting the synovial space that is composed of small and tightly bound cells orientated along the major direction of movement, contains scant and isotropic matrix, and produces joint lubricants; a thick intermediate/deep zone that contains oval large chondrocytes aligned in columns vertical to the joint surface and flanked by abundant and anisotropic matrix rich in collagen II, aggrecan and other macromolecules; and a calcified zone below the tidemark with round chondrocytes that faces, and is linked to, the subchondral bone [81]. It is widely recognized that such structural organization and features are important for long-term tissue biomechanical function, endurance, and homeostasis through life and that their derangements due to acute injury or chronic conditions can lead to irreversible changes and pathologies, such as OA [82]. A central question facing developmental biologists for a long time has been not only how articular cartilage forms within the developing joints but also how it acquires its functional structure, composition, and unique multizone organization. As we and others showed [83,84], the tissue does not initially display a stratified organization in neonatal or young animals but acquires it during postnatal growth. Fig. 4.2A shows articular cartilage in the medial portion of proximal tibia in a neonatal mouse identified by expression of typical articular gene markers such as *tenascin-C* and *Erg* [44] and lack of expression of shaft growth plate chondrocyte markers such as *matrilin-1* [85], with all its cells being *Gdf5*+ progeny [44]. Clearly, the neonatal articular tissue was quite thin at that stage and composed by 6−7 layers of small and variously shaped cells and scant isotropic matrix (Fig. 4.2A, vertical green bar). During the following 2−3 weeks, the tissue underwent remarkable growth in thickness (Fig. 4.2B, vertical green bar). With the exception of the small flat cells composing its superficial zone, the bulk of chondrocytes were large and separated by an abundant cartilage matrix. By 6−8 weeks of age

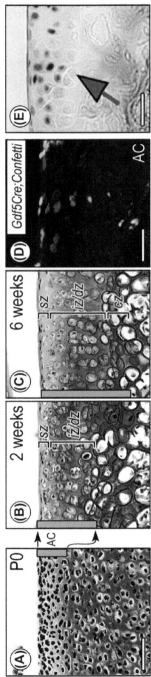

FIGURE 4.2 Histological and gene reporter analyses of knee articular cartilage development and maturation from neonatal to adult stages. (A–C) Longitudinal sections through the medial portions of tibia in P0 (A), P14 (B), and 6-weeks-old (C) mice stained with safranin O and fast green. The nascent and growing articular (AC) tissue at each stage (green vertical bars) was defined and recognized by the expression of typical articular molecular markers including *tenascin-C* and *Erg* and absence of growth plate markers such as *matrilin-1* (see Ref. [56]). Indicated are the main articular cartilage zones: *sz*, superficial zone; *iz/dz*, intermediate/deep zones; and *cz*, calcifying zone. The latter was not readily distinguishable at P14 because some of the cartilaginous tissue will be incorporated into the subchondral bone. (D) Fluorescent image of reporter-expressing articular chondrocytes in a section through medial tibial articular cartilage in 6-week-old dual *Gdf5-Cre;Confetti* transgenic mice. Note that the chondrocyte stacks typical of mature cartilage at this age were composed of cells expressing different reporters, indicating that the stacks likely arose from repositioning and rearrangement of neighboring cells rather than by columnar apposition of daughter cells (that would express the same-color reporter). (E) Bright field image of a section of tibia articular cartilage in adult *Gli1-LacZ* (hedgehog signaling) reporter mice. Note that the highest level of reporter activity was appreciable in the central portion of the tissues, but levels were lower in superficial and bottom (calcifying) zones (arrow). Bar in (A) for (A–C): 40 μm; bar in (D): 60 μm; and bar in (E): 35 μm.

(Fig. 4.2C), the tissue finally displayed its typical mature multizonal organization with a clear superficial zone (*sz*), an intermediate zone and a thick deep zone (*iz/dz*) with chondrocyte columns flanked by anisotropic matrix, and a calcified zone (*cz*) at its bottom. So, how does articular cartilage go from a thin and matrix-poor tissue with flat, compacted cells at neonatal stages to a thick, highly structured, and zonal tissue with chondrocyte columns by adult age?

This seemingly innocent question has actually been hard to answer. One idea was originally provided by the identification of cells with a progenitor character in postnatal articular cartilage [86]. The authors used sequential enzymatic treatments to isolate cells from the superficial, intermediate, and deep zones from P7 bovine articular cartilage and characterized the cells using a variety of in vitro and in vivo assays. Cells from the superficial zone exhibited high affinity binding to fibronectin and efficient colony unit−forming ability and also expressed the progenitor marker gene *Notch1* both in vitro and in vivo. The cells displayed phenotypic and developmental plasticity in that they engrafted into several tissues once transplanted into chick embryos, including bone, tendon, and perimysium. These and other data led the authors to conclude that cells in the superficial zone are endowed with a progenitor/stem character and potentials. Over postnatal time, the cells would be responsible for the formation of the thick multizone articular cartilage seen in adults by a mechanism of apposition. The cells would proliferate and produce vertical columns of overlapping daughter chondrocytes spanning the entire tissue thickness from superficial to deep zone, a notion also proposed in their previous studies [87]. Cells with a progenitor/stem character expressing typical markers such as CD90 and STRO-1 have also been identified in human articular cartilage [88].

A recent study of mouse limb articular cartilage development has also reached the conclusion that articular cartilage mainly develops and grows by apposition [89]. The authors created knockin (heterozygous null) *Prg4-CreER^{T2}* mice and mated them with *LacZ* reporter mice. The logic of this genetic approach was that *Prg4* is strongly expressed in developing limb joints by late embryogenesis [90]. After a single tamoxifen injection at E17.5, reporter-positive (*Prg4*+) cells were found to be present exclusively within a single cell layer at the very surface of incipient articular tissue at P0. Over time, the *Prg4*+ cells became more numerous and were present in the form of vertical chondrocyte columns spanning the whole tissue thickness in adult mice. When tamoxifen was first injected at postnatal times, the reporter-positive chondrocyte columns were partial and did not span the whole tissue thickness. Tests of cell proliferation by BrdU incorporation indicated that more than 70% of superficial cells were proliferative even in 1-month-old mice. The authors concluded that the *Prg4*-expressing cells located at the neonatal surface serve as an active and dynamic progenitor cell population responsible for generation of all zones in adult articular cartilage by columnar apposition of daughter cells over postnatal time.

Appositional growth is an important mechanism by which tissues and organs can grow [91], and it is thus possible that it may also account for articular cartilage growth from neonatal to adult stages. However, this mechanism raises a number of questions in the case of articular cartilage. It would require fairly high rates of cell turnover and proliferation when in fact (and at variance of what was observed in the above study) proliferation of cells within neonatal and juvenile articular cartilage is relatively low [56]. Appositional growth does not provide an obvious explanation for why the cells have different phenotypes, morphologies, and volumes in the different zones in adult tissue. It would also require considerable matrix turnover to go along with dynamic cell renewal when in fact the matrix—and collagen in particular—is very stable [92,93]. Given the importance of this issue and relevance to both basic science and translational medicine, we recently reexamined it using new BAC-based transgenic mouse lines $Prg4\text{-}CreER^{T2}$ and $Dkk3\text{-}creER^{T2}$ in addition to $Gdf5\text{-}CreER^{T2}$ and $Gdf5\text{-}Cre$ mentioned above, in conjunction with single and *Confetti* reporter mice [56]. As mentioned above, our cell tracing-tracking cell lineage analyses showed that knee joint progenitors produced small nonmigratory progenies and distinct local tissues over prenatal and postnatal time. Proliferation was prominent at prenatal stages but dropped considerably during neonatal and juvenile stages. Fluorescent imaging of articular cartilage at successive postnatal ages indicated that the chondrocyte columns present in adult tibial articular cartilage consisted of non—daughter cells expressing different *Confetti* reporters (see Fig. 4.2D for an example). The lack of daughter cell columns was not due to technical limitations of the *Confetti* approach because typical columns of same reporter-expressing chondrocytes were present in the subadjacent tibial growth plate [56]. Further stereological imaging and 3D reconstruction throughout articular cartilage thickness in late juvenile and adult mice actually showed that the articular chondrocytes were not perfectly aligned and did not overlap each other fully, thus forming stacks rather than columns. Determination of cell volume in each articular cartilage zone and monitoring and quantification at successive postnatal stages indicated that zone-selective increases in cell volume were major drivers of overall tissue growth and thickening, a mechanism reminiscent of chondrocyte hypertrophy in growth plates that is the prime driver of skeletal elongation and growth [94]. Interestingly and importantly, the maximum volume displayed by deep zone chondrocytes did not exceed about 60% of that of typical hypertrophic chondrocytes in the growth plate, indicating that chondrocytes are able to set and maintain distinct volumes in different settings and that articular chondrocytes normally avoid full hypertrophy. Second harmonic generation with two-photon microscopy showed that the collagen matrix went from being isotropic and scattered at neonatal stages to being anisotropic and aligned along the chondrocyte stacks in adults. Together, our observations do not wholly sustain a model of appositional growth for articular cartilage. Rather, they indicate that the tissues grow and thicken over postnatal life mainly by formation of non—daughter cell stacks, with limited contribution by cell proliferation and a major role played by zone-specific cell volume increases and differential matrix deposition, accumulation, and/or remodeling. The formation of chondrocyte stacks may involve a process of realignment and reorientation of neighboring cells, possibly brought about by mechanisms such as convergent extension [95]. Our data agree quite well with those in a very recent study on articular cartilage development using similar genetic approaches, including *Confetti* mice [96].

By being at the interface and boundary between articular cartilage and synovial cavity and fluid, the superficial zone exerts a very important function in joint function and endurance. Indeed, this zone can be affected at early stages of OA as indicated by loss of tensile strength and water content and changes in gene expression [97] and may thus represent a significant culprit in disease progression. As indicated above, the bulk of superficial zone cells normally maintain an elongated fibrocartilaginous morphology, are orientated along the main axis of joint movement likely to minimize friction and cell abrasion, are characterized by tight intercellular connections and scanty matrix, and elaborate and secrete lubricants [81,98]. Although the cells derive from *Gdf5+* progenitors [44], it is not at all clear how they acquire their distinct phenotypic traits and maintain them throughout life. To address this significant gap in current information, Jia et al. recently determined the consequences of EGFR signaling deficiency on superficial zone and articular cartilage function and responses to experimental OA in mice [99]. The rational for the study was previous work from the same group showing that conditional loss of *Egfr* expression in cartilage deranged growth plate organization and secondary ossification [100]. The authors made use of a similar conditional gene ablation approach and found that juvenile mutant mice displayed a markedly reduced number of superficial cells in their knees and lower production of joint lubricants, including Prg4. Mutant articular cartilage contained disorganized collagen fibrils, was characterized by reduced biomechanical properties, and became more sensitive over age to experimental surgery-induced OA consisting of destabilization of the medial meniscus (DMM) [101]. The effects of *Egfr* deficiency on cellular phenotype and behavior appeared to be direct as indicated by experiments with isolated superficial cells and long bone explants in culture. These interesting and novel data indicate that EGFR signaling is important in the maintenance of superficial zone cell function and the protection of cartilage structure and suggest that this signaling pathway may offer a target to boost joint function and dampen functional loss and aberration during OA.

Zhang et al. have recently made an additional important contribution to the field by examining the consequences of experimental cell death of surface cells on joint function and response to injury [102]. They used knockin *Prg4-CreER^{T2}* mice to conditionally induce expression of diphtheria toxin (DTA) after tamoxifen treatment in the joints of juvenile mice, resulting in autonomous cell death of targeted superficial cells and underlying chondrocytes. Surprisingly, this did not lead to major cartilage surface damage in the knees over time up to 8 months, but there was two- to threefold higher chondrocyte proliferation within 2 months from DTA induction compared with controls. In addition, when the DTA-ablated mice were subjected to DMM-induced OA, they actually displayed less cartilage damage after 12 weeks compared with companion-operated controls. Although there was significant and expected chondrocyte loss in operated controls, there was no further decrease in cell number in operated DTA-ablated joints. The authors concluded that loss of superficial cells and chondrocytes after DTA-induced death does not initiate tissue damage, implying that chondrocyte survival and maintenance are not needed to protect cartilage at least within the time frames examined. The authors propose that chondrocytes may adopt a catabolic phenotype during the early stages of OA, responsible for overall tissue damage and destruction over time. Thus, the reduced number of chondrocytes after DTA induction may have been beneficial in that it would protect the tissue by reducing the degree of catabolic response. Strategies to reduce or even prevent chondrocyte-induced catabolism may slow down degenerative joint disease. These are interesting and certainly intriguing observations, insights, and conclusions, and it will be important to further test them using different, and possibly more efficient, *CreER* drivers. The apparent endurance of DTA-ablated joints likely reflects the fact that (1) articular cartilage in juvenile and adult animals is a rather stable tissue both structurally and functionally and is largely postmitotic and (2) a partial acute loss of its cells would thus not overtly alter its character and organization [103]. It will be interesting to see whether effective chemical or biological agents could be identified to specifically modulate superficial cell function and phenotype, possibly including kartogenin and resveratrol [104,105], and assess benefits on joint endurance and repair responses during natural age-dependent OA or acute conditions such as trauma. We showed that acute transgenic activation of Wnt/β-catenin signaling leads to thickening of the superficial zone and increased *Prg4* expression in mice [106], offering an additional potential therapeutic tool to target its cells.

The *Confetti* approach used by our group [56] and by Chagin and coworkers [96] has provided additional insights into the behavior and properties of superficial cells. The multiple reporters in this genetic approach allow for assessment of parental relationships among cells expressing the same reporter and also their mitotic activity. The data in both studies showed that some superficial cells were present as same-color doublets/quadruplets oriented horizontally along the surface axis of articular cartilage and parallel to the synovial cavity in neonatal and juvenile tissues, indicative of recent mitotic activity along a preferential orientation. Chondrocytes in deeper zones appeared instead to divide more randomly at those stages, before forming stacks of cells with different reporters by adulthood (Fig. 4.2D) [56]. Thus, superficial zone cells may be able to renew themselves and produce progenies that maintain histological position, orientation, and function. Such active renewal properties could also reflect the fact that superficial cells may generally be more labile and short-lived compared with deeper chondrocytes, given that they are subjected to continuous fluid flow, abrasion, and biomechanical prodding during joint motion. Their renewal abilities (possibly ascribable to local progenitors) would thus be beneficial for long-term tissue homeostasis and could also become a therapeutic target. Although these possibilities need further experimentation, the current data point to the interesting idea that cells within the different zones of articular cartilage have distinct longevity and different needs for cell renewal and maintenance.

HIP JOINT FORMATION AND GROWTH

The pelvis comprises the ilium, ischium, and pubic bones and is connected to the axial and appendicular skeleton via the sacroiliac and hip joints, respectively. The hip joint is an architecturally complex structure that exerts a major weight-bearing function and is characterized by the fossa-shaped acetabulum on the pelvis side articulating with the reciprocally shaped and ball-like head of the proximal femur [107]. The acetabulum is contributed by the developmental confluence of the ilium, ischium, and pubis and normally covers about half of the femoral head in adults. Interactions and articulation between the acetabulum and femur are expanded, reinforced, and stabilized by the following: the labrum, a circular fibrocartilaginous rim-like structure increasing coverage; the teres, an intrajoint ligament that interconnects the acetabulum and femoral head by insertions into central depressions in each element; and a large capsule and three extracapsular ligaments [107,108]. The femoral head and the acetabulum are orientated toward each other with specific geometries and angles that are needed for proper and long-term function. These defined angles are used prognostically and diagnostically in the clinic, including Sharp's angle [109]. In addition, the femoral neck itself is orientated at a specific, species-specific angle with respect to the main longitudinal axis of the femoral shaft and measures about 128 degrees in adult humans, reflecting distinct body structure, postures, and movement in different species [110,111]. Deviations in this angle can be pathogenic and cause coxa vara and coxa valga. Shallowness of the acetabulum characterizes the pediatric condition developmental dysplasia of the hip (DDH) that if untreated can lead to joint malfunction and juvenile OA [112]. These and many other basic and important biomedical aspects of hip joint anatomy, organization, and function are well established and appreciated, but relatively little is known about their developmental biology, a reflection of the sheer complexity of the hip joint and the multiplicity of features and parameters characterizing it. Progress has been provided by several important studies that are highlighted next, followed by our new data on hip joint interzone specification and joint growth in mouse.

Elegant quail-to-chick embryo transplantation and tissue extirpation studies by Malashichev et al. originally showed that the entire pelvic girdle derives from the lateral plate mesoderm [113,114]. Development of pubis and ischium was found to require signals from the local limb field ectoderm, whereas ilium development required both ectodermal and somitic signals, with differential roles played by transcription factors *Pax1*, *Alx4*, and *Emx2*. Alcian blue staining and 3D reconstruction analyses in those studies indicated that separate mesenchymal condensations emerging within the lateral plate mesenchyme were responsible for ilium, ischium, and pubis development. Ilium chondrification was the first to occur and became appreciable at embryonic Hamburger Hamilton stage 28 (HH28). The neocartilaginous ilium anlage encircled—but was not continuous with—the developing femoral head and presaged the location of the acetabular fossa. At that stage, the development of the knee joint was far more advanced than the hip joint, indicating that each limb joint develops according to local clocks and not on the basis of the general proximal-to-distal limb developmental pattern. Ilium chondrification was followed by that of pubis and ischium.

Important and complementary observations were obtained by Pomikal and Streicher in their highly detailed histochemical, 3D and 4D (temporal) reconstruction studies in mouse embryos [115]. The authors found that as in chick, the pelvic skeletal elements derive from a single broad field of mesenchymal cells positioned in close proximity to the incipient appendicular skeleton and appreciable by E12.5. Separate and distinct preskeletal cell condensations formed over time, with the first one to chondrify being the ilium by E13.5 followed by those of pubis and ischium. Initially, the pelvic elements developed very close to the limb field but with time, they vigorously grew and elongated toward the axial skeleton and also underwent a dramatic reorientation with respect to the body axis. The authors closely examined the developing hip joint and found that the cartilaginous femoral head was already intimately surrounded by some condensed pelvic mesenchymal cells by E12.5, reaffirming the notion that the femoral head plays an important role in regulating the position, shaping, and geometry of the acetabular fossa [116]. This notion was further substantiated in subsequent 3D imaging studies in chick embryos [117]. The data indicated that by being already cartilaginous, the femoral head not only exerted a preponderant role in acetabulum development but also influenced the overall hip joint morphology before both cavitation and emergence of a fluid-filled synovial cavity. Extending these interesting observations and questions to human development, a recent mechanobiological dynamic simulation study made use of previously published embryonic human hip developmental data [118] and found that normal translational and rotational movements are required to maintain femoral head sphericity and acetabular depth, particularly at early developmental stages [119]. Reduced or abnormal prenatal movements— such as that occurring in embryos malpositioned in utero—were predicted to lead to substandard sphericity and acetabular depth and coverage, increasing the risk of pathologies such as DDH.

Given the availability of genetic cell lineage—tracing approaches, we carried out studies to further analyze the contributions and genealogical relationships of lateral plate and limb cells in mouse hip join development. In line with the histological studies above [115], we found in E12.5 embryos that the prospective head of the cartilaginous femur was surrounded by interzone-like mesenchymal progenitors vigorously expressing *Sox9* and filling the entire space between the femoral head and ilium cartilaginous primordium (Fig. 4.3A and B, red arrow). The latter already had an elongated shape, and its cells strongly expressed *Sox9* and the typical chondrogenic marker *Ucma* (Fig. 4.3A, B, and D, arrowhead). *Ucma* and *Sox9* expression characterized also the cartilaginous femur (Fig. 4.3D, *fe*). Interestingly, the interzone cell marker *Gdf5* was exclusively expressed by 5—7 layers of cells intimately associated with the developing femoral head at this stage (Fig. 4.3C, yellow arrow). These data revealed for the first time that *Gdf5*-expressing cells in the developing hip joint did not comprise the entire mesenchymal population occupying the whole putative joint formation as *Gdf5*-expressing interzone cells do in other joints, such as the knee. The differential distribution of hip *Gdf5*-expressing cells point to unique interactions and interplays with both the femur and limb field. Higher magnification analysis (Fig. 4.3E—H) confirmed that while the cells filling the prospective joint area between ilium and femur displayed a mesenchymal and fibroblastic morphology (Fig. 4.3E and F, yellow bracket), they were readily distinguishable into *Gdf5*-expressing cells surrounding the femoral head and *Gdf5*-negative cells projecting toward the ilium (Fig. 4.3G and H).

By E13.5, the hip joint area had undergone significant changes (Fig. 4.3I—L). The *Gdf5*-expressing cells had become very compacted and intimately connected to the femoral head and thus resembling interzone cells (Fig. 4.3J—L, arrow). In contrast, those facing the developing pelvis remained variably shaped and had actually become sparse and diffused, possibly and likely reflecting the onset of cavitation (Fig. 4.3J, arrowheads). *Gdf5*-positive cells located around the basilar portion of the femoral head remained compacted and appeared to bridge and interconnect the femur and pelvis primordia (Fig. 4.3I and J, double arrowhead). Notably, the opposing pelvic and femoral sides of the joint at this stage clearly displayed their reciprocal fossa-like and ball-like shapes, reiterating the notion that the femoral head has a major role in determining contour and geometry of the acetabulum [116].

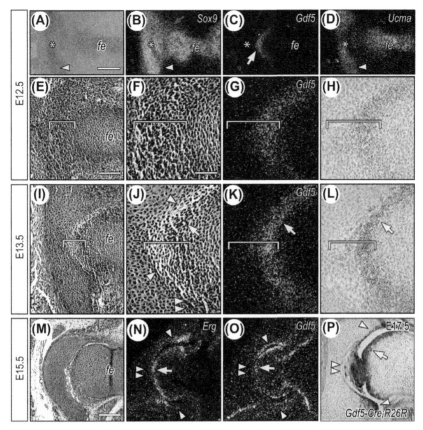

FIGURE 4.3 Analysis of hip joint development in mouse embryos. Serial sections through the developing hip joint and pelvis at indicated embryonic stages were processed for histological staining, in situ hybridization or histochemical detection of *LacZ*, as indicated. Location of prospective acetabular fossa is marked by an asterisk. (A–D) In E12.5 embryos, the incipient hip joint was characterized by the cartilaginous femur (*fe*) expressing *Sox9* and *Ucma* (B and D) and the ilium primordium (A, B, and D, arrowhead) that were interconnected by a large number of mesenchymal cells occupying the prospective joint area and all expressing *Sox9* (B, red arrow). *Gdf5* expression was restricted to cells adjacent to femoral head (C, yellow arrow). (E–H) Higher magnification images showing that while the mesenchymal cell population interdispersed between femoral head and ilium primordium was uniform and fibroblastic histologically (E and F, yellow bracket), it was readily distinguishable into a *Gdf5*-expressing cohort adjacent to femur and *Gdf5*-negative cells toward the ilium (G and H). (I–L) In E13.5 embryos, the hip joint displayed incipient traits of cavitation (bracketed area) as indicated by reduced cell density and irregular cell arrangement in central fossa region (I and J, arrowheads). The *Gdf5*-expressing cells around the femoral head were more compacted at this stage (J and I, arrow) as were those located around the femoral head's basal perimeter (J, double arrowhead) that will likely give rise to acetabular articular cartilage over time. (M–O) At E15.5, the joint was undergoing full cavitation. Expression of both *Gdf5* and *Erg* characterized the articular tissue on femoral head (N and O, arrow) and acetabular perimeter (N and O, arrowheads), but expression was low to negligible in the central fossa region (N and O, double arrowhead). (P) Bright field histochemical image of hip joint in E17.5 *Gdf5-Cre;R26R* mice showing that *Gdf5+* lineage cells were restricted to articular cartilage on the femoral head (arrow) and peripheral portion of acetabulum (arrowheads) but were largely absent from the fossa region (double arrowhead). Bar in (A) for (A–D): 200 μm; bar in (E) for (E and I): 125 μm; bar in (F) for (F–H and J–L): 60 μm; and bar in (M) for (M–P): 150 μm.

By E15.5, cavitation was well underway (Fig. 4.3M), and it was clear that *Gdf5* expression characterized the entire articulating layers on the femoral head (Fig. 4.3O, arrow), but only the peripheral half of the acetabulum (Fig. 4.3O, arrowheads), and was undetectable in the central portion of the fossa (Fig. 4.3O, double arrowhead) where the teres was attaching (Fig. 4.3P, double arrowhead). Indeed, the expression pattern of *Erg*, a transcription factor characterizing developing articular chondrocytes [120], overlapped with that of *Gdf5* expression (Fig. 4.3N). To fully delineate the fate map of *Gdf5*-expressing cells, we used double *Gdf5-Cre;R26R* transgenic mouse embryos and processed serial sections of developing hips for *LacZ* staining. Very clearly, *Gdf5+* cells comprised the entire articulating layers of femur (Fig. 4.3P, arrows), the peripheral half of the acetabulum (Fig. 4.3P, arrowhead) and the teres at E17.5, but were not detectable in the central acetabular fossa (Fig. 4.3P, double arrowhead).

By 3 weeks postnatally, hip joint development had advanced substantially, although it was not yet complete. The femoral head still contained a very substantial amount of growth cartilage (Fig. 4.4A, *gc*) beneath articular cartilage (Fig. 4.4A, *ac*), and the same could be seen beneath articular cartilage around the peripheral acetabular half (Fig. 4.4A, arrows). In addition, the still growing ilium and pubis were bridged by a characteristic synchondrosis (mirror-image growth plate) sustaining skeletal elongation in both directions (Fig. 4.4A, arrowhead). Two additional synchondroses comprised the triradiate growth cartilage that interconnects the growing ilium, ischium, and pubis and overlays the acetabulum [112], best appreciated when viewed from the spinal side (Fig. 4.4B and C, arrowheads). By 6 weeks, the triradiate cartilage and its synchondroses were no longer appreciable indicating that pelvis growth was largely completed (Fig. 4.4D), but the femoral head still had a conspicuous growth cartilage (Fig. 4.4D, *gc*). High magnification showed that whereas acetabular articular cartilage was compact, made of uniform articular chondrocytes and attached to bone (Fig. 4.4E), articular cartilage on the femoral head (Fig. 4.4F, *ac*) was still flanked by growth cartilage with its large hypertrophying chondrocytes. By 6 months of age, the hip joint finally displayed its fully mature features, with femoral and acetabular articular cartilage overlaying subchondral bone and lack of growth cartilage (Fig. 4.4G). The conspicuous nature of the fat pad and teres (Fig. 4.4G, *fp* and *te*) was also evident.

Our data relate quite well with previous studies and provide additional and novel insights into hip joint development. They also point to intriguing differences in the development of hip and more distal limb joints and lead us to the following summary and model (Fig. 4.5). In limb joints such as the knee, *Gdf5+* progenitor cells fully and symmetrically bridge the flanking femur and tibia cartilaginous anlagen and contribute in largely equal manners to the formation of joint tissues on both elements, including articular cartilage and ligaments (Fig. 4.5D) [44]. These interzone cells would be composed of

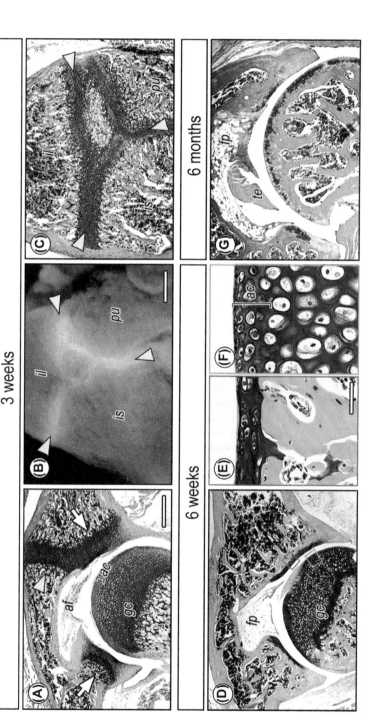

FIGURE 4.4 Hip joint development at postnatal stages. (A–C) In 3 week-old mice, hip development was advanced but not yet complete, as revealed by histochemical inspection of safranin O/fast green-stained sections. In (A), note the presence of abundant growth cartilage (*gc*) beneath both femoral articular cartilage (*ac*) and acetabular articular cartilage (white arrows) and the presence of an active synchondrosis between ilium and pubis (yellow arrowhead). In (B and C), the synchondroses forming the tripartite growth cartilage structure (yellow arrowheads) and interconnecting ilium (*il*), pubis (*pu*), and ischium (*is*) were well appreciable when viewed from the spinal side (B) and also histologically (C). (D–F) In 6-week-old mice, the compacted femoral articular cartilage (D, F, *ac*) still overlaid growth cartilage (D, *gc*), while articular cartilage along the acetabular peripheral half (D) overlaid bone (E). The red and green squared areas in (D) are shown at higher magnification in (E) and (F), respectively. (G) In 6-month-old mice, the hip joint displayed a fully mature structure, including the presence of articular cartilage attached to subchondral bone in every location and typical additional components such as fat pad (*fp*) and teres (*te*). Bar in (A, D, and G): 200 μm; bar in (B) for (B–C); 120 μm; and bar in (E) for (E–F): 40 μm.

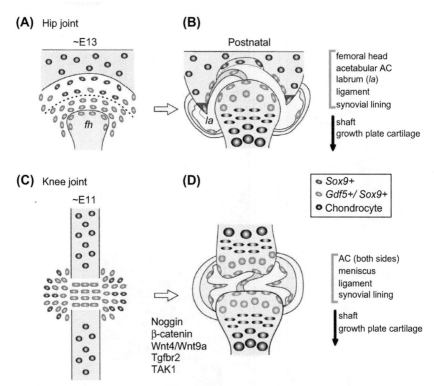

(A) Hip joint   **(B)**

~E13   Postnatal

fh   la

femoral head
acetabular AC
labrum (*la*)
ligament
synovial lining

shaft
growth plate cartilage

(C) Knee joint   **(D)**

~E11

Sox9+
Gdf5+/ Sox9+
Chondrocyte

Noggin
β-catenin
Wnt4/Wnt9a
Tgfbr2
TAK1

AC (both sides)
meniscus
ligament
synovial lining

shaft
growth plate cartilage

FIGURE 4.5 Schematic model summarizing and comparing major features and players in hip and knee joint development and growth. (A) At midgestation stages around approximately E13.0, the developing hip joint is characterized by the following: a prominent *Gdf5/Sox9*-expressing interzone-like mesenchymal cell population (green-colored cells) surrounding the nascent femoral head (*fh*); and a population of *Sox9*-expressing progenitor cells (red-colored cells) located in closer proximity to the ilium cartilaginous primordium (blue-colored cells). (B) During postnatal stages, *Gdf5/Sox9* lineage cells (green) would give rise to the entire femoral articular cartilage, teres, labrum (*la*) synovial lining, and inner capsule, while they would produce only articular cartilage located around the peripheral half of acetabulum. The tissues located in the central portion of acetabular fossa—and possibly including fat pad and accessory components—would originate from *Sox9* lineage cells (red). (C) In the incipient embryonic knee joints starting around E11.0, *Gdf5/Sox9*-expressing cells would constitute the interzone and associated population (green-colored cells), while additional *Sox9*-expressing cells (red-colored cells) would be induced in, and recruited from, the immediate surroundings. The formation, functioning, fate, and roles of these cells would be dictated by an array of critical signaling proteins and transcription factors, including *Noggin*, *β-catenin*, *Wnt4/Wnt9a*, *Tgfbr2*, and *TAK1* in addition to important others such as *Gdf5*, *Gdf6*, *BMPs*, *Erg*, and *c-Jun* (see text for details). (D) During postnatal development and growth, those progenitor populations would give rise to all the tissues present on the opposing femoral and tibial sides of knee joints, including articular cartilage, intrajoint ligaments, synovial lining, and inner portions of capsule and meniscus.

progenitors descending from local dedifferentiated chondrocytes, as well as *Gdf5+/Sox9+* and *Sox9+* cells induced in and recruited from the immediate joint surroundings. However, in the developing hip joint, it appears that *Gdf5*-expressing interzone-like progenitors (with a *Sox9* origin) do not fill the entire prospective joint area but are preferentially associated with the developing femoral head; the area toward the ilium primordium is filled by less dense *Sox9*-expressing cells (Fig. 4.5A and 4.3C). Based on our *Gdf5-Cre;R26R* genetic cell tracing data, the *Gdf5+* cells are responsible for generation of the entire femoral articular cartilage, synovial lining, and teres over time but would produce only articular cartilage present along the peripheral half of the acetabulum, with the central fossa tissues deriving from *Gdf5*-negative and likely *Sox9*-positive progenitors (Fig. 4.5A). It is thus possible that the femoral head is not only responsible for the convex fossa-like shape of the acetabulum [116] but may also be instructing and directing the *Gdf5* progenitors to produce and establish acetabular and femoral articular cartilage, synovial lining, and inner capsule as well as the physical link (the teres) with the pelvis. By its close association with the femur, onset of hip joint development could itself be part of processes under the jurisdiction of, and influenced by, the limb field and local ectoderm [113,114].

CONCLUSIONS AND IMPLICATIONS FOR CARTILAGE REPAIR

As the above summary and analysis of most recent literature show, significant advances have been made toward the identification and clarification of mechanisms subtending limb joint development. It is now clear that the origin, behavior, and fate of *Gdf5+* cells emerging at each prospective joint site are more complex and fluid than previously realized. The cells' dynamism, diverse origin, and sequential temporal genesis provide insights into possible mechanisms by which the cells become determined and committed to different fates and differentiation paths and produce distinct tissues at appropriate times and locations within each joint. The mesen-chymal character of the interzone has long been a puzzle given that at least in medial and distal limb joints, the interzone expresses prochondrogenic factors such as BMPs [67] and interzone cells derive in part from chon-drocytes. The recent study by Klammert et al. [76] does provide a plausible and interesting answer to this riddle indicating that GDF5 can antagonize BMP signaling in certain contexts and by being expressed at the very onset of interzone formation, GDF5 could promote or permit the establishment of interzone's mesenchymal phenotype. As importantly, the study also provides a more probing way to analyze and interpret the biology of gain-of-function *GDF5* mutations linked to abnormalities and even fusion of autopod joints in

SYM1, SYNS, and related syndromes. As described earlier, such mutants could alter the delicate balance between anti- and prochondrogenic mechanisms normally operating within the mesenchymal interzone and tilt it toward chondrogenesis, thus deranging the interzone properties and causing joint fusion or defects. We should add that the most recent literature has also provided novel and illuminating insights into the following: the morphogenesis of joint tissues, and in particular articular cartilage; the multiple regulatory mechanisms operating within, and influencing the phenotype of, the superficial zone; the distinct developmental program of the hip joint; the presence and persistence of embryonic progenitors and their progenies within adult joints (see below); and the pathogenesis of common chronic joint conditions including OA.

In contrast to their innate development, growth, and morphogenetic abilities delineated above, limb joint tissues have long been know to have a regrettably poor capacity for self-repair and regeneration. This has been a continuous source of frustration, puzzlement, and inquiry, and recent authoritative and comprehensive reviews have taken the pulse of the field and pointed to possible and plausible steps forward [23,24,121]. Articular cartilage is central to joint function and is affected by both acute and chronic conditions and thus has received particular attention in repair and regeneration research. Bioengineering approaches have been sought out to compensate for, and supplant, the poor innate repair capacity of articular cartilage. Common themes and goals of such efforts have been the identification of an ideal mixture of scaffolds, growth factors, and stem cells to repair or even regenerate articular cartilage. As the above reviews describe, a very large number of formulations and delivery techniques have been conceived and tested in animal models and implemented in patients. These interventions can certainly be helpful and effective in terms of restoring some joint function and reduce pain, but regrettably, the outcomes are far from ideal, particularly long term. The neocartilaginous tissue resulting from those strategies does not usually resemble native articular cartilage, lacks its structural and phenotypic complexity and zonal organization, has suboptimal biomechanical properties, and does not maintain joint function long term. Possible reasons have been identified to explain these outcomes and limitations [23]. Certainly, a likely reason is that current bioengineering approaches do not incorporate key principles of developmental biology that would mimic or even reproduce the mechanisms by which articular cartilage normally forms and matures. Needless to say, it would be hard to precisely recreate all those mechanisms within the context of joint therapy, but the developmental studies summarized above can offer some general—though admittedly limited and conjectural—suggestions.

Given that joint tissues derive from *Gdf5+* progenitors, this population could theoretically be ideal for joint tissue repair and superior to generic stem cells commonly used in cartilage repair strategies, including bone marrow—derived mesenchymal stem cells (BM-MSCs) and induced pluripotent stem cells (iPSCs) [11,23,24]. Toward establishing a proof-of-principle strategy, we isolated the interzones from the digit joints of E13.0 *Gdf5-GFP* reporter mouse embryos, dispersed them into single-cell population enzymatically, mixed them with hyaluronate hydrogels [122], and implanted them into osteochondral defects created by needle puncture in the knee of adult constitutive *RFP*-expressing host mice. As shown in Fig. 4.6A, the cells were readily deliverable and remained confined within the scaffold-filled defect, at least in the short term. Interzone cells isolated from companion dual-reporter *Gdf5-GFP;Prg4-RFP* mouse embryos were seeded in monolayer. At seeding, most of the cells were *GFP*-positive reflecting their origin and purity (Fig. 4.6B). By 48 h of culture, many had already turned on expression of *Prg4* (Fig. 4.6C) that normally characterizes incipient embryonic limb joint cells and postnatal articular chondrocytes [44,90], indicating their preparedness, responsiveness, and developmental potency. Given that interzone cells can be isolated in sufficient quantities to carry out such in vivo and in vitro mouse experiments, the current data indicate that similar, but long term, experiments could be carried out to compare the repair and regenerative capacities of interzone cells to those of BM-MSCs and iPSCs. Such experiments are in progress in our labs. Should interzone cells turn out to be able to regenerate articular cartilage and thus be superior and ideal compared with other stem cell populations, the task would then be to find conditions to expand the interzone cells in vitro while retaining their native properties [106] and test them in the larger animal models. Alternatively and possibly, transcriptome analyses being carried out on embryonic joint cells [123,124] could be directed to specifically identify master genes imparting and controlling the "interzone phenotype." Such genes could be introduced into iPSCs or embryonic stem cells [125] and induce them to acquire an interzone phenotype, followed by a testing of the reprogrammed cells in animal models.

A related and important issue to consider is the multizone functional structure that adult articular cartilage normally has and in particular, its superficial zone containing flat, tightly bound, and lubricant-producing fibrocartilaginous cells in a scarce isotropic matrix, the intermediate/deep zone with thick stacks of enlarged and round chondrocytes and an abundant, anisotropic, and stable cartilage matrix, and the underlying calcified zone. Given that these structural and compositional features are important not only for resilience but also for endurance, an effective cartilage repair strategy would need to recreate them, at least to a large extent. Ideally, a next generation of combinatorial scaffolds [126] could be designed to contain three to four distinct zones in which each zone would mimic the biomechanical features, architecture, layout, and relative height of the superficial, intermediate, deep, and calcifying zones of mature articular cartilage. Each zone should permit the seeded cells to reach appropriate cell densities,

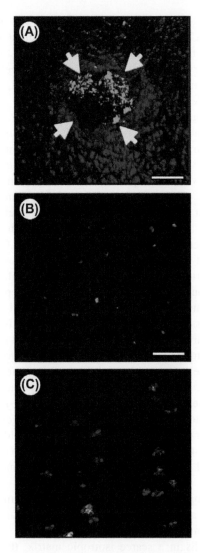

FIGURE 4.6 Fluorescent images of operated mouse knee joint and interzone cells. (A) Whole-mount view of a knee joint osteochondral defect (arrows) that had been filled with GFP-expressing interzone cells isolated from embryonic digit joints and mixed with an HA scaffold. Note that the host mice constitutively expressed *RFP* and thus, all the cells and tissues surrounding the implanted green (*GFP*) interzone cells are red (*RFP*). (B and C) Interzone cells were isolated from dual-reporter *Gdf5-GFP;Prg4-RFP* mouse embryos and seeded in monolayer. Note that soon after seeding, most cells expressed *GFP* in line with their origin and purity (B). By 48 h from seeding, many of the cells became *RFP*-positive, reflecting activation of *Prg4* gene expression and providing evidence of their developmental fidelity and potency (C). Bar in (A): 150 μm; and bar in (B) for (B−C): 60 μm.

architecture, and phenotypes over time and should thus contain sets of inductive or repressive growth factors and cues to promote the development of such zone-selective phenotypes, acting for adequate amounts of time possibly through slow-release technologies [18]. For instance, moderate levels of Wnt/β-catenin and/or EGFR signaling built into the top zone of the construct could promote a superficial zone fibrocartilaginous cell phenotype and *Prg4* expression [99,106]. Given its apparent ability to inhibit BMP signaling in certain contexts, GDF5 could help as well. In comparison, a moderate level of hedgehog or BMP signaling within the intermediate and/or deep zones of the construct could promote a fully cartilaginous phenotype with vigorous cartilage matrix production, deposition, and anisotropic remodeling [66]. As revealed by *Gli1-LacZ* mice, hedgehog signaling does appear to be high in the intermediate/deep zones but low in superficial and bottom zones (Fig. 4.2E). Insulin growth factor signaling has been found to be an important regulator of chondrocyte cell volume enlargement [127], and this pathway could also be built into the construct's deep zone to favor cell volume expansion and in turn, overall tissue thickening. In the bottom zone, reagents such as hydroxyapatite could favor bone integration.

An additional strategy would be to stimulate the endogenous repair capacity of joint tissues and trigger an effective response, and this approach could be tested on its own or in combination with bioengineering interventions above. This is of course a rather difficult goal, but recent developmental studies provide ground for hope. In a just published study, De Bari and coworkers found that embryonic *Gdf5+* lineage cells persist in adult mouse knee synovium and display phenotypic traits of MSCs [128]. The cells vigorously proliferated after acute surgical osteochondral joint injury, and some were even recruited into a *Nestin+* perivascular population. Synovial *Gdf5+* cell hyperplasia was dependent on transcription cofactor *Yap* known to regulate MSC proliferation, and importantly, the cells produced new cartilaginous tissue at the injury site over time. The authors reached the interesting conclusion that synovium is a reservoir of MSCs descending from the embryonic joint interzone and may be able to maintain joint tissues in adults. The data and insights in this study relate quite well with those we reported and described in our own recent study [56]. We found that after osteochondral surgical injury in adult mouse knee joints, the synovium underwent very rapid and prominent hyperplasia associated with strong local cell proliferation and became filled with cells descending from embryonic *Prg4+* progenitors. A similar vigorous response was observed when *Prg4+* progeny cells were first induced in juvenile mice and injury was carried out in adults. At joint locations, where the hyperplastic synovium was near or in contact with the injury site, *Prg4+* cells continuous with the synovial population had filled the injury site itself, providing a clear hint that the synovial cells were migrating en masse into, and filling, the injury site. Interestingly, chondrocytes located within articular cartilage flanking the injury site did not appear to respond in terms of proliferation, migration or change in cell phenotype, at least within the acute 7 day postinjury period examined. These and other data led us to conclude that synovial *Prg4+* lineage cells descending from embryonic or juvenile founders have a strikingly fast ability to respond to joint injury and migrate to/into the injured site, thus behaving as pioneers in injury response. It is likely that the *Prg4+* cells are closely related to, and largely overlapping with, the *Gdf5+* cells [128]. Together, these two just published studies provide clear evidence that cells with a progenitor or even embryonic character persist in the joints of adults. Such cell populations could be isolated, expanded, and incorporated into bioengineered constructs for tissue repair. Additionally, large-scale experiments could be carried out to identify pharmacological, biological, or other means to cox, recruit, induce, or otherwise force those resident progenitors to perform beyond their natural, but unfortunately poor, repair capacity, mount an effective endogenous response to injury, and in turn be able to repair and regenerate joint tissues more fully. Should such strategies become apparent, they could also be tested as a measure to alleviate or reverse the progressive loss of joint structure and function that affects and afflicts the elderly population or even children with early onset and congenital forms of joint disease.

ACKNOWLEDGMENTS

The original studies in the authors' laboratories on which this chapter is based were supported by NIAMS grant R01AR062908, NIAMS grant F32AR064071, and NICDR grant RO1DE023841. The authors express their gratitude to the many colleagues participating in those studies and to our collaborator Dr. David Rowe at the University of Connecticut, Farmington, for providing mouse reporter lines described earlier. Owing to the concise nature of this review chapter, not all relevant and deserving literature and authors could be cited.

REFERENCES

[1] Archer CW, Dowthwaite GP, Francis-West P. Development of synovial joints. Birth Defects Res C 2003;69:144−55.

[2] Pacifici M, Koyama E, Iwamoto M. Mechanisms of synovial joint and articular cartilage formation: recent advances, but many lingering mysteries. Birth Defects Res C 2005;75:237−48.

[3] Pitsillides AA, Ashhurst DE. A critical evaluation of specific aspects of joint development. Dev Dyn 2008;237:2284−94.

[4] Garretson RB, Williams GR. Clinical evaluation of injuries to the acromioclavicular and sternoclavicular joints. Clin Sports Med 2003;22:239−54.

[5] Kulowski J. Flexion contracture of the knee. Clin Orth Relat Res 2007;464:4−10.

[6] Nelson CE, Morgan BA, Burke AC, Laufer E, DiMambro E, Murtaugh LC, et al. Analysis of Hox gene expression in the chick limb bud. Development 1996;122:1449−66.

[7] Yokouchi Y, Sasaki H, Kuroiwa A. Homeobox gene expression correlated with the bifurcation process in limb cartilage development. Nature 1991;353:443−5.

[8] Koyama E, Yasuda T, Wellik DM, Pacifici M. Hox11 paralogous genes are required for formation of wrist and ankle joints and articular surface organization. Ann NY Acad Sci 2010;1192:307−16.

[9] Koyama E, Yasuda T, Minugh-Purvis N, Kinumatsu T, Yallowitz AR, Wellik DM, et al. Hox11 genes establish synovial joint organization and phylogenetic characteristics in developing mouse zeugopod skeletal elements. Development 2010;137:3795−800.

[10] Archer CW, Caterson B, Benjamin M, Ralphs JR. The biology of the synovial joint. London: Harwood Academics; 1999. p. 1−444.

[11] Hunziker EB. Articular cartilage repair: basic science and clinical progress. A review of the current status and prospects. Osteoarthr Cart 2002;10:432−63.

[12] Bhosale AM, Richardson JB. Articular cartilage: structure, injuries and review of management. Br Med Bull 2008;87:77−95.

[13] Kurth TB, Dell' Accio F, Crouch V, Augelio A, Sharpe PT, De Bari C. Functional mesenchymal stem cell niches in adult mouse knee joint synovium in vivo. Arthr Rheum 2011;63:1289−300.

[14] Levick JR, McDonald JN. Fluid movement across synovium in healthy joints: role of synovial fluid macromolecules. Ann Rheum Dis 1995;54:417−23.

[15] Jones BA, Pei M. Synovium-derived stem cells: a tissue-specific stem cell for cartilage engineering and regeneration. Tissue Eng Part B 2012;18:301−11.

[16] Ellison A, Berg E. Embryology, anatomy and function of the anterior cruciate ligament. Ortho Clin North Am 1985;16:3−14.

[17] Schtte MJ, Dabezies EJ, Zimny ML, Happel LT. Neural anatomy of the human anterior cruciate ligament. J Bone Joint Surg Am 1987;69:243−7.

[18] Decker RS, Koyama E, Pacifici M. Genesis and morphogenesis of limb synovial joints and articular cartilage. Matrix Biol 2014;39:5−10.

[19] Salazar VS, Gamer LW, Rosen V. BMP signaling in skeletal development, disease and repair. Nat Rev Endocrinol 2016;12:203−21.

[20] Longobardi L, Li T, Tagliaferro L, Temple JD, Willicockson HH, Ye P, et al. Synovial joints: from development to homeostasis. Curr Osteoporos Rep 2015;13:41−51.

[21] Decker RS. Articular cartilage and joint development from embryogenesis to adulthood. Sem Cell Dev Biol 2017;62:50−6.

[22] Parvizi J, Pour AE, Hillibrand AS, Goldeberg G, Sharkey PF, Rothman RH. Back pain and total hip arthroplasty: a prospective natural history study. Clin Orth Relat Res 2010;468:1325−30.

[23] Johnstone B, Alini M, Cucchiarini M, Dodge GR, Eglin D, Guilak F, et al. Tissue engineering for articular cartilage repair. The state of the art. Eur Cells Mater 2013;25:248−67.

[24] Makris EA, Gomoll AH, Malizos KN, Hu JC, Athanasiou KA. Repair and tissue engineering techniques for articular cartilage. Nat Rev Rheumatol 2015;11:21−34.

[25] Capdevilla J, Izpisua Belmonte JC. Patterning mechanisms controlling vertebrate limb development. Annu Rev Cell Dev Biol 2001;17:87−132.

[26] Hinchliffe JR, Johnson DR. The development of the vertebrate limb. New York: Oxford University Press; 1980. p. 72−83.

[27] Zeller R, Lopez-Rios J, Zuniga A. Vertebrate limb bud development: moving towards integrative analysis of organogenesis. Nature Rev Genet 2009;10:845−58.

[28] Woods A, Wang G, Dupuis H, Shao Z, Beier F. Rac1 signaling stimulates N-cadherin expression, mesenchymal condensation, and chondrogenesis. J Biol Chem 2007;282:23500−8.

[29] Lim J, Tu X, Choi K, Akiyama H, Mishina Y, Long F. BMP-Smad4 signaling is required for precartilaginous mesenchymal condensation independent of Sox9 in the mouse. Dev Biol 2015;400:132−8.

[30] Drushel RF, Pechak DG, Caplan AI. The anatomy, ultrastructure and fluid dynamics of the developing vasculature of the embryonic chick wing bud. Cell Diff 1985;16:13−28.

[31] Feinberg RN, Latker CH, Beebe DC. Localized vascular regression during limb morphogenesis in the chicken embryo. I. Spatial and temporal changes in the vascular pattern. Anat Rec 1986;214:405−9.

[32] Yin M, Pacifici M. Vascular regression is required for mesenchymal condensation and chondrogenesis in the developing limb. Dev Dyn 2001;222:522−33.

[33] Provot S, Zinyk DL, Gunes Y, Kathri R, Le Q, Kronenberg HM, et al. Hif-1alpha regulates differentiation of limb bud mesenchyme and joint development. J Cell Biol 2007;177:451−64.

[34] Amarilio R, Viukov SV, Sharir A, Eshkar-Oren I, Johnson RS, Zelzer E. HIF1alpha regulation of Sox9 is necessary to maintain differentiation of hypoxic prechondrogenci cells during early skeletogenesis. Development 2007;134:3917−28.

[35] Hamrick MW. Primate origins: evolutionary change in digital ray patterning and segmentation. J Human Evol 2001;40:339−51.

[36] Holder N. An experimental investigation into the early development of the chick elbow joint. J Embryol Exp Morphol 1977;39:115−27.

[37] Mitrovic D. Development of the diathrodial joints in the rat embryo. Am J Anat 1978;151(4):475−85.

[38] Shubin NH, Daeschler EB, Jenkins FA. The pectoral fin of Tiktaalik roseae and the origin of the tetrapod limb. Nature 2006;440:764−71.

[39] George D, Blieck A. Rise of the earliest tetrapods: an early Devonial origin from marine environment. PLoS One 2011;6:e22136.

[40] Cosden-Decker RS, Bickett MM, Lattermann C, MacLeod JN. Structural and functional analysis of intra-articular interzone tissue in axolotl salamanders. Osteoarthr Cart 2012;20:1347−56.

[41] Lee J, Gardiner DM. Regeneration of limb joints in the axolot (*Ambystoma mexicanum*). PLoS One 2012;7:e50615.

[42] Rountree RB, Schoor M, Chen H, Marks ME, Harley V, Mishina Y, et al. BMP receptor signaling is required for postnatal maintenance of articular cartilage. PLoS Biol 2004;2:1815−27.

[43] Storm EE, Kingsley DM. GDF5 coordinates bone and joint formation during digit development. Dev Biol 1999;209:11−27.

[44] Koyama E, Shibukawa Y, Nagayama M, Sugito H, Young B, Yuasa T, et al. A distinct cohort of progenitor cells participates in synovial joint and articular cartilage formation during mouse limb skeletogenesis. Dev Biol 2008;316:62−73.

[45] Craig FM, Bentley G, Archer CW. The spatial and temporal pattern of collagens I and II and keratan sulphate in the developing chick metatarsophalangeal joint. Development 1987;99:383−91.

[46] Nalin AM, Greenlee TK, Sandell LJ. Collagen gene expression during development of avian synovial joints: transient expression of type II and XI collagen genes in the joint capsule. Dev Dyn 1995;203:352−62.

[47] Soeda T, Deng JM, de Crombrugghe B, Behringer RR, Nakamura T, Akiyama H. Sox9-expressing precursors are the cellular origin of the cruciate ligament of the knee joint and the limb tendons. Genesis 2010;48:635–44.

[48] Zhang Q, Cigan AD, Marrero L, Lopreore C, Liu S, Ge D, et al. Expression of doublecortin reveals articular chondrocyte lineage in mouse embryonic limbs. Genesis 2010;49:75–82.

[49] Spagnoli A, O'Rear L, Chandler RL, Granero-Molto F, Mortlock DP, Gorska AE, et al. TGF-β signaling is essential for joint morphogenesis. J Cell Biol 2007;177:1105–17.

[50] Li TF, Longobardi L, Myers TJ, Temple JD, Chandler RL, Ozkan H, et al. Joint TGF-β type II receptor-expressing cells: ontogeny and characterization as joint progenitors. Stem Cells Dev 2013;22:1342–59.

[51] Longobardi L, Li T, Myers TJ, O'Rear L, Ozkan H, Li Y, et al. TGF-β type II receptor/MCP-5 axis: at the crossroad between joint and growth plate development. Dev Cell 2012;23:71–81.

[52] Shwartz Y, Viukov S, Krief S, Zelzer E. Joint development involves a continuous influx of Gdf5-positive cells. Cell Rep 2016;15:1–11.

[53] Koyama E, Ochiai T, Rountree RB, Kingsley DM, Enomoto-Iwamoto M, Iwamoto M, et al. Synovial joint formation during mouse limb skeletogenesis. Roles of Indian hedgehog signaling. Ann NY Acad Sci 2007;1116:100–12.

[54] Niedermaier M, Schwabe GC, Fees S, Helmrich A, Brieske N, Seeman P, et al. An inversion involving the mouse Shh locus results in brachydactyly through dysregulation of Shh expression. J Clin Invest 2005;115:900–9.

[55] Ray A, Ingh PNP, Sohaskey ML, Harland RM, Bandyopadhyay A. Precise spatial restriction of BMP signaling is essential for articular cartilage differentiation. Development 2015;142:1169–79.

[56] Decker RS, Um HB, Dyment NA, Cottingham N, Usami Y, Enomoto-Iwamoto M, et al. Cell origin, volume and arrangement are drivers of articular cartilage formation, morphogenesis and response to injury in mouse limbs. Dev Biol 2017;426:56–68.

[57] Snippert HJ, van der Flier LG, Sato T, van Es JH, van den Born M, Kroon-Veeboer C, et al. Intestinal crypt homeostasis results from neutral competition between symmetrically dividing Lgr5 stem cells. Cell 2010;143:134–44.

[58] Decker RS, Koyama E, Pacifici M. Articular cartilage: structural and developmental intricacies and questions. Curr Osteoporos Rep 2015;13:407–14.

[59] Brunet LJ, McMahon JA, McMahon AP, Harland RM. Noggin, cartilage morphogenesis, and joint formation in the mammalian skeleton. Science 1998;280:1455–7.

[60] Gunnell LM, Jonason JH, Loiselle AE, Kohn A, Schwarz EM, Hilton MJ, et al. TAK1 regulates cartilage and joint development via the MAPK and BMP signaling pathways. J Bone Min Res 2010;25:1784–97.

[61] Spater D, Hill TP, O'Sullivan RJ, Gruber M, Conner DA, Hartmann C. Wnt9a signaling is required for joint integrity and regulation of Ihh during chondrogenesis. Development 2006;133:3039–49.

[62] Gao Y-H, Lan Y, Liu H, Jiang R. The zinc finger transcription factors Osr1 and Osr2 control synovial joint formation. Dev Biol 2011;352:83–91.

[63] Kan A, Tabin CJ. c-Jun is required for specification of joint cell fates. Genes Dev 2013;27:514–24.

[64] Dy P, Smits P, Silvester A, Penzo-Mendez A, Dumitriu B, Han Y, et al. Synovial joint morphogenesis requires the chondrogenic action of Sox5 and Sox6 in growth plate and articular cartilage. Dev Biol 2010;341:346–59.

[65] St-Jacques B, Hammerschmidt M, McMahon AP. Indian hedgehog signaling regulates proliferation and differentiation of chondrocytes and is essential for bone formation. Genes Dev 1999;13:2076—86.

[66] Rockel JS, Yu C, Whetstone H, Craft AM, Reilly K, Ma H, et al. Hedgehog inhibits β-catenin activity in synovial joint development and osteoarthritis. J Clin Invest 2016;126:1649—63.

[67] Bitgood MJ, McMahon AP. Hedgehog and Bmp genes are coexpressed at many diverse sites of cell-cell interaction in the mouse embryo. Dev Biol 1995;172:126—38.

[68] Lefebvre V, Smits P. Transcriptional control of chondrocyte fate and differentiation. Birth Defects Res C 2005;75:200—12.

[69] Storm EE, Kingsley DM. Joint patterning defects caused by single and double mutations in members of the bone morphogenetic protein (BMP) family. Development 1996;122:3969—79.

[70] Seemann P, Schwappecher R, Kjaer KW, Krakow D, Lehmann K, Dawson K, et al. Activating and deactivating mutations in the receptor interaction site of GDF5 cause symphalangism or brachydactyly type A2. J Clin Invest 2005;115:2373—81.

[71] Schwaerzer GK, Hiepen C, Schrewe H, Nickel J, Pioeger F, Sebald W, et al. New insights into the molecular mechanism of multiple synostoses syndorme (SYNS): mutation within the GDF5 knuckle epitope causes noggin-resistance. J Bone Min Res 2012;27:429—42.

[72] Chen H, Capellini TD, Schoor M, Mortlock DP, Reddi AH, Kingsley DM. Heads, shoulders, elbows, knees, and toes: modular Gdf5 enhancers control different joints in the vertebrate skeleton. PLoS Genet 2016;12:e1006454.

[73] Loughlin J. Genetic contribution to osteoarthritis development: current state of evidence. Curr Opin Rheumatol 2015;27:284—8.

[74] Hotten GC, Matsumoto T, Kimura M, Bechtold RF, Kron R, Ohara T, et al. Recombinant human growth/differentiation factor 5 stimulates mesenchyme aggregation and chondrogenesis responsible for the skeletal development of limbs. Growth Factors 1996;13:65—74.

[75] Seemann P, Brehm A, Konig J, Reissner C, Stricker S, Kuss P, et al. Mutations in GDF5 reveal a key residue mediating BMP inhibition by NOGGIN. PLoS Genet 2009;5:e1000747.

[76] Klammert U, Mueller TD, Hellmann TV, Wuerzler KK, Kotzsch A, Schliermann A, et al. GDF-5 can act as a context-dependent BMP-2 antagonist. BMC Biol 2015;13:77.

[77] Heinecke K, Seher A, Schmitz W, Mueller TD, Sebald W, Nickel J. Receptor oligomerization and beyond: a case study in bone morphogenetic proteins. BMC Biol 2009;7:59.

[78] Nickel J, Kotzsch A, Sebald W, Mueller B. A single residue of GDF-5 defines binding specificity to BMP receptor IB. J Mol Biol 2005;349:933—47.

[79] Shen B, Bhargav D, Wei A, Williams LA, Tao H, Ma DD, et al. BMP-13 emerges as a potential inhibitor of bone formation. Int J Biol Sci 2009;5:192—200.

[80] Wang W, Rigueur D, Lyons KM. TFGβ signaling in cartilage development and maintenance. Birth Defects Res C 2014;102:37—51.

[81] Hunziker EB, Kapfinger E, Geiss MD. The structural architecture of adult mammalian articular cartilage evolves by a synchronized process of tissue resorption and neoformation during postnatal development. Osteoarthr Cart 2007;15:403—13.

[82] Mow VC, Ratcliffe A, Poole AR. Cartilage and diartrodial joints as paradigms for hierarchical materials and structures. Biomaterials 1992;13:67—97.

[83] Gannon AR, Nagel T, Bell AP, Avery NC, Kelly DJ. Postnatal changes to the mechanical properties of articular cartilage are driven by the evolution of its collagen network. Eur Cells Mater 2015;29:105−21.

[84] Julkunen P, Harijula T, Iivarinen J, Marjanen J, Seppanen K, Narhi T, et al. Biomechanical, biochemical and structural correlations in immature and mature rabbit articular cartilage. Osteoarthr Cart 2009;17:1628−38.

[85] Hyde G, Dover S, Aszodi A, Wallis GA, Boot-Handford RP. Lineage tracing using matrilin-1 gene expression reveals that articular chondrocytes exist as the joint interzone forms. Dev Biol 2007;304:825−33.

[86] Dowthwaite GP, Bishop JC, Redman SN, Khan IM, Rooney P, Evans DJR, et al. The surface of articular cartilage contains a progenitor cell population. J Cell Sci 2004;117:889−97.

[87] Hayes AJ, MacPherson S, Morrison H, Dowthwaite GP, Archer CW. The development of articular cartilage: evidence for an appositional growth mechanism. Anat Embryol 2001;203:469−79.

[88] Williams R, Khan IM, Richardson K, Nelson I, McCarthy HE, Analbelsi T, et al. Identification and clonal characterirization of a progenitor cell sub-population in normal human articular cartilage. PLoS One 2010;5:e13246.

[89] Kozhemyakina E, Zhang MQ, Ionescu A, Kobayashi A, Kronenberg HM, Warman ML, et al. Identification of a Prg4-expressing articular cartilage progenitor cell population in mice. Arthr Rheum 2015;67:1261−73.

[90] Rhee DK, Marcelino J, Baker M, Gong Y, Smits P, Lefebvre V, et al. The secreted glycoprotein lubricin protects cartilage surfaces and inhibits synovial cell outgrowth. J Clin Invest 2005;115:622−31.

[91] Robling AG, Dujvelaar KM, Geevers JV, Chashi N, Turner CH. Modulation of appositional and longitudinal bone growth in the rat ulna by applied static and dynamic force. Bone 2001;29:105−13.

[92] Eyre D. Review: collagen of articular cartilage. Arthr Res 2002;4:30−5.

[93] Poole AR, Kojima T, Yasuda T, Mwale F, Kobayashi M, Laverty S. Composition and structure of articular cartilage: a template for tissue repair. Clin Orth Relat Res 2001;391:526−33.

[94] Breur GJ, VanEnkevort BA, Farnum CE, Wilsman NJ. Linear relationship between the volume of hypertrophic chondrocytes and the rate of longitudinal bone growth in growth plates. J Ortho Res 1991;9:348−59.

[95] Tada M, Heisenberg C-P. Convergent extension: using collective cell migration and cell intercalation to shape embryos. Development 2012;139:3897−904.

[96] Li L, Newton PT, Bouderlique T, Sejnohova M, Zikmund T, Kozhemyakina E, et al. Superficial cells are self-renewing chondrocyte progenitors, which form the articular cartilage in juvenile mice. FASEB J 2016;31:1067−84.

[97] Aigner T, Vornehm SI, Zetler C, Dudhia J, von der Mark K, Bayliss MI. Suppression of cartilage matrix gene expression in upper zone chondrocytes of osteoarthritic cartilage. Arthr Rheum 1997;40:562−9.

[98] Becerra J, Andrades JA, Guerado E, Zamora-Navas P, Lopez Puertas JM, Reddi AH. Articular cartilage: structure and regeneration. Tissue Eng Part B 2010;16:617−27.

[99] Jia H, Ma X, Tong W, Doyran B, Sun Z, Wang L, et al. EGFR signaling is critical for maintaining the superficial layer of articular cartilage and preventing osteoarthritis initiation. Proc Natl Acad Sci USA 2016;113:14360−5.

[100]   Zhang X, Siclari VA, Lan S, Zhu J, Koyama E, Dupuis HL, et al. The critical role of the epidermal growth factor receptor in endochondral ossification. J Bone Min Res 2011;26:2622−33.

[101]   Glasson SS, Blanchet TJ, Morris EA. The surgical destabilization of the medial meniscus (DMM) model of osteoarthritis in the 129/SvEv mouse. Osteoarthr Cart 2007;15:1061−9.

[102]   Zhang M, Mani SB, He Y, Hall AM, Xu L, Zurakowski D, et al. Induced superficial chondrocyte death reduces catabolic cartilage damage in murine posttraumatic osteoarthritis. J Clin Invest 2016;126:2893−902.

[103]   Henry SP, Liang S, Akdemir KC, de Crombrugghe B. The postnatal role of Sox9 in cartilage. J Bone Min Res 2012;27:2511−25.

[104]   Decker RS, Koyama E, Enomoto-Iwamoto M, Maye P, Rowe P, Zhu S, et al. Mouse limb skeletal growth and synovial joint development are coordinately enhanced by Kartogenin. Dev Biol 2014;395:255−67.

[105]   Maepa M, Razwinani M, Motaung S. Effects of resveratrol on collagen type II protein in the superficial and middle zone chondrocytes of porcine articular cartilage. J Ethnopharmacol 2016;178:25−33.

[106]   Yasuhara R, Ohta Y, Yuasa T, Kondo N, Hoang T, Addya S, et al. Roles of β-catenin signaling in phenotypic expression and proliferation of articular cartilage superficial zone cells. Lab Invest 2011;91:1739−52.

[107]   Bencardino JT, Palmer WE. Imaging of hip disorders in athletes. Radiol Clin North Am 2002;40:267−87.

[108]   Grant AB, Sala DA, Davidovitch RI. The labrum: structure, function, and injury with femero-acetabular impingment. J Child Orthop 2012;6:357−72.

[109]   Brownbill RA, Ilich JZ. Hip geometry and its role in fracture: what do we know so far? Curr Osteoporos Rep 2003;1:25−31.

[110]   Boese CK, Jostmeier J, Oppermann J, Dargel J, Chang D, Eysel P, et al. The neck shaft angle: CT reference values of 800 adult hips. Skeletal Radiol 2016;45:455−63.

[111]   Millan MS, Kaliontzopoulou A, Rissech C, Turbon D. A geometric morphometric analysis of acetabular shape of the primate hip joint in relation to locomotor behaviour. J Human Evol 2015;83:15−27.

[112]   Ponseti IV. Morphology of the acetabulum in congenital displocation of the hip. J Bone Joint Surg Am 1978;60:586−99.

[113]   Malashichev Y, Borkhvardt V, Christ B, Scaal M. Differential regulation of avian pelvic girdle development by the limb field ectoderm. Anat Embryol 2005;210:187−97.

[114]   Malashichev Y, Christ B, Prols F. Avian pelvis originates from lateral plate mesoderm and its development requires signals from both ectoderm and paraaxial mesoderm. Cell Tissue Res 2008;331:595−604.

[115]   Pomikal C, Streicher J. 4D-analysis of early pelvic girdle development in the mouse (*Mus musculus*). J Morphol 2010;271:116−26.

[116]   Harrison TJ. The influence of the femoral head on pelvic growth and acetabular form in the rat. J Anat 1961;95:12−24.

[117]   Nowlan NC, Sharpe J. Joint shape morphogenesis precedes cavitation of the developing hip joint. J Anat 2014;224:482−9.

[118]   Ralis Z, McKibbin B. Changes in shape of the human hip joint during its development and their relation to its stability. J Bone Joint Surg Br 1973;55:780−5.

[119]   Giorgi M, Carriero A, Shefelbine SJ, Nowlan NC. Effects of normal and abnormal loading conditions on morphogenesis of the prenatal hip joint: application to hip dysplasia. J Biomech 2015;48:3390−7.

[120] Iwamoto M, Tamamura Y, Koyama E, Komori T, Takeshita N, Williams JA, et al. Transcription factor ERG and joint and articular cartilage formation during mouse limb and spine skeletogenesis. Dev Biol 2007;305:40−51.

[121] Correa D, Lietman SA. Articular cartilage repair: current needs, methods and research directions. Sem Cell Dev Biol 2017;62:67−77.

[122] Kim IL, Mauck RL, Burdick JA. Hydrogel design for cartilage tissue engineering: a case study with hyaluronic acid. Biomaterials 2011;32:8771−82.

[123] Jenner F, Ijpma A, Cleary MA, Heijsman D, Narcis R, van der Spek PJ, et al. Differential gene expression of the intermediate and outer interzone layers of developing articular cartilage in murine embryos. Stem Cells Dev 2014;23:1883−98.

[124] Wu LN, Bluguermann C, Kyupelyan L, Latour B, Gonzalez S, Shah S, et al. Human developmental chondrogenesis as a basis for engineering chondrocytes from pluripotent stem cells. Stem Cell Reports 2013;1:575−89.

[125] Craft AM, Ahmed N, Rockel JS, Baht GS, Alman BA, Kandel RA, et al. Specification of chondrocytes and cartilage tissues from embryonic stem cells. Development 2013;140:2597−610.

[126] Steele JAM, McCullen SD, Callanan A, Autefage H, Accardi MA, Dini D, et al. Combinatorial scaffold morphologies for zonal articular cartilage engineering. Acta Biomater 2014;10:2065−75.

[127] Cooper KL, Oh S-H, Sung Y, Dasari RR, Kirschner MW, Tabin CJ. Multiple phases of chondrocyte enlargement underlie differences in skeletal proportions. Nature 2013;495:375−8.

[128] Roelofs AJ, Zupan J, Riemen AHK, Kania K, Ansboro S, White N, et al. Joint morphogenetic cells in the adult mammalian synovium. Nat Commun 2017. epub 15040.

Chapter 5

Stem Cell-Based Approaches for Cartilage Tissue Engineering: What Can We Learn From Developmental Biology

Roberto Narcisi[1], April M. Craft[2, 3]

[1]Erasmus MC, Rotterdam, The Netherlands; [2]Boston Children's Hospital, Boston, MA, United States; [3]Harvard Stem Cell Institute, Cambridge, MA, United States

CURRENT STATE OF THE CARTILAGE REPAIR FIELD

General Introduction

Patients who have damaged their cartilage are extremely likely to experience joint pain. At that time, health care providers and orthopaedic surgeons can currently only provide treatments with limited long term efficacy, such as pain medication or surgery. Depending on the extent of the lesions, surgical options to treat cartilage injuries include autologous chondrocyte implantation (ACI), marrow-stimulating techniques, or total joint replacement. The most common treatment options for focal cartilage defect repair are marrow-stimulating techniques such as microfracture and Pridie-drilling. Both are used to generate channels through the subchondral bone in order to permit bone marrow-derived cells to enter the defect and initiate and contribute to the healing process. The newly generated tissue does alleviate pain in patients for some time, however, the tissue typically does not contain significant amounts of normal articular cartilage matrix proteins, and is thus mechanically inferior. The realistic aim for cartilage repair strategies is to provide patients with long-lasting pain relief and improved mobility. Each year that a joint replacement surgery can be delayed is associated with a tremendous amount of benefits for patients, their families, their communities, and the overall health care system. Those include reductions in pain management costs and risk of addictions, reduced healthcare costs associated with multiple joint replacements and rehabilitation and disability, reduced burdens on families and local community health and social services, and the overall ability of individuals to return to the workforce and their normal daily activities.

Once damaged, cartilage progressively deteriorates because it lacks the ability to repair itself efficiently, which increases a patient's predisposition to develop severe degenerative pathologies such as osteoarthritis. Investigators have begun testing cell and tissue replacement therapies in patients with damaged cartilage in order to prevent this degenerative progression. ACI is a procedure in which a patient's own cartilage cells (chondrocytes) are harvested, expanded ex vivo, and then transplanted back into the damaged cartilage area. Although this procedure transiently alleviates joint pain, the effect is short-lived because the transplanted cells do not regenerate bona fide articular cartilage but instead produce a fibrocartilage-like tissue that lacks the structural integrity of articular cartilage, and can ultimately fail [1]. Investigators have been developing these types of repair strategies with advances resulting in products such as DeNovo NT [2,3], which is an implant of particulate allograft cartilage (Zimmer, Inc.), and NeoCart (Histogenics), a tissue implant comprised of autologous chondrocytes currently in Phase II clinical trials [4,5]. Due to the differences in mechanical properties of autologous implants that are still observed after several years, there remains an increased chance of delaminating from the surface and causing further damage in the joint. Procedures that utilize autologous cells also have increased risk and costs, as the patient is required to undergo two procedures, the first to retrieve tissue and cells from a healthy intact region of articular cartilage, and the second to implant the expanded cells and/or tissue into the first or original damaged cartilage area. As such, the integrity of an otherwise healthy region of cartilage has been violated in order to harvest autologous cells to treat the primary defect, introducing the possibility of donor site morbidity. While none of these approaches is ideal, they do not preclude a patient from 1 day undergoing a joint replacement surgery. Encouragingly, they have already shown the potential to delay the need for joint replacement surgery, which again, provides a tremendous amount of benefit to the patient and to overall healthcare systems.

In order to better recreate or regenerate articular cartilage tissues, we need to understand how it develops normally and identify cell sources that have the potential to recapitulate these processes.

CARTILAGE DEVELOPMENT

Embryonic Origins of Cartilage

One of the main challenges with cartilage repair and regeneration is that articular cartilage forms prenatally and regeneration does not normally occur after birth. Thus, by understanding the processes that result in articular cartilage formation during embryonic development, we can gain insights into how to better approach cartilage tissue engineering strategies with appropriately staged cell populations, whether embryonic/pluripotent stem cells or post-natal stem cells are used.

During development, cartilage is formed from both mesoderm and ectoderm/neural crest lineages. Mesoderm derived cartilage lays the template for both the axial skeleton and the skeletal elements of the limbs. The cartilage that lines axial skeletal elements is derived from paraxial mesoderm and somites, while lateral plate mesoderm gives rise to the skeletal elements in our limbs, including the articular cartilage lining these bone surfaces in joints. Neural crest-derived cartilage found in the craniofacial regions (i.e., nasal chondrocytes) provides shape and important biological functions, however, they do not function to support compressive loads like the cartilage found in our joints. While nasal chondrocytes can be isolated and have elegantly been shown to repair damaged cartilage [6], for the purpose of this chapter, we will focus on the development of cartilage derived from mesoderm, as these tissues are most affected by degenerative joint disease and injuries.

We will start in the early vertebrate embryo prior to the emergence of the three germ layers: endoderm, mesoderm and ectoderm. A fertilized egg will shortly develop into a blastocyst, which contains a cluster of cells known as the inner cell mass (Fig. 5.1A). The cells of the inner cell mass will give rise to the embryo proper, and are the immediate source of ex vivo derived embryonic stem cells [7,8] (Fig. 5.1E). Once the embryo attaches to the uterine lining, epiblast cells will begin to proliferate and organize in a very dynamic manner. While primitive ectoderm lining the embryo exists by this development timepoint, both mesoderm and endoderm first emerge during gastrulation, in an area of the embryo known as the primitive streak (Fig. 5.1B). In mice, epiblast cells form a cup-shaped structure and cells will begin to transverse an area in this structure called the primitive streak. The germ layers of the developing human embryo are organized as a trilaminar disc at this stage, with the primitive streak occupying the medial region along the rostro-caudal axis (Fig. 5.2). Although these two physical structures differ between humans and mice, the dynamic movement of the cells and the signals they receive during gastrulation are quite similar. As continuous waves of epiblast cells move through the primitive streak in a temporal and spatially distinct manner, they are induced to differentiate into specific mesendoderm lineages by gradients of signaling factors. Nodal, a member of the transforming growth factor-β (TGF-β) superfamily, is highly expressed in distal/anterior region (i.e., the node), bone morphogenetic protein-4 (BMP4) is expressed in the proximal region of the embryo in the yolk sac, and canonical WNT signaling is active throughout the primitive streak region in the embryo and is required for this process (Fig. 5.1B). The combination, concentration and duration of these factors dictate not only whether an epiblast cell becomes endoderm or mesoderm,

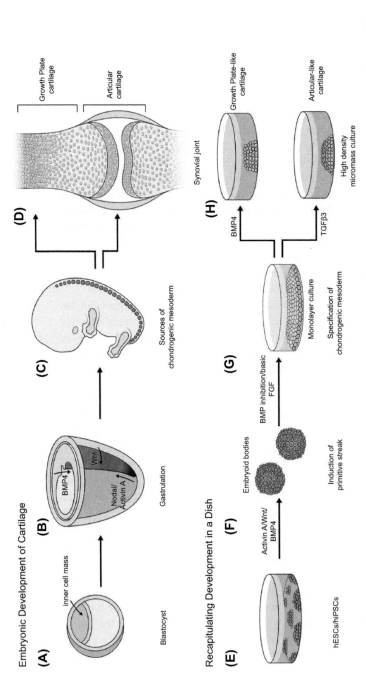

FIGURE 5.1 Stages of embryonic cartilage development are recapitulated in vitro using pluripotent stem cells. The cells of the inner cell mass (A, green) of the blastocyst staged embryo give rise to the embryo proper and are also the source of embryonic stem cells (E). Mesoderm progenitors arise during gastrulation as epiblast cells traverse the primitive streak (B). Here, the cells will receive signals such as BMP4, Nodal, and Wnt. Cells that give rise to cartilage are found in the anterior most part of the posterior primitive streak (B, purple). Paraxial mesoderm/somites (C, dark blue) give rise to the axial skeleton and lateral plate mesoderm gives rise to chondrogenic and skeletal elements in the limbs (C, light blue). (D) Illustrated in a synovial joint, articular cartilage lines the surfaces of the long bones while growth plate cartilage lays the template for new bone to form. Human (or mouse) embryonic stem cells, as well as induced pluripotent stem cells, can either be maintained in self-renewing culture conditions (E) or induced to become primitive streak mesoderm progenitors in vitro when cultured as embryoid bodies (or in monolayer) in the presence of Activin A (a surrogate for Nodal), BMP4, and Wnt (F, purple). Chondrogenic mesoderm is specified by inhibiting the BMP signaling pathway and by promoting FGF signaling with basic-FGF/FGF-2 in monolayer culture (G). Cartilaginous tissues are generated from chondrogenic progenitors though high density micromass culture in the presence of TGFβ3 to produce articular chondrocytes and articular-like cartilage or BMP4 to produce hypertrophic chondrocytes and growth plate-like cartilage (H). Protocol described in detail in Craft et al. [37]. *Illustrations by Tasha McAbee, M.S. CMI (Boston Children's Hospital).*

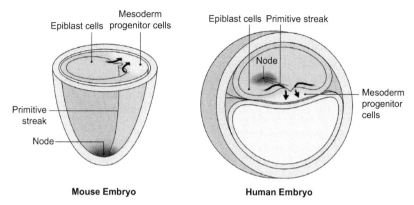

FIGURE 5.2 Differences and similarities between mouse and human embryos during gastrulation. Epiblast cells (green) traverse the primitive streak region as they acquire either a mesoderm (blue) or endoderm fate. The location of the embryonic node (purple) is indicated. *Illustrations by Tasha McAbee, M.S. CMI (Boston Children's Hospital).*

but distinct mesodendoerm lineages are also specified at this earliest divergence point. All cells that traverse the primitive streak initiate expression of the transcription factor Brachyury [9,10]. The expression of cell surface markers such as Flk-1/KDR (VEGF2R) and Platelet-derived growth factor receptor alpha (PDGFRa), among others, can however, be used to distinguish specific mesoderm subsets with differential developmental potential [11,12]. Lineage tracing studies in animal models (e.g., xenopus, mice and chick) confirmed that cells found in specific regions of the primitive streak give rise to distinct tissues later in development. A subset of anterior primitive streak cells are fated to become lateral plate mesoderm which gives rise to skeletal elements that form in the limb described in detail below (Fig. 5.1C). As the embryo further develops, the primitive streak itself elongates and forms 2 strips of paraxial mesoderm lining the notochord that coordinately segment into paired somites (Fig. 5.1C). The axial skeleton derives from the ventral part of the somite called the sclerotome, which received signals such as hedgehog agonists and BMP antagonists from the notochord and floor plate regions. In contrast, the dorsal part of the somite containing the myotome and dermomyotome receives signals such as WNT from the overlying ectoderm, and gives rise to skeletal muscle and the dermis. While much of this information may seem irrelevant for the repair of adult cartilage tissue, these stages of differentiation are imperative when the goal is to generate cartilage from an embryonic stem cell or an induced pluripotent stem cell source.

Regulation of Cartilage Differentiation During Development

Developmental processes regulating articular cartilage formation have been extensively studied in the last years, first to increase the knowledge in basic cartilage biology, but also in order to provide new insights for the development of new tissue engineer-based approaches. The development of synovial joints is discussed in detail in a previous chapter, however, the following section will focus on the generation of articular cartilage specifically. During cartilage development, cells in the joint undergo a series of precise structural and functional changes, finely regulated by environmental factors. Among them, growth factors and mechanical stimulation play a critical role.

Growth Factors

Early limb growth is regulated by a WNT-FGF signaling loop and experiments demonstrated that an ectopic expression of those growth factors can induce limb development [13]. As limb development proceeds, signals from the apical ectodermal ridge (AER) maintain the underlying mesodermal cells in a proliferative and undifferentiated state [14,15], mainly under the synergistic effect of WNT3a and FGF-8 and via the suppression of SOX9 expression [16]. Non-canonical WNT signals (β-catenin independent) also play an important role since, for example, cell proliferation directly correlate with the presence of WNT5a in the limb mesoderm [17]. TGF-β was also found to be expressed in the AER with a negative effect on proliferation through interference with the FGF signaling pathway and the down-regulation of the BMP inhibitor Gremlin [18], known to sustain the proliferation via the maintenance of the expression of FGF-8 in the AER [19].

Later, during chondrocyte specification and maturation, WNT, FGF and TGF-β play a central role, often working differently depending on the differentiation stage. The chondrogenic differentiation process proceeds via three main stages: condensation, differentiation and hypertrophy/terminal differentiation (Fig. 5.1D). Each of these three stages is characterized by the expression of specific markers, such as N-cadherin for condensing chondrogenic mesenchyme, aggrecan and type II collagen (COL2A1) for chondrogenic differentiation, and alkaline phosphatase and type X collagen (COL10A1) for the hypertrophy/terminal differentiation stage [20]. WNT signals and the typical FGF downstream molecule mitogen-activated protein kinase (MAPK) are crucial to initiate chondrogenic differentiation by repressing N-cadherin [21]. During chondrocyte specification, the regulation of WNT signaling by TGF-β molecules is particularly complex (reviewed by Cleary et al. [22]). Depending on the differentiation stage, chondrogenic cells can, for example, respond to TGF-β by either enhancing or repressing WNT signaling. Regulation of FGF pathways during late developmental stages is poorly understood, although the presence of FGF receptors in the different developmental stages has been characterized (reviewed by Cleary et al. [22]). Stimulation of TGF-β and/or BMP signaling is required for cartilage formation, particularly for the production of extracellular matrix proteins.

As described and explored in detail in a previous chapter, in mice and other vertebrate model species, it is thought that articular chondrocytes and chondrocytes found in the growth plate derive from distinct progenitor cells during development. The cells that are hypothesized to give rise to articular cartilage are first observed in a region of the developing joint called the interzone [23,24]. Interzone cells, characterized by the expression of *Gdf5*, *chordin* and *autotaxin*, also give rise to other joint tissues such as ligaments and menisci, but importantly, they do not contribute significantly to the development of growth plate cartilage and the long bones [25,26]. Since interzone cells may be the direct embryonic precursors to important joint structures, there has been great interest in understanding how they arise and differentiate into specific joint tissues. Developmentally relevant progenitor cell populations would be ideally suitable for generating the large numbers of cells that will be required for engineering articular cartilage tissues as well as ligaments and menisci, which also fail to heal efficiently after injury and go on to cause degenerative joint disease in many individuals.

Mechanical Loading

Mechanical stimulation is known to promote chondrogenesis. Compression of embryonic limb bud mesenchymal cells stimulates expression of SOX9 [27], indicating that mechanical stimulation induce chondrogenic differentiation by directly regulating gene expression in progenitor cells. However, is particularly difficult to experimentally control biomechanics during embryogenesis, and therefore embryos with altered limb mobility by chemical/physical strategies are often used to study the effect of lack or excessive mechanical stimulation [28,29]. In all these cases, the interference with the mechanical forces during cartilage development leads to disrupted skeletogenesis or failure to form a proper synovial joint [29,30]. Analysis in vitro at single cell level also revealed the importance of mechanical stimulation to address chendrogenic differenciation and to define different differentiation stages. Chondrogenically differentiated mouse stem cells, for example, show increased stiffness compared to undifferentiated cell [31]. Similarly, the mechanical properties of single cells can be employed to evaluate differentiation states of embryonic [32] and marrow-derived [33] stem cells. Moreover, applying compressive loading in 3-dimensions enhances chondrogenesis of progenitor cells, leading to an increase of cartilage specific matrix protein synthesis [27].

EMBRYONIC STEM CELLS AND INDUCED PLURIPOTENT STEM CELLS GENERATE CARTILAGE TISSUES

Human pluripotent stem cells, including both embryonic (ESCs) and induced pluripotent stem cells (iPSCs), have the unique capacity to produce any cell type in our body, including heart cells, liver cells, pancreatic cells, nerve cells and chondrocytes [34–38]. It is possible to faithfully recapitulate early embryonic development in the ESC system [10,39,40] by providing stage specific signals to the cells as they would experience them in vivo (Fig. 5.1E–H). The challenge has been to develop strategies to direct them to make only the cell lineage of interest, which for cartilage regeneration and repair are articular chondrocytes capable of generating stable cartilaginous tissue. Serum-free directed differentiation approaches are more efficient due to the chemically defined nature of the culture medias and the ability to modulate signaling pathways with utmost control. This approach is more readily translated into the clinic compared to those that rely on animal-derived products that vary among reagent lots. However, endogenous signals produced by the cells themselves can vary, however, and must be taken into consideration when monitoring for efficiency of each developmental stage and when using each individual pluripotent stem cell line.

Chondrocytes and cartilage tissues derived from lateral plate mesoderm and paraxial/somitic mesoderm, both of which emerge during gastrulation in the primitive streak staged embryo (Fig. 5.1B—C). Lateral plate mesoderm progenitors in particular have diverse potential in that they can also give rise to the heart and the hematopoietic system in addition, to the musculoskeletal system in the limbs. Despite these challenges, several groups have successfully generated chondrogenic mesoderm progenitors from embryonic stem cells using a directed differentiation approach. As BMP signaling is required during primitive streak induction to generate Flk-1/KDR expressing cardiac and hematopoietic mesoderm [40,41], we and others hypothesized that inhibiting the BMP pathway in the presence of Activin/Nodal and WNT signaling would induce a subset of mesoderm with chondrogenic potential (Fig. 5.1F). This was a successful approach to generate primitive streak mesoderm that expressed Brachyury and PDGFRa, but lacked Flk-1/KDR expression as well as cardiac and hematopoietic potential [11,12,42,43]. Paraxial mesoderm/somite progenitors with skeletal muscle potential have also been derived from hESC/hIPSCs through BMP inhibition and WNT activation [44,45].

As in the embryo, FGF signaling is required during the next phase of differentiation to specify these early mesoderm cells towards a mesenchymal-like progenitor fate with chondrogenic potential (Fig. 5.1G). FGF-treated mesoderm progenitors derived from ESCs or IPSCs express transcription factors found in the somite [12,37,42,43] such as *Meox1/MEOX1*, *Pax1/PAX1* and *Nkx3.2/NKX3.2*, and are marked by the expression of cell surface markers (e.g., CD73, CD105) akin to those found on adult MSCs [37].

We and others have been successful in generating cartilage tissues in vitro from human ESCs and iPSCs using a high-density micromass culture, as well as through encapsulation or seeding onto a variety of natural and synthetic biomaterials [37,43,46,47]. Ideally, directed differentiation culture conditions are highly efficient and reproducible such that when ESC and iPSC-derived chondrogenic mesoderm is placed in high cell density conditions such as micromass culture in the presence of one or more TGF-β agonists, they begin to generate cartilaginous matrix [37,43,48] (Fig. 5.1H). Cartilage tissues

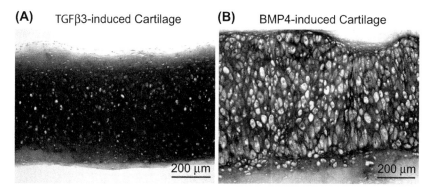

(A) TGFβ3-induced Cartilage   **(B)** BMP4-induced Cartilage

200 μm   200 μm

FIGURE 5.3 Cartilaginous tissues comprised of articular or hypertrophic chondrocytes are generated from human pluripotent stem cells. Representative images of human embryonic stem cell-derived cartilage tissues following 12 weeks of micromass culture in the presence of TGFβ3 (A) or BMP4 (B). Human chondrocytes specified in this developmental biology-based manner produce matrix and glycosaminoglycans that are visualized by metachromatic toluidine blue staining. Scale bar (200 μm).

specified from human pluripotent stem cells (hPSCs) in a developmentally relevant manner are rich in proteoglycans and express important proteins that function to support compressive loads (type II collagen, aggrecan) [37] (Fig. 5.3). Interestingly, molecularly and functionally distinct chondrogenic phenotypes can be generated via long term exposure to specific members of the TGF-β or BMP signaling pathways in the hPSC-derived micromass cultures. Articular cartilage-like tissues that express proteins that function to lubricate joint surfaces (lubricin/*PRG4*) are specified by prolonged TGF-β3 stimulation (Fig. 5.3A), while BMP4 treatment induces chondrocyte hypertrophy and the upregulation of growth plate chondrocyte genes such as type X collagen (*COL10A1*) [37] (Fig. 5.3B). Interestingly, hPSC-derived articular chondrocytes may be transitioning through a joint interzone progenitor population in vitro, as the expression of interzone-specific genes such as *GDF5* occurs prior to the upregulation of lubricin/*PRG4*, consistent with the development of this tissue in vivo [37].

A gold standard of cartilage stability is determined by its fate in vivo after extended periods of time. By understanding the signaling pathways that are responsible for articular and growth plate cartilage formation, it is possible to specify these two chondrocyte lineages derived from hPSCs and show that they have different fates in vivo, similar to those of fetal articular and growth plate cartilage tissues. Articular cartilage tissue generated from hPSCs through TGF-β stimulation resisted ossification and maintained characteristics of primary articular cartilage in vivo for extended periods of time when transplanted subcutaneously (under the skin) of immunodeficient mice [37]. Conversely, hypertrophic chondrocyte populations induced with BMP4 generated a cartilage tissue that initiated the endochondral ossification process in vivo [37]. These distinct in vitro and in vivo phenotypes and functions of chondrocytes and cartilage tissues derived from human ESC and iPSCs provides us with an unlimited source of material for regenerative medicine applications.

POST-NATAL STEM CELLS (E.G., ADULT HUMAN MSCS) FOR CARTILAGE TISSUE ENGINEERING

History and General Characteristics

Stem cells from adult (post-natal) tissues are known as mesenchymal stem/stromal cells (MSCs). These cells were originally identified and studied by Friedenstein et al. and Tavassoli, for their capacity to form colonies in vitro and to give rise to bone when heterotopically transplanted in vivo [49–51]. The unique characteristic of these fibroblastic-like cells make them of particular interest and from 2012, more than 2000 new PubMed entries each year on MSCs are available. However, two main and opposing descriptions of MSCs can be found in the literature (reviewed by Bianco [52]). The first definition describes MSCs as postnatal, self-renewing and multipotent stem cells able to form all skeletal tissues. This definition is strongly linked with the original view of Friedenstein and it has evolved during the last years allowing researchers to further characterize these clonogenic MSCs in vivo, to recognize their localization as the already known perivascular cells and to identify CD146 as the main characteristic surface marker expressed by these cells [53]. The second view identifies MSCs as all the non-hematopoietic cells from bone marrow capable of plastic adherence ("bulk" MSCs), independent of their clonogenicity, differentiation capacity, or their in vivo function(s) [54,55].

For practical reasons, "bulk" MSCs are most widely used by researchers since they are easily isolated from many tissues including bone marrow, synovium and adipose tissues. They can be expanded in vitro while maintaining, to a certain extent, their differentiation capacity. However, MSCs isolated simply by plastic adherence are much more heterogeneous in terms of their surface marker expression. For example, all of the cells most likely express CD105, CD73 and CD90, however, only a small percentage express CD146. They also can vary with respect to clonogenic capacity and differentiation potential (Fig. 5.4). This may be due to the fact that "bulk" MSCs may contain other cells capable of adhering to tissue culture plastic, such as fibroblasts, adipocytes and smooth muscle cells. The purpose of this chapter is not to convince you about the validity of each definition of MSCs, but simply to make the reader aware of the existence of these different views. There is, however, a general consensus about two other concepts:

1. MSCs need a certain environment, or a "niche," to preserve their characteristics. This natural in vivo environment was recently described as perivascular in multiple adult tissues [53,56,57]. This reflects what we observe

Morphology
during expansion

Chondrogenic
induction

High chondrogenic/
low passage MSC

Poor Chondrogenic/
high passage MSC

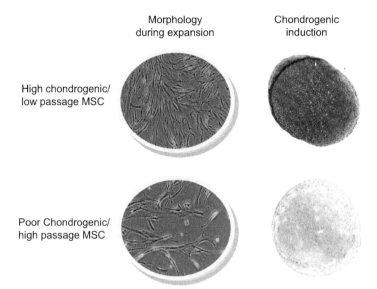

FIGURE 5.4 Representative images of human mesenchymal stem/stromal cells (MSCs) during expansion (left panels) and after chondrogenic induction (right panels). Chondrogenic induction was performed following the standard protocol by Johnstone et al. [86] for 35 days. Glycosaminoglycans are visualized by thionine staining.

during development, where skeletal progenitors residing in the adventitial/perivascular region contribute to cartilage and bone formation [58,59].

2. MSCs secrete factors to maintain their "niche," and also do so in response to tissue injury in order to modulate trophic and immune responses of other cells [60,61]. Identifying the signals that MSCs need to maintain their "niche" (both in vitro and in vivo), and the paracrine factors/cytokines secreted by them under specific conditions (mimicking for example pathological conditions) remains one of the biggest challenges for researchers in this field.

Expansion In Vitro

Post-natal human MSCs have a great potential for tissue engineering cell based approaches, especially due to their differentiation capacity, but also due to ease of accessibility and low donor site morbidity. Although originally isolated from bone marrow [62], cells sharing many of the morphological and phenotypic characteristics of MSCs have been isolated from the superficial layer of the cartilage [63], periosteum [64], synovium [65], infrapatellar fat pad [66], adipose tissue [67] and from fetal tissues [68] (e.g., placenta, amniotic fluid, umbilical cord). Following isolation, MSCs can be expanded in vitro, and the role of several growth factors in maintaining their stem cell characteristics (or "in vitro niche") have been investigated. Since From now on, we will mainly refer to MSCs as those isolated from bone marrow, because they are the most widely used, unless otherwise specified.

Fibroblast growth factor-2 (FGF-2; also known as basic-FGF) sustains the growth of progenitor cells during development and thus was also one of the first growth factors tested and successfully used to enhance the expansion of MSCs ex vivo [69]. FGF-2 also maintains the differentiation capacity of MSCs, and thus remains one of the most commonly used exogenous factors for MSC expansion today. Clear evidence of the mechanism of action by which FGF-2 retains the differentiation capacity of MSCs is lacking, although it seems that FGF-2 works via the stimulation of the expression of SOX9 and RUNX2 [70,71], two well-known differentiation-specific transcription factors. Moreover, it has been demonstrated that increasing the differentiation potential in vitro via FGF-2 administration does not always predict their differentiation capacity in vivo [53]. Originally, other growth factors were studied in combination with FGF-2 in order to further increase the expansion and differentiation capacity of MSCs, but there was no significant improvements compared to the use of FGF-2 alone. Only recently and upon taking inspiration from cartilage development [16], it was found that the combination of FGF-2 and WNT3a synergistically improved the proliferation and subsequent chondrogenic differentiation capacity of MSCs both in vitro and in vivo [72]. Interestingly, the effects of adding WNT proteins alone during expansion (not in combination with FGF-2) can vary from inhibiting proliferation (non-canonical WNT proteins) to inhibiting chondrogenic differentiation (canonical WNT proteins). Proliferation of MSCs is also enhanced by epidermal growth factor [73], although there are conflicting results regarding its ability to retain differentiation capacity [74].

In some studies, TGF-β was shown to promote senescence of MSCs [75], while other groups reported that TGF-β stimulates proliferation [76]. In line with the pro-proliferative role of TGF-β, the depletion of TGF-β from the expansion media decreases the proliferation rate of human chondrocytes [77] and, moreover, a chemically defined medium mainly composed of TGF-β, FGF-2 and platelet-derived growth factor (PDGF) was proposed as a standardized expansion media for MSCs [78]. PDGF is also one of the key components of platelets and plasma derivatives, products used as serum substitutes for cell expansion. Platelets and plasma derivatives are in fact known for their proliferative-inductive proprieties and for sustaining the differentiation capacity of MSC [79,80]. Recently, the combination of platelet rich plasma with a scaffold made of degradable nanofibers was proposed as an efficient method to expand and chondrogenically differentiate human MSCs [81].

Other systems have been developed in the attempt to better mimic in vitro the conditions that MSCs experience in their natural "niche." Two of these systems are of particular interest: hypoxic and three dimensional (3D) culture conditions. It was demonstrated by several reports that lowering oxygen from atmospheric level ($\sim$20%) to a more physiological in vivo level (1%–5%) enhances proliferation and subsequent differentiation capacity of MSCs from different origins [82,83]. This phenomenon was always associated with a direct or indirect regulation of the hypoxia inducible factor proteins. Culturing MSCs in a 3D environment (with or without scaffold) is thought to help maintain the cells in a more natural environment (reviewed by Hoch et al. [84]). In particular, the culture of MSCs in three dimensions enhances cell seeding efficiency (more homogeneous distribution of the cells in the construct) and their chondrogenic capacity [85].

Chondrogenic Differentiation In Vitro

Differentiation of MSCs towards the chondrogenic lineage is performed in a high density culture, generally through the formation of cell aggregates (such as pellet cultures) in the presence of TGF-β, as described for the first time by Johnstone et al. [86]. For hypertrophic cartilage formation from MSCs, the protocol is relatively efficient such that many different groups developed strategies to initiate ectopic bone formation in vivo by implanting chondrocytes derived from MSCs into recipient mice [87,88]. Researchers believe the bone formation via a cartilage intermediate step recapitulates the endochondral ossification process observed during bone development. Only one report recently suggests that MSCs form bone in vivo via a distinct and direct osteogenic process [89].

For stable cartilage formation, adaptations to the protocol of Johnstone et al. [86] are needed in order to reduce the tendency of MSCs to undergo hypertrophic maturation. During cartilage development there are three main signaling pathways involved in the zonal specification of the limb and therefore implicated in regulating cartilage formation: TGF-β, WNT and FGF (Fig. 5.5). The cross-talk among these pathways is particularly complex and the same signaling molecules can have different roles when target different cells or when acting in different developmental stages (reviewed by Cleary et al. [22]). Interestingly, interfering with WNT signaling [72], TGF-β pathways [90], or providing different FGF ligands during the different chondrogenic differentiation stages [91] has been shown to improve the stability of the cartilage generated by MSCs. Other methods to limit the hypertrophic maturation of MSCs, such as inhibition of the parathyroid hormone signaling pathway [92], mechanical stimulation [93,94] and differentiation of MSCs under hypoxic conditions, which is the physiological condition in the joint, also seem promising [95,96].

More recently, gathering a growing area of interest are alternative methods to induce chondrogenesis in MSCs via the silencing of anti-chondrogenic factors or microRNA (reviewed by Lolli et al. [97] and Clark et al. [98]) or by blocking pro-angiogenic molecules [99], often in absence of exogenous

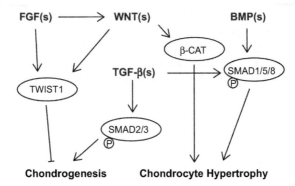

FIGURE 5.5 Main interactions between TGF-β WNT and FGF, signaling pathways during chondrogenic differentiation of adult human bone marrow-derived mesenchymal stem/stromal cells.

growth factors. All these approaches have in common the direct or the indirect activation of the chondrogenic transcription factor SOX9 and/or interference with WNT signaling. The validity of some of these in vitro studies was confirmed in vivo. An example is MSCs silenced for mir-221 expression, which were successfully used to repair cartilage tissue in an osteochondral defect model [100].

An overview of the main regulatory signals influencing differentiation of human MSCs over time is illustrated in Fig. 5.6.

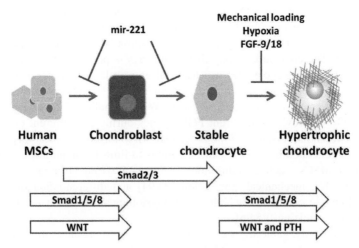

FIGURE 5.6 Schematic overview of the most relevant factors and interventions that regulate the chondrogenic differentiation of human mesenchymal stem/stromal cells (MSCs) in a stage-specific manner.

CONCLUSIONS

The potential of stem cells for clinical applications is uncontested and has undoubtedly inspired several new research avenues focused on regenerative medicine. Translating into clinical practice the growing amount of knowledge provided by basic research remains a difficult challenge. In the cartilage field, two particularly important challenges remain: variability of autologous and allogeneic donor cells and the ability to generate and maintain a stable (hyaline) articular cartilage phenotype. In this chapter we have highlighted novel developments in the field of cartilage repair, and how the application of developmental biology-inspired concepts, both new and old, have pushed research efforts towards innovative and fascinating discoveries. We should build on this foundation of knowledge now, and still consider additional aspects of how basic research can propel the tissue engineering field forward. In our opinion, the following approaches should be considered as the central core for future investigations:

1. Development of new "complex" systems where multiple aspects regulating cartilage specification, differentiation and maturation are considered and studied at the same time, such as mechanical loading, hypoxia, temperature and cell-cell interactions between different cell types normally present in the in vivo environment.
2. Develop new strategies to isolate or generate stem or progenitor cells with homogeneous (re)generative capacity.
3. Appreciate the role of timing during cartilage differentiation. The stage at which stem cells are exposed to a certain factor during differentiation, and the duration of such treatment(s), will contribute to the resulting effects on cell fate. While stages of chondrogenesis are seemingly distinct in vivo, this is not the case for the majority of in vitro efforts to recapitulate them.

ACKNOWLEDGMENTS

Our work and the writing of this chapter were supported by the Dutch Arthritis Foundation (16-1-201) and a VENI grant from NWO (13659; R.N.), and the McEwen Centre for Regenerative Medicine, the Krembil Foundation and Boston Children's Hospital (A.M.C.).

REFERENCES

[1] Genzyme. Carticel for Implantation. www.carticel.com. Genzyme Corporation; 2014. Available from: http://www.carticel.com/#footer-safetyInfo.
[2] Stevens HY, Shockley BE, Willett NJ, Lin AS, Raji Y, Guldberg RE, et al. Particulated juvenile articular cartilage implantation in the knee: a 3-year EPIC-microCT and histological examination. Cartilage 2014;5(2):74−7.
[3] Yanke AB, Tilton AK, Wetters NG, Merkow DB, Cole BJ. DeNovo NT particulated juvenile cartilage implant. Sports Med Arthrosc Rev 2015;23(3):125−9.

[4] Crawford DC, DeBerardino TM, Williams 3rd RJ. NeoCart, an autologous cartilage tissue implant, compared with microfracture for treatment of distal femoral cartilage lesions: an FDA phase-II prospective, randomized clinical trial after two years. J Bone Joint Surg 2012;94(11):979−89.

[5] Anderson DE, Williams 3rd RJ, DeBerardino TM, Taylor DC, Ma CB, Kane MS, et al. Magnetic resonance imaging characterization and clinical outcomes after NeoCart surgical therapy as a primary reparative treatment for knee cartilage injuries. Am J Sports Med 2017;45(4):875−83.

[6] Mumme M, Steinitz A, Nuss KM, Klein K, Feliciano S, Kronen P, et al. Regenerative potential of tissue-engineered nasal chondrocytes in goat articular cartilage defects. Tissue Eng Part A 2016;22(21−22):1286−95.

[7] Martin GR. Isolation of a pluripotent cell line from early mouse embryos cultured in medium conditioned by teratocarcinoma stem cells. Proc Natl Acad Sci USA 1981;78(12):7634−8.

[8] Cowan CA, Klimanskaya I, McMahon J, Atienza J, Witmyer J, Zucker JP, et al. Derivation of embryonic stem-cell lines from human blastocysts. N Engl J Med 2004;350(13):1353−6.

[9] Clements D, Taylor HC, Herrmann BG, Stott D. Distinct regulatory control of the Brachyury gene in axial and non-axial mesoderm suggests separation of mesoderm lineages early in mouse gastrulation. Mech Dev 1996;56(1−2):139−49.

[10] Fehling HJ, Lacaud G, Kubo A, Kennedy M, Robertson S, Keller G, et al. Tracking mesoderm induction and its specification to the hemangioblast during embryonic stem cell differentiation. Development 2003;130(17):4217−27.

[11] Darabi R, Gehlbach K, Bachoo RM, Kamath S, Osawa M, Kamm KE, et al. Functional skeletal muscle regeneration from differentiating embryonic stem cells. Nat Med 2008;14(2):134−43.

[12] Craft AM, Ahmed N, Rockel JS, Baht GS, Alman BA, Kandel RA, et al. Specification of chondrocytes and cartilage tissues from embryonic stem cells. Development 2013;140(12):2597−610.

[13] Kawakami Y, Capdevila J, Buscher D, Itoh T, Rodriguez Esteban C, Izpisua Belmonte JC. WNT signals control FGF-dependent limb initiation and AER induction in the chick embryo. Cell 2001;104(6):891−900.

[14] Saunders Jr JW. The proximo-distal sequence of origin of the parts of the chick wing and the role of the ectoderm. J Exp Zool 1948;108(3):363−403.

[15] Todt WL, Fallon JF. Development of the apical ectodermal ridge in the chick wing bud. J Embryol Exp Morphol 1984;80:21−41.

[16] ten Berge D, Brugmann SA, Helms JA, Nusse R. Wnt and FGF signals interact to coordinate growth with cell fate specification during limb development. Development 2008;135(19):3247−57.

[17] Chimal-Monroy J, Montero JA, Ganan Y, Macias D, Garcia-Porrero JA, Hurle JM. Comparative analysis of the expression and regulation of Wnt5a, Fz4, and Frzb1 during digit formation and in micromass cultures. Dev Dyn 2002;224(3):314−20.

[18] Lorda-Diez CI, Montero JA, Garcia-Porrero JA, Hurle JM. Tgfbeta2 and 3 are coexpressed with their extracellular regulator Ltbp1 in the early limb bud and modulate mesodermal outgrowth and BMP signaling in chicken embryos. BMC Dev Biol 2010;10:69.

[19] Merino R, Rodriguez-Leon J, Macias D, Ganan Y, Economides AN, Hurle JM. The BMP antagonist Gremlin regulates outgrowth, chondrogenesis and programmed cell death in the developing limb. Development 1999;126(23):5515−22.

[20] Barry F, Boynton RE, Liu B, Murphy JM. Chondrogenic differentiation of mesenchymal stem cells from bone marrow: differentiation-dependent gene expression of matrix components. Exp Cell Res 2001;268(2):189−200.

[21] Oh CD, Chang SH, Yoon YM, Lee SJ, Lee YS, Kang SS, et al. Opposing role of mitogen-activated protein kinase subtypes, erk-1/2 and p38, in the regulation of chondrogenesis of mesenchymes. J Biol Chem 2000;275(8):5613−9.

[22] Cleary MA, van Osch GJ, Brama PA, Hellingman CA, Narcisi R. FGF, TGFβ and Wnt crosstalk: embryonic to in vitro cartilage development from mesenchymal stem cells. J Tissue Eng Regen Med 2015;9(4):332−42.

[23] Holder N. An experimental investigation into the early development of the chick elbow joint. J Embryol Exp Morphol 1977;39:115−27.

[24] Archer CW, Dowthwaite GP, Francis-West P. Development of synovial joints. Birth Defects Res C Embryo Today 2003;69(2):144−55.

[25] Pacifici M, Koyama E, Shibukawa Y, Wu C, Tamamura Y, Enomoto-Iwamoto M, et al. Cellular and molecular mechanisms of synovial joint and articular cartilage formation. Ann NY Acad Sci 2006;1068:74−86.

[26] Shwartz Y, Viukov S, Krief S, Zelzer E. Joint development involves a continuous influx of Gdf5-positive cells. Cell Reports 2016;15(12):2577−87.

[27] Takahashi I, Nuckolls GH, Takahashi K, Tanaka O, Semba I, Dashner R, et al. Compressive force promotes sox9, type II collagen and aggrecan and inhibits IL-1beta expression resulting in chondrogenesis in mouse embryonic limb bud mesenchymal cells. J Cell Sci 1998;111(Pt 14):2067−76.

[28] Dolan SB, Jentes ES, Sotir MJ, Han P, Blanton JD, Rao SR, et al. Pre-exposure rabies vaccination among US international travelers: findings from the global TravEpiNet consortium. Vector Borne Zoonotic Dis 2014;14(2):160−7.

[29] Roddy KA, Prendergast PJ, Murphy P. Mechanical influences on morphogenesis of the knee joint revealed through morphological, molecular and computational analysis of immobilised embryos. PLoS One 2011;6(2):e17526.

[30] Solem RC, Eames BF, Tokita M, Schneider RA. Mesenchymal and mechanical mechanisms of secondary cartilage induction. Dev Biol 2011;356(1):28−39.

[31] Pillarisetti A, Desai JP, Ladjal H, Schiffmacher A, Ferreira A, Keefer CL. Mechanical phenotyping of mouse embryonic stem cells: increase in stiffness with differentiation. Cell Reprogram 2011;13(4):371−80.

[32] Ofek G, Willard VP, Koay EJ, Hu JC, Lin P, Athanasiou KA. Mechanical characterization of differentiated human embryonic stem cells. J Biomech Eng 2009;131(6):061011.

[33] Chiang H, Hsieh CH, Lin YH, Lin S, Tsai-Wu JJ, Jiang CC. Differences between chondrocytes and bone marrow-derived chondrogenic cells. Tissue Eng Part A 2011;17(23−24):2919−29.

[34] Kattman SJ, Witty AD, Gagliardi M, Dubois NC, Niapour M, Hotta A, et al. Stage-specific optimization of activin/nodal and BMP signaling promotes cardiac differentiation of mouse and human pluripotent stem cell lines. Cell Stem Cell 2011;8(2):228−40.

[35] Kriks S, Shim JW, Piao J, Ganat YM, Wakeman DR, Xie Z, et al. Dopamine neurons derived from human ES cells efficiently engraft in animal models of Parkinson's disease. Nature 2011;480(7378):547−51.

[36] Ogawa S, Surapisitchat J, Virtanen C, Ogawa M, Niapour M, Sugamori KS, et al. Three-dimensional culture and cAMP signaling promote the maturation of human pluripotent stem cell-derived hepatocytes. Development 2013;140(15):3285−96.

[37] Craft AM, Rockel JS, Nartiss Y, Kandel RA, Alman BA, Keller GM. Generation of articular chondrocytes from human pluripotent stem cells. Nat Biotechnol 2015.

[38]   Nostro MC, Sarangi F, Yang C, Holland A, Elefanty AG, Stanley EG, et al. Efficient generation of NKX6-1(+) pancreatic progenitors from multiple human pluripotent stem cell lines. Stem Cell Reports 2015;4(4):591—604.

[39]   Kattman SJ, Huber TL, Keller GM. Multipotent flk-1+ cardiovascular progenitor cells give rise to the cardiomyocyte, endothelial, and vascular smooth muscle lineages. Dev Cell 2006;11(5):723—32.

[40]   Gadue P, Huber TL, Paddison PJ, Keller GM. Wnt and TGF-beta signaling are required for the induction of an in vitro model of primitive streak formation using embryonic stem cells. Proc Natl Acad Sci USA 2006;103(45):16806—11.

[41]   Nostro MC, Cheng X, Keller GM, Gadue P. Wnt, activin, and BMP signaling regulate distinct stages in the developmental pathway from embryonic stem cells to blood. Cell Stem Cell 2008;2(1):60—71.

[42]   Tanaka M, Jokubaitis V, Wood C, Wang Y, Brouard N, Pera M, et al. BMP inhibition stimulates WNT-dependent generation of chondrogenic mesoderm from embryonic stem cells. Stem Cell Res 2009;3(2—3):126—41.

[43]   Umeda K, Zhao J, Simmons P, Stanley E, Elefanty A, Nakayama N. Human chondrogenic paraxial mesoderm, directed specification and prospective isolation from pluripotent stem cells. Sci Rep 2012;2:455.

[44]   Chal J, Oginuma M, Al Tanoury Z, Gobert B, Sumara O, Hick A, et al. Differentiation of pluripotent stem cells to muscle fiber to model Duchenne muscular dystrophy. Nat Biotechnol 2015;33(9):962—9.

[45]   Xi H, Fujiwara W, Gonzalez K, Jan M, Liebscher S, Van Handel B, et al. In vivo human somitogenesis guides somite development from hPSCs. Cell Reports 2017;18(6):1573—85.

[46]   Toh WS, Lee EH, Guo XM, Chan JK, Yeow CH, Choo AB, et al. Cartilage repair using hyaluronan hydrogel-encapsulated human embryonic stem cell-derived chondrogenic cells. Biomaterials 2010;31(27):6968—80.

[47]   Yamashita A, Morioka M, Yahara Y, Okada M, Kobayashi T, Kuriyama S, et al. Generation of Scaffoldless hyaline cartilaginous tissue from human iPSCs. Stem Cell Reports 2015;4(3):404—18.

[48]   Tsumaki N, Nakase T, Miyaji T, Kakiuchi M, Kimura T, Ochi T, et al. Bone morphogenetic protein signals are required for cartilage formation and differently regulate joint development during skeletogenesis. J Bone Miner Res 2002;17(5):898—906.

[49]   Friedenstein AJ, Gorskaja JF, Kulagina NN. Fibroblast precursors in normal and irradiated mouse hematopoietic organs. Exp Hematol 1976;4(5):267—74.

[50]   Friedenstein AJ, Piatetzky II S, Petrakova KV. Osteogenesis in transplants of bone marrow cells. J Embryol Exp Morphol 1966;16(3):381—90.

[51]   Tavassoli M, Crosby WH. Transplantation of marrow to extramedullary sites. Science 1968;161(3836):54—6.

[52]   Bianco P. "Mesenchymal" stem cells. Annu Rev Cell Dev Biol 2014;30:677—704.

[53]   Sacchetti B, Funari A, Michienzi S, Di Cesare S, Piersanti S, Saggio I, et al. Self-renewing osteoprogenitors in bone marrow sinusoids can organize a hematopoietic microenvironment. Cell 2007;131(2):324—36.

[54]   Dominici M, Le Blanc K, Mueller I, Slaper-Cortenbach I, Marini F, Krause D, et al. Minimal criteria for defining multipotent mesenchymal stromal cells. The International Society for Cellular Therapy position statement. Cytotherapy 2006;8(4):315—7.

[55]   Pittenger MF, Mackay AM, Beck SC, Jaiswal RK, Douglas R, Mosca JD, et al. Multilineage potential of adult human mesenchymal stem cells. Science 1999;284(5411):143—7.

[56]   Caplan AI. All MSCs are pericytes? Cell Stem Cell 2008;3(3):229—30.

[57] Crisan M, Yap S, Casteilla L, Chen CW, Corselli M, Park TS, et al. A perivascular origin for mesenchymal stem cells in multiple human organs. Cell Stem Cell 2008;3(3):301−13.

[58] Diaz-Flores L, Gutierrez R, Varela H, Rancel N, Valladares F. Microvascular pericytes: a review of their morphological and functional characteristics. Histol Histopathol 1991;6(2):269−86.

[59] Maes C, Kobayashi T, Selig MK, Torrekens S, Roth SI, Mackem S, et al. Osteoblast precursors, but not mature osteoblasts, move into developing and fractured bones along with invading blood vessels. Dev Cell 2010;19(2):329−44.

[60] Bernardo ME, Fibbe WE. Mesenchymal stromal cells: sensors and switchers of inflammation. Cell Stem Cell 2013;13(4):392−402.

[61] Caplan AI, Correa D. The MSC: an injury drugstore. Cell Stem Cell 2011;9(1):11−5.

[62] Prockop DJ. Marrow stromal cells as stem cells for nonhematopoietic tissues. Science 1997;276(5309):71−4.

[63] Dowthwaite GP, Bishop JC, Redman SN, Khan IM, Rooney P, Evans DJ, et al. The surface of articular cartilage contains a progenitor cell population. J Cell Sci 2004;117(Pt 6):889−97.

[64] Fukumoto T, Sperling JW, Sanyal A, Fitzsimmons JS, Reinholz GG, Conover CA, et al. Combined effects of insulin-like growth factor-1 and transforming growth factor-beta1 on periosteal mesenchymal cells during chondrogenesis in vitro. Osteoarthritis Cartilage 2003;11(1):55−64.

[65] De Bari C, Dell'Accio F, Tylzanowski P, Luyten FP. Multipotent mesenchymal stem cells from adult human synovial membrane. Arthritis Rheum 2001;44(8):1928−42.

[66] Khan WS, Adesida AB, Tew SR, Longo UG, Hardingham TE. Fat pad-derived mesenchymal stem cells as a potential source for cell-based adipose tissue repair strategies. Cell Prolif 2012;45(2):111−20.

[67] Zuk PA, Zhu M, Ashjian P, De Ugarte DA, Huang JI, Mizuno H, et al. Human adipose tissue is a source of multipotent stem cells. Mol Biol Cell 2002;13(12):4279−95.

[68] Wang HS, Hung SC, Peng ST, Huang CC, Wei HM, Guo YJ, et al. Mesenchymal stem cells in the Wharton's jelly of the human umbilical cord. Stem Cell 2004;22(7):1330−7.

[69] Martin I, Muraglia A, Campanile G, Cancedda R, Quarto R. Fibroblast growth factor-2 supports ex vivo expansion and maintenance of osteogenic precursors from human bone marrow. Endocrinology 1997;138(10):4456−62.

[70] Handorf AM, Li WJ. Fibroblast growth factor-2 primes human mesenchymal stem cells for enhanced chondrogenesis. PLoS One 2011;6(7):e22887.

[71] Ito T, Sawada R, Fujiwara Y, Seyama Y, Tsuchiya T. FGF-2 suppresses cellular senescence of human mesenchymal stem cells by down-regulation of TGF-beta2. Biochem Biophys Res Commun 2007;359(1):108−14.

[72] Narcisi R, Cleary MA, Brama PA, Hoogduijn MJ, Tuysuz N, ten Berge D, et al. Long-term expansion, enhanced chondrogenic potential, and suppression of endochondral ossification of adult human MSCs via WNT signaling modulation. Stem Cell Reports 2015;4(3):459−72.

[73] Mastrogiacomo M, Cancedda R, Quarto R. Effect of different growth factors on the chondrogenic potential of human bone marrow stromal cells. Osteoarthritis Cartilage 2001;9(Suppl. A):S36−40.

[74] Tamama K, Kawasaki H, Wells A. Epidermal growth factor (EGF) treatment on multipotential stromal cells (MSCs). Possible enhancement of therapeutic potential of MSC. J Biomed Biotechnol 2010;2010:795385.

[75] Sawada R, Ito T, Tsuchiya T. Changes in expression of genes related to cell proliferation in human mesenchymal stem cells during in vitro culture in comparison with cancer cells. J Artif Organs 2006;9(3):179−84.

[76] Jian H, Shen X, Liu I, Semenov M, He X, Wang XF. Smad3-dependent nuclear translocation of beta-catenin is required for TGF-beta1-induced proliferation of bone marrow-derived adult human mesenchymal stem cells. Genes Dev 2006;20(6):666−74.

[77] Narcisi R, Signorile L, Verhaar JA, Giannoni P, van Osch GJ. TGFbeta inhibition during expansion phase increases the chondrogenic re-differentiation capacity of human articular chondrocytes. Osteoarthritis Cartilage 2012;20(10):1152−60.

[78] Chase LG, Lakshmipathy U, Solchaga LA, Rao MS, Vemuri MC. A novel serum-free medium for the expansion of human mesenchymal stem cells. Stem Cell Res Ther 2010;1(1):8.

[79] Doucet C, Ernou I, Zhang Y, Llense JR, Begot L, Holy X, et al. Platelet lysates promote mesenchymal stem cell expansion: a safety substitute for animal serum in cell-based therapy applications. J Cell Physiol 2005;205(2):228−36.

[80] Muraglia A, Todeschi MR, Papait A, Poggi A, Spano R, Strada P, et al. Combined platelet and plasma derivatives enhance proliferation of stem/progenitor cells maintaining their differentiation potential. Cytotherapy 2015;17(12):1793−806.

[81] Diaz-Gomez L, Alvarez-Lorenzo C, Concheiro A, Silva M, Dominguez F, Sheikh FA, et al. Biodegradable electrospun nanofibers coated with platelet-rich plasma for cell adhesion and proliferation. Mater Sci Eng C Mater Biol Appl 2014;40:180−8.

[82] Drela K, Sarnowska A, Siedlecka P, Szablowska-Gadomska I, Wielgos M, Jurga M, et al. Low oxygen atmosphere facilitates proliferation and maintains undifferentiated state of umbilical cord mesenchymal stem cells in an hypoxia inducible factor-dependent manner. Cytotherapy 2014;16(7):881−92.

[83] Fotia C, Massa A, Boriani F, Baldini N, Granchi D. Hypoxia enhances proliferation and stemness of human adipose-derived mesenchymal stem cells. Cytotechnology 2015;67(6):1073−84.

[84] Hoch AI, Leach JK. Concise review: optimizing expansion of bone marrow mesenchymal stem/stromal cells for clinical applications. Stem Cells Transl Med 2015;4(4):412.

[85] Papadimitropoulos A, Piccinini E, Brachat S, Braccini A, Wendt D, Barbero A, et al. Expansion of human mesenchymal stromal cells from fresh bone marrow in a 3D scaffold-based system under direct perfusion. PLoS One 2014;9(7):e102359.

[86] Johnstone B, Hering TM, Caplan AI, Goldberg VM, Yoo JU. In vitro chondrogenesis of bone marrow-derived mesenchymal progenitor cells. Exp Cell Res 1998;238(1):265−72.

[87] Farrell E, Both SK, Odorfer KI, Koevoet W, Kops N, O'Brien FJ, et al. In-vivo generation of bone via endochondral ossification by in-vitro chondrogenic priming of adult human and rat mesenchymal stem cells. BMC Musculoskelet Disord 2011;12:31.

[88] Scotti C, Tonnarelli B, Papadimitropoulos A, Scherberich A, Schaeren S, Schauerte A, et al. Recapitulation of endochondral bone formation using human adult mesenchymal stem cells as a paradigm for developmental engineering. Proc Natl Acad Sci USA 2010;107(16):7251−6.

[89] Serafini M, Sacchetti B, Pievani A, Redaelli D, Remoli C, Biondi A, et al. Establishment of bone marrow and hematopoietic niches in vivo by reversion of chondrocyte differentiation of human bone marrow stromal cells. Stem Cell Res 2014;12(3):659−72.

[90] Hellingman CA, Davidson EN, Koevoet W, Vitters EL, van den Berg WB, van Osch GJ, et al. Smad signaling determines chondrogenic differentiation of bone-marrow-derived mesenchymal stem cells: inhibition of Smad1/5/8P prevents terminal differentiation and calcification. Tissue Eng Part A 2011;17(7−8):1157−67.

[91] Correa D, Somoza RA, Lin P, Greenberg S, Rom E, Duesler L, et al. Sequential exposure to fibroblast growth factors (FGF) 2, 9 and 18 enhances hMSC chondrogenic differentiation. Osteoarthritis Cartilage 2015;23(3):443–53.

[92] Weiss S, Hennig T, Bock R, Steck E, Richter W. Impact of growth factors and PTHrP on early and late chondrogenic differentiation of human mesenchymal stem cells. J Cell Physiol 2010;223(1):84–93.

[93] Gardner OF, Fahy N, Alini M, Stoddart MJ. Joint mimicking mechanical load activates TGFbeta1 in fibrin-poly(ester-urethane) scaffolds seeded with mesenchymal stem cells. J Tissue Eng Regen Med 2016.

[94] Steward AJ, Kelly DJ, Wagner DR. Purinergic signaling regulates the transforming growth factor-beta3-induced chondrogenic response of mesenchymal stem cells to hydrostatic pressure. Tissue Eng Part A 2016;22(11–12):831–9.

[95] Anderson DE, Markway BD, Bond D, McCarthy HE, Johnstone B. Responses to altered oxygen tension are distinct between human stem cells of high and low chondrogenic capacity. Stem Cell Res Ther 2016;7(1):154.

[96] Markway BD, Cho H, Anderson DE, Holden P, Ravi V, Little CB, et al. Reoxygenation enhances tumour necrosis factor alpha-induced degradation of the extracellular matrix produced by chondrogenic cells. Eur Cell Mater 2016;31:425–39.

[97] Lolli A, Penolazzi L, Narcisi R, van Osch G, Piva R. Emerging potential of gene silencing approaches targeting anti-chondrogenic factors for cell-based cartilage repair. Cell Mol Life Sci 2017.

[98] Clark EA, Kalomoiris S, Nolta JA, Fierro FA. Concise review: microRNA function in multipotent mesenchymal stromal cells. Stem Cell 2014;32(5):1074–82.

[99] Marsano A, Medeiros da Cunha CM, Ghanaati S, Gueven S, Centola M, Tsaryk R, et al. Spontaneous in vivo chondrogenesis of bone marrow-derived mesenchymal progenitor cells by blocking vascular endothelial growth factor signaling. Stem Cells Transl Med 2016;5(12):1730–8.

[100] Lolli A, Narcisi R, Lambertini E, Penolazzi L, Angelozzi M, Kops N, et al. Silencing of antichondrogenic MicroRNA-221 in human mesenchymal stem cells promotes cartilage repair in vivo. Stem Cell 2016;34(7):1801–11.

Chapter 6

Endochondral Ossification: Recapitulating Bone Development for Bone Defect Repair

Caoimhe Kiernan, Callie Knuth, Eric Farrell
Erasmus MC, University Medical Centre Rotterdam, Rotterdam, The Netherlands

INTRODUCTION

Bone defect repair is one of the most common regenerative procedures with more than 2 million bone grafts performed worldwide annually. The major causes of large bone defects are trauma, congenital anomalies, and tissue resection due to cancer. Although bone has an excellent inherent repair capacity, its ability to bridge very large defects remains limited. Despite the many advances in medicine, the gold standard for the treatment of these defects is still autologous bone transplantation. This involves harvesting bone from another anatomical location in the patient for use as a transplant material to place at the defect site. Interestingly, little has changed in the approach to large bone defect repair in 500 years. A book written by surgeon Ibrahim Bin Abdullah in 1505 AD contains the first mention of cranioplasty in which he describes the technique of using goat or dog skull bones to heal a human cranial defect [1]. This is believed to be the first detailed description of the cranioplasty procedure. Today the three approaches of autologous (from the patient), allogeneic (from another human donor), and xenogeneic (animal derived) bone graft transplantation would not seem so different to the approach described in his book. Fast forward 500 years and autologous bone transplantation, rather than the use of xenogeneic bone, represents the current "gold standard" for the treatment of large bone defects that do not sponta-neously heal. Obviously, there would be many areas (such as hygiene and reproducibility, connection to existing vasculature) where significant im-provements have been made, but the painful and costly harvesting of patient bone does not represent the ideal situation for patient or surgeon.

Developmental Biology and Musculoskeletal Tissue Engineering. https://doi.org/10.1016/B978-0-12-811467-4.00006-1
125

Furthermore, despite being a gold standard procedure, autologous bone transplantation still does not have a success rate above 90% in most reported cases and can be reported to be as low as 80% [2]. Although we do not know the success rate surgeon Ibrahim achieved and cannot compare the repair rate his patient cohort (a small number of young healthy soldiers) with that of aged diabetic smokers or cancer patients that are routinely treated nowadays, the comparison is still interesting. If we were to compare the progress that has been achieved in the area of large bone defect repair with progress in the fields of computing, transport, or telecommunications even in the last 100 years, it is clear that advancement in the development of new treatment options has been slow.

Currently, no other material has a better overall success rate in relation to graft survival and integration with the surrounding tissues than autologous bone. Free nonvascularized bone grafts are usually taken from the iliac crest [3,4]. In these instances there are unwanted surgical side-effects and complications at the harvest site, including pain, infection, scarring, hypersensitivity, and instability. This is the predominant problem associated with bone grafting. However, the success rate of such procedures must also be taken into consideration. Vascularized free flaps (autologous tissue transplantation complete with attached vasculature) can have success rates of up to 90% (as reviewed by Hayden et al.) with nonvascularized bone grafts more likely to have success rates around 80% [2]. Of course, these outcomes are much related to indication for surgery, anatomical site, and postoperative radiotherapy. Given the increased complexity and cost of vascularized bone grafts, nonvascularized bone grafts are often used [5]. With a failure rate of between 10% and 20% and the many problems associated with harvesting of autologous bone, there is a clear need for alternative solutions for replacement of bone in large defects.

At present there is a huge focus on the field of tissue engineering and regenerative medicine (TERM) to provide new and alternative approaches to autologous bone grafting. Approaches vary greatly. One strategy is to develop new materials, bioactive or inert that should augment bone repair by "guided tissue regeneration" as occurs in the oral and maxillofacial surgery field. Materials include resorbable polymers such as polylactic-polyglycolic acids and collagens that are used as barrier membranes to improve bone formation and integration without interference from the oral mucosa [6]. Obviously, these approaches still involve the use of harvested iliac crest bone and only serve to improve the outcomes using autologous bone. Another approach is to use growth factors such as bone morphogenetic proteins (BMPs) combined with novel biomaterials to replace the use of iliac crest bone. These factors (most commonly BMP2) have shown variable results, sometimes performing as well as or better than iliac crest bone but sometimes less so [7]. Two drawbacks in this approach are that these factors are extremely expensive and are used at supraphysiological levels that are not without risk of causing uncontrolled bone growth. Platelet-rich plasma (PRP) has also been used in combination with iliac crest bone to augment bone formation, again with mixed results. Compared with iliac crest bone, PRP generally does not improve bone formation/repair/integration [7]. It is important to note that the majority of these TERM-based approaches are most often combined with iliac crest bone, which then does not address the issues of donor site morbidity, risk of infection, cost of a second surgery, etc. The use of various adult progenitor cells to aid in the formation of new bone and repair of critical-sized bone defects has received much attention in recent decades. However, to understand and consider the approaches taken, it is first necessary to briefly discuss how bone is formed developmentally.

DEVELOPMENTAL BONE FORMATION; ENDOCHONDRAL VERSUS INTRAMEMBRANOUS OSSIFICATION

During development, bone forms through the processes of intramembranous and endochondral ossification [8,9]. In both processes, preexisting mesen-chymal tissue is converted into bone tissue. Craniofacial bones and the clavicle form by intramembranous ossification, which is the direct differ-entiation of mesenchymal cells to osteoblasts. Until approximately 2006, the main focus of the field of TERM with regard to bone formation was on the direct formation of bony tissues, usually using mesenchymal stem cells/marrow stromal cells (MSCs), mimicking the developmental process of intramembranous ossification. In this process, bone is formed by the concentration of mesenchymal cells in a highly vascularized region. These cells are stimulated to undergo osteogenic differentiation becoming osteoblasts. These newly formed osteoblasts lay down the organic matrix of bone, the osteoid, comprised mainly of collagen type I. Over time, this osteoid becomes mineralized through the binding of calcium salts leading to the formation of mature bone. Some osteoblasts retreat to the edge of this newly forming bone and play a role in further bone formation and formation of the periosteum. However, some cells also become trapped in the newly formed matrix and further differentiate into osteocytes. This process is controlled mainly by Runt-related transcription factor 2 (Runx2) and the BMP family of growth factors. From a TERM perspective, this is a very logical physiological process to try to emulate. Direct differentiation of MSCs into osteoblasts was shown to be quite straightforward in vitro [10]. However, in vivo and on a larger scale, this process appeared to be less successful. The main issue was lack of vascularization and necrosis at the core of large tissue-engineered "bone" [11,12]. Another issue was that the tissue that was formed in vitro was not complete bone and further bone tissue formation did not occur in vivo, resulting in small amounts of poor-quality bone being formed.

The process of bone formation via endochondral ossification on the other hand is a more complex route to the generation of new bone tissue. There are multiple stages of this process that have been described extensively [13—18]. First and most importantly, a cartilaginous template is formed from condensing mesenchymal cells. The cells of this template continue to proliferate and differentiate into chondrocytes. At this early stage, the transcription factor SRY-related high-motility group box family of proteins (Sox9) is expressed and is required for chondrogenic differentiation [19]. Collagen type II, aggrecan, and glycosaminoglycan (GAG) deposition by differentiating chondrocytes leads to the formation of the cartilage extracellular matrix (ECM) [20,21]. A dense layer of connective tissue called the perichondrium condenses around the periphery of the cartilage template. After proliferation, chondrocytes undergo hypertrophy and selectively express matrix metalloproteinase 13 (MMP13) [22]. MMP13 is a collagenase, which degrades collagen type II fibrils. Therefore, hypertrophic chondrocytes begin to downregulate collagen type II expression [23]. Hypertrophic chondrocytes express various proteins such as collagen type X (COLX), bone morphogenetic protein 6 (BMP6), and the transcription factors indian hedgehog (Ihh), Runx2, and osterix promoting further differentiation. At the same time, blood vessels invade the cartilage template delivering osteoblasts and chondroclasts/osteoclasts, which leads to erosion of the cartilage and development of the primary ossification center [24,25]. The breakdown of the cartilage ECM by osteoclasts in combination with the deposition of bone by osteoblasts on the cartilage fragments leads to the extension of the primary ossification centre. Under the control of Runx2, hypertrophic chondrocytes upregulate vascular endothelial growth factor expression to promote vascular invasion [13] and together with COLX and Ihh induce osteoblast differentiation [26]. High-motility group box 1 is also secreted by hypertrophic chondrocytes to attract osteoblasts, osteoclasts, and endothelial cells [27]. Deposition of hydroxyapatite crystals (mainly calcium and phosphate) in the ECM surrounding hypertrophic chondrocytes assists in the mineralization of the cartilage template. Concurrent with hypertrophy, osteoblast differentiation takes place within the perichondrium. Osteoid matrix mainly made up of collagen type I is secreted by differentiating osteoblasts to form the bone collar [28]. Mature osteoblasts become embedded in the matrix and develop into osteocytes [29]. Long bones in the body are formed through this process. In long bones, at the end of the cartilage model, a secondary ossification center also develops to form the growth plate [13,30]. The growth plate leads to longitudinal growth of these bones.

ENDOCHONDRAL OSSIFICATION AND BONE DEFECT REPAIR

A number of studies have attempted to use adult stem cells, such as MSCs in combination with biomaterials (tricalcium phosphate or collagen sponges) [11]. Unfortunately, to date, using these cells has not yielded the promising results expected. This is usually because of the absence of vascularization of the implanted construct causing necrosis [11,12]. There have also been interesting single-patient case studies reporting the generation of vascularized bone by the use of the patient as a bioreactor [31]. This involves implanting a construct containing a combination of biomaterials, growth factors, and cells into a subcutaneous pocket in the patient to allow for vascularization and bone formation and then removal and placement of the construct in the defect site. This type of approach, although interesting, still involves multiple surgical sites and would be prohibitively expensive as a mass treatment option. Clearly, there is exciting research being performed in the field of bone TERM. As mentioned, one of the major issues facing the treatment of large bone defects is that of vascularization. Any implant or construct needs to be able to induce or support blood vessel invasion and an associated seamless integration with the surrounding bone. Ideally, rapid vascularization or even prevascularization of tissue-engineered bone is desired and would greatly improve treatment outcomes. One strategy to improve vascularization and overall bone quality focuses on endochondral ossification as a route to bone repair and replacement therapies and is the focus of this chapter. Several groups have shown that, not only is mature vascularized bone formed, but the entire marrow niche is recapitulated when chondrogenically primed adult MSCs are implanted subcutaneously either as pellets or seeded in scaffolds [11−15]. This seems to mirror developmental bone formation via endochondral ossification involving generation of bone via cartilage anlagen intermediate and furthermore shows excellent integration with the host as regard vascularization. This mode of bone formation is in contrast to the direct formation of bone by osteogenically differentiated MSCs, known as intramembranous ossification. Critically, endochondral ossification has also been shown to be necessary for marrow niche formation [32] and is closely linked to haematopoiesis/lymphopoiesis. Before this work, the standard approach to tissue engineering bone was via intramembranous ossification by osteogenically differentiating MSCs in vitro or using osteoblasts, approaches that are still being researched extensively. Bone tissue engineering by direct intramembranous ossification has not led to an improvement over the current bone repair technique of autologous bone transplantation and is still not suitable for large bone defects [11,12]. The important difference between these approaches is that during development, endochondral ossification induces blood vessel invasion into the cartilage template, whereas intramembranous ossification requires that the mesenchymal tissue is already well vascularized before bone formation.

For several years, many groups have demonstrated the ability to regenerate or at least partially regenerate critical-sized bone defects using chondrogenically primed MSCs. In 2006, Huang et al. demonstrated partial yet impressive reconstruction of the rabbit lunate using chondrogenically primed MSCs loaded into a hyaluronan/gelatin scaffold. In this paper, the authors hypothesized that the chondrogenic priming of cells of mesodermal origin and replacement of a whole joint might lead to the formation of cartilage and bone in the appropriate areas in response to environmental cues such as mechanical and paracrine stimuli. Although they did not completely regenerate the joint, the results were impressive, with both bone and cartilage formation occurring. In the following years, several research groups also became aware of the potential of "stem cells" of different origin (embryonic, adult marrow, etc.) to generate bone via the endochondral route. Very often this arose from the observation that tissue-engineered cartilage, on subcutaneous implantation in mice, was not phenotypically stable. Farrell and van der Jagt observed in 2009 that chondrogenically differentiated MSCs seeded into collagen-GAG scaffolds became vascularized and expressed COLX (a commonly used marker of hypertrophic/growth plate chondrocytes destined) [33]. They hypothesized that this process might be the onset of endochondral ossification occurring instead of the formation of stable cartilage. At the same time, Jukes et al. hypothesized that in vitro chondrogenically primed embryonic stem cells (ESCs) might form bone via the endochondral ossification route based on observations from teratomas formed by ESCs [34]. In these tumors, they observed hypertrophic cartilaginous tissue frequently associated with bone. Furthermore, although they were able to osteogenically differentiate ESCs in vitro, they did not go on to form bone in vivo. In contrast, chondrogenically differentiated ESCs formed bone in vivo once they were differentiated for a minimum of 14 days in vitro. Similarly, the group of Richter repeatedly demonstrated the phenotypic instability of chondrogenically differentiated adult stem cells from various tissues (adipose, marrow, and synovium) and the tendency of these tissues either to be degraded and resorbed or to mineralize [35–39]. Capitalizing on this, they demonstrated that chondrogenically priming MSCs seeded onto β-tricalcium phosphate scaffolds led to bone formation in vivo by endochondral ossification. Interestingly, it has also been shown that chondrocytes derived from articular cartilage or other sites will not undergo endochondral ossification in vivo nor do they express COLX [40–42]. In 2010, Scotti et al. introduced the term "developmental engineering" to name this approach of tissue engineering bone by mimicking the developmental process of endochondral ossification, although the paradigm of mimicking development in regenerative medicine was, in and of itself, not new [43].

Following interest in the reproducibility of chondrogenically primed MSCs to form bone via endochondral ossification, several groups assessed the ability of such tissues to repair critical-sized defects in small animals. In 2014, van der Stok and Bahney separately demonstrated the ability of chondrogenically primed MSC pellets to partially repair a critical-sized defect in the long bones without the need for a biomaterial [44,45]. The issue with each of these reports, as discussed by the authors themselves is one of scale. Each pellet usually contains 200,000 to 500,000 cells and measures approximately 1−2 mm in diameter. Although culturing MSC pellets chondrogenically and implanting them to form bone is quite reproducible, it is not a viable approach to generate thousands of these pellets to treat the kinds of defects seen in patients. This would be too time-consuming, not cost-effective, difficult for a surgeon to handle, and ultimately not scalable to large patient numbers (Fig. 6.1).

In 2014, Harada et al. also investigated the possibility of using chondrogenically primed MSCs to heal a critical-sized defect [46]. In this

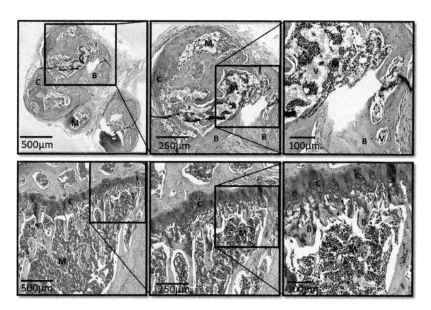

FIGURE 6.1 (Top row) Example of mesenchymal stem cell/marrow stromal cell (MSC)−mediated endochondral ossification. Chondrogenically primed human MSC pellets were subcutaneously implanted in immunodeficient mice for 8 weeks. On retrieval, samples were fixed, decalcified, sectioned and stained with hematoxylin and eosin. Visible is the newly formed bone (B), marrow (M), and blood vessels or sinusoids (V) as well as some of the original cartilage-like material (C). (Bottom row) Example of physiological endochondral ossification occurring in the growth plate of a young adult mouse tibia. As the cartilage (C) progresses toward bone formation (B) the chondrocytes become hypertrophic before mineralization and bone formation occurs.

manuscript, the researchers demonstrated the chondrogenic capacity of chondrogenically primed rat MSCs to express the usual markers of chondrogenesis when cultured in a poly(D,L-lactic-co-glycolic acid) scaffold. As would be expected, the cells produced a GAG-rich matrix and expressed collagen type II. Strangely, however, they did not observe COLX expression, which is always seen in chondrogenically differentiated MSCS. Nevertheless, the authors demonstrated extremely impressive long bone defect repair in animals with a 5-mm femur defect. So much so that they proceeded to demonstrate the same result in animals with a 15-mm long defect in the femur, almost the total diaphysial length. Interestingly, no other group has yet demonstrated repair of this magnitude in a comparably sized defect. Nor has this research been effectively translated to a larger animal model. However, it has been shown in direct comparisons that an endochondral approach to large bone defect repair is superior to the intramembranous approach. Thompson et al. demonstrated that the endochondral ossification approach was better than intramembranous in a calvarial defect, despite this bone having an intramembranous developmental origin [47].

THE ROLE OF COLLAGEN TYPE X

When looking at the series of events that take place preceding mineralization, it is clear that chondrocytes undergo hypertrophy, the process of chondrocyte enlargement and what was previously described as terminal differentiation. During this phase COLX accounts for more than 45% of the total collagen produced by hypertrophic chondrocytes [48]. Although its expression is known to be specific to chondrocytes undergoing hypertrophy, the stage preceding bone formation developmentally during endochondral ossification (EO), the function of COLX is still poorly understood. This highly evolutionarily conserved collagen is the classical marker used to identify hypertrophic chondrocytes and forms hexagonal lattices in vitro, which have been hypothesized to add stability to the pericellular network of maturing chondrocytes and structural support to the ECM, as it is remodeled during bone formation [49,50].

COLX plays a role not only in development and fracture repair but also in disease situations, which result in spontaneous and detrimental bone formation. For instance, COLX can consistently be detected in newly formed osteophytes in rheumatoid and osteoarthritis patients. In these osteophytes, COLX is found as the once healthy cartilage transitions into bone [51,52]. Numerous different mutations in the protein can lead to genetic disorders, which present with skeletal abnormalities such as schmid metaphyseal chondrodysplasia, a disorder resulting not only in dwarfism but several other musculoskeletal abnormalities including *coxa vara* and *genu varum* [53]. Understanding the role COLX plays in all of these situations can lead to a better understanding of its function in bone formation, which will ultimately improve our understanding of the process of endochondral ossification and can improve how we approach the construction of tissue-engineered constructs. To date, it has been hypothesized that COLX plays several roles including stabilizing the remodeling ECM [50], regulating the distribution of collagens and proteoglycans throughout the ECM [54], regulating the zonal distribution of matrix vesicles [55], and more recently shown to play a role in maintaining or supporting the hematopoietic stem cell niche [56]. What role COLX plays in MSC-mediated endochondral ossification in a TERM setting is still unclear. Most, if not all researchers, use it as a marker of terminal differentiation or hypertrophy or as an indicator of bone-forming capacity in vivo, yet it is unknown if any of these traits are accurately attributable to COLX. For example, Hellingman et al. demonstrated COLX expression in nasal and auricular chondrocytes, but these cells did not undergo endochondral ossification on implantation [41]. Further as discussed below, recent research has shown that chondrocytes do not necessarily undergo terminal differentiation, in which case COLX is not a marker for it. Given its very specific expression in hypertrophic chondrocytes and its large contribution to the protein content of this tissue, it is clear that it is an important matrix molecule. What its role in TERM-based approaches to bone defect repair via endochondral ossification is less clear.

HOST CELL RECRUITMENT AND THE FATE OF THE CHONDROCYTE

An issue with this approach to bone defect repair is that we still do not fully understand the relative role of the donor and host cells in orchestrating the repair and regenerative processes regardless of the type of implanted construct. This makes it impossible to fully control the process in vivo. Vascular invasion, formation of a marrow cavity, and presence of osteoclast activity observed in tissue-engineered constructs demonstrates that endogenous host cells have a role in coordinating the ultimate bone formation [57–60]. This can be assumed because of the nonmesenchymal lineage of these cells but has also been shown using specific cell labeling and tracking. Using immunocompetent transgenic rats overexpressing human placental alkaline phosphatase, Farrell et al. were able to track donor cell fate on implantation into syngeneic wild-type rats [57]. Staining for alkaline phosphatase (ALP) activity allowed the identification of the origin of the osteoblasts in the bone regeneration process. In this study, osteocytes were shown to be positive for ALP activity and therefore were shown to be of donor origin. There was a mixed population of osteoblasts that stained both positively and negatively for ALP activity and were shown to be both donor- and host-derived. These results again demonstrate that both the donor and host cells are involved in the generation of the bone and would suggest that the donor cells are slowly replaced by host cells over time. In a later study by Scotti and colleagues, they reported using a hypertrophic cartilage template that the formation of the bone in the peripheral regions of the construct is driven by host cells, whereas the implanted donor cells were found to guide the bone formation in the central regions [61]. They suggested that because the implanted cells are capable of surviving in an avascular environment, they could support the bone formation in the central regions. However, on the peripheral regions of the cartilage constructs, blood vessel invasion allows host cells to invade the surface of the construct and guide bone tissue formation. Over time, it is possible that the donor cells might eventually be replaced. The role of the donor and host cells during stem cell−mediated endochondral bone formation was also investigated by Bahney and colleagues; however, unlike the other studies they concluded that the majority of the bone regenerate was donor-derived [45]. The difference here was that the authors used the cartilage portion of a seven-day-old stabilized fracture callus from a rat as the implanted tissue rather than a true tissue-engineered construct. Most likely, there is an initial contribution from the donor cells that are implanted, and these cells are slowly replaced over time. This would be most desirable when thinking of the possibility of using allogeneic cells to generate tissue-engineered constructs. Current research approaches are mostly focused on the use of autologous cells. Although there is a clear interaction between the donor and host cells for successful bone formation, in an allogeneic setting, this situation could be completely different. There is very little known about the potential of allogeneic cell-based bone regeneration therapies and how the donor and host cells interact in coordinating the bone regeneration. Furthermore, even with the use of autologous cells, the fate of the chondrocyte in the process of stem cell−mediated endochondral bone formation is controversial. Earlier research involving chondrogenically differentiated cells for endochondral bone formation presumed that on implantation the cells would form bone via the developmental process, whereby hypertrophic chondrocytes undergo apoptosis and are replaced by host-derived osteoblasts. Interestingly, in the study by Bahney and colleagues, their lineage-tracing studies reported that not only the bone regenerate was donor-derived but also the chondrocytes did not undergo apoptosis but transformed into osteoblasts.

Until recently, all texts describing the process of endochondral ossification document the terminal fate of the chondrocyte. Chondrocytes are either not clearly mentioned in the final stages of the process or described to undergo apoptosis or even necrosis as the invading blood vessels bring osteoblasts and osteoclasts/chondroclasts to remodel the matrix. However, it has become clear that this is not the case. Since 2014, several groups have used lineage tracing and other techniques to demonstrate that not all chondrocytes die and are replaced by osteoblasts. At the very least, a subset of these cells appear to survive and transdifferentiate into bone-forming osteoblasts [45,62−64]. Ultimately, knowing the fate of the implanted cells, including the hypertrophic chondrocytes and the contributions of various host cell populations, including immune cells, will be critical in developing new therapies based on this developmental approach to bone defect repair.

THE ROLE OF THE IMMUNE SYSTEM AND USE OF THIRD-PARTY CELLS

Large-scale tissue damage, such as that leading to a large bone defect, will inevitably lead to a strong immune response coupled with inflammation. This must be an important consideration in any regenerative medicine—based approach to bone healing. Repair of damaged bone is a complex process involving cells of the skeletal system and the immune system. Inflammation is a prerequisite of fracture repair, which leads to a sequence of events crucial for successful repair [65—67]. Within the first 24 h of injury, there is an influx of proinflammatory cytokines (such as TNFα, interleukin (IL)-6, and IL-1β), which induce angiogenesis and initiate the immune response [65,68]. The first cells to respond to these signals are those of the innate immune system (monocytes, macrophages, dendritic cells, neutrophils, and natural killer cells), which secrete a variety of cytokines and growth factors to recruit cells of the adaptive immune system (T and B lymphocytes) [69]. Secretion of bone-specific growth factors resulting from the inflammatory response, e.g., TGF-β and BMP, attract osteoprogenitor cells (such as MSCs) to the site of injury [69]. Together, the inflammatory mediators and factors expressed lead to the proliferation and differentiation of the osteoprogenitors to osteoblasts [70—72]. The differentiation of osteoprogenitors during fracture repair can occur either through the developmental process of intramembranous ossification or more commonly through the process of endochondral ossification [66,73—75]. Secretion of TGF-β2 or TGF-β3, BMPs, or other molecular signals induces the formation of the cartilaginous callus, which is ultimately replaced by woven bone [68,76]. Studies from fracture repair can give vital insight into the immune cells participating in bone repair and consequently those cells that may be required for bone regeneration, for example, by active recruitment by exogenously applied cells or proteins. Lymphocytes have also been shown to be necessary for fracture repair. T and B cells have been shown to infiltrate the fracture callus to participate in bone repair during the bone remodeling stage [77]. Fracture healing has also been shown to be accelerated in mice lacking T and B cells [78]. CD8 T cells have been shown to negatively impact fracture repair [79], whereas CD4 T cells have varying effects on bone regeneration, depending on the subtype [80]. These complex interactions between the skeletal cells and immune cells during fracture repair are critical for the bone regeneration/repair outcome.

Although the use of autologous cells for bone regeneration purposes will not lead to serious immune rejection of the implanted cells, a fracture will still elicit a major response from the immune system. The issue with autologous cells is that they can be limited in the quantity that can be obtained. Furthermore, the use of autologous MSCs from elderly or diseased patients can have reduced differentiation and regenerative potential compared with healthy donors [81]. In addition, there will be a major lag time in treating a patient with their own cells because of the requirements to expand the cells in vitro and perform the necessary quality controls on these cells before administration to the patient. Considering this, it is important to take the immune response into account when designing new regenerative medicine—based approaches to repair bone defects. Research investigating the use of MSCs for bone repair most commonly involves autologous cells, cells from syngeneic donors, or in the context of animal-based models, human cells implanted into immunosuppressed animals. However, in the context of the clinic, this would lead to potentially long waiting periods to harvest and culture autologous cells, in addition to increasing expenses related to testing individual batches of cells for approval, limiting their clinical application. From a practical and ultimately regulatory point of view, it would be preferable to be able to use allogeneic cells that are immediately available from approved stocks or even to have a treatment that does not require cells at all. As a result, there has been much interest in the use of allogeneic cells for regenerative medicine to make the approach more clinically viable. Because of the low expression levels of major histocompatibility complex (MHC) class II and other immune stimulatory molecules (e.g., CD80 and CD86), MSCs have been shown to be somewhat "immunoprivileged" and can be used in an allogeneic setting [82—84]. Furthermore, considering that in fracture repair, cells will be transplanted into an inflammatory environment, MSCs are advantageous in that they can modulate the immune system [85—97]. Similar to fracture repair, the innate immune system would activate the adaptive immune system, in this case to reject cells/tissues. Immunomodulation by allogeneic MSCs is important to allow specific immune responses to be regulated and avoid rejection. Th1 T cells are known to negatively impact bone regeneration by allogeneic MSCs through the inhibition of the expression of osteogenesis-specific genes (osteocalcin, Runx2, and ALP) [98]. Th2 T cells and regulatory T cells have been shown to promote osteogenesis [69,99,100]. What is much less clear is the interaction between the immune system and MSCs that have been pre-differentiated into another tissue type before implantation. Allogeneic undifferentiated MSCs are generally known to be nonimmunogenic in that they do not provoke immune responses; however, they are immunomodulatory and can modulate the immune system for their desired purposes [82,84,101—103]. Immunogenicity is defined as the ability of cells/tissues to provoke an immune response and is generally considered to be an undesirable physiological response. Reports on the immunomodulatory properties of allogeneic chondrogenically differentiated MSCs are few and somewhat contradictory. There are conflicting studies of the immunogenicity of allogeneic chondrogenically differentiated MSCs with reports showing that they can be both nonimmunogenic [104−107] and immunogenic [108,109] depending on the culture system used in the investigations. In addition, very little is known about what takes place in the in vivo setting in immuno-competent animals. In 2011, Liu et al. showed that bone formation by undifferentiated MSCs could not take place in fully immune-competent syngeneic animals when these were implanted subcutaneously on calcium phosphate ceramics [11]. However, they were able to demonstrate that modulation of the T-cell response could enable bone formation. What is unknown is what the outcome would be if MSCs were chondrogenically primed before implantation into an allogeneic setting (nongenetically similar as with human donors instead of patients' own cells). Farrell et al. have shown that chondrogenically primed rat MSCs can form bone in immuno-competent animals of a similar genetic background [57]. What will be necessary to investigate is the use of allogeneic chondrogenically primed cells in a similar setting. Furthermore, it is unclear what the role of the biomaterial would be in such a situation. It is well known that biomaterials themselves elicit varied and often unwanted responses from the immune system, and this is also critical for the understanding of how chondrogeni-cally primed MSCs behave and in the allogeneic setting and what responses they induce from the host.

SCALE UP ENDOCHONDRAL OSSIFICATION APPROACHES: TOWARD THE CLINIC

This leads us to discuss options to actually translate this approach to the clinic. The main issue here is one of scale, both in the context of the size of the defect that needs to be healed and the number of patients to be treated. Both of these issues are very important when considering how to bring a promising new therapy from proof of concept in small animal studies to large animal experiments and ultimately clinical trials. With regard to the defect size, this is important from the perspective of vascularization and required cell numbers, growth factor doses, or quantities of biomaterials. The requirements here are orders of magnitude above what are necessary for rodent studies. Given that most cells of the body are rarely more than 100−200 μm from a capillary, ensuring proper vascularization in tissue-engineered constructs is critical. One of the proposed advantages of endochondral ossification is that chondrocytes naturally reside in a low-oxygen environment and will survive the initial hypoxic insult on implantation into an avascular region. They will then orchestrate the attraction of surrounding vasculature concurrent with the remodeling of the construct matrix and the formation of new bone. This appears to be the case in small animal models, at least with regard to the ultimate formation of vascularized bone. In addition, at least a proportion of the implanted cells appear to persist in this new tissue [57]. However, there is a limit to how fast blood vessels can invade a larger tissue and probably also a limit to how long chondrogenically primed MSCs can survive in an oxygen and nutrient-poor and inflamed environment. This is likely to present significant issues when directly translating this approach to a clinically relevant-sized defect. Ultimately, this will depend on the ability of the chondrogenically primed cells to survive in low-oxygen and low-nutrient environments until vascularization and remodeling of the construct into vascularized bone occurs.

There are possible ways to circumvent this issue. The most obvious is to remove the requirement for cells in the first place. However, ideally one would like to retain the positive effects observed by using chondrogenically primed MSC constructs. Several groups have attempted to devitalize in vitro–generated chondrogenically primed MSC constructs before implantation. The rationale was that the ECM and various secreted proteins bound to the matrix were most important in instructing endogenous cells to regenerate the bone. Given the prevailing dogma that the hypertrophic chondrocytes die during endochondral ossification this was a logical approach. We and others observed that this approach did not result in de novo bone formation (unpublished work and personal communications). The group of Martin et al. also observed this phenomenon when they used a freeze thaw approach to devitalize cell-seeded chondrogenically primed constructs [110]. Interestingly, when they induced apoptosis in the cells in these matrices by transducing cells with an apoptosis-inducible cassette, they observed comparable bone formation with viable cell containing constructs. This would suggest that the cells, or at least living cells, are not absolutely critical for bone formation to take place. However, although the ability to produce devitalized matrices that will induce bone formation is very desirable, there is no consensus on the scalability of such an approach. Work from the group of Kelly has also taken the approach of using devitalized matrices to generate new bone in vivo. Cunniffe et al. generated devitalized scaffolds from pig growth plates that were turned into slurry and subsequently freeze-dried as scaffolds of a desired shape and size to implant into calvarial defects. This approach was shown to be superior to empty defect controls with regard to bone healing and vascularization of the defect area. However, a comparison of another commonly used approach was not performed. The same group also generated "hypertrophic constructs" from cultured MSCs, which were then homogenized, devitalized, and freeze-dried as scaffold materials for femoral defect repair [111]. In this interesting approach, the authors observed some defect bridging in the treated conditions, but this was quite variable and not significantly better than the empty defects. This might again suggest the requirement for a source of exogenous cells to instruct defect repair, at least within the first weeks after implantation. Although the scalability of such an approach is debatable, the idea is sound.

Combining materials engineering with stem cell biology and new technologies, such as 3D printing, also represents a new area of interest for the generation of whole bone organs or even joints [112,113]. These fields require entire chapters or books within their own rights and will not be expanded on here sufficed to mention that such approaches exist and are viable methods to reproducibly scale up the production of constructs for the repair of large bone defects. It is the combination of reproducibility and flexibility provided by hydrogels and 3D printing that suggests such an approach, ideally cell-free, could become a reality for the generation of bone repair constructs in the coming years. For this to become a reality, certain advances will be required. The ability to print stronger tissues and composites will be necessary to withstand physiological load bearing. Inclusion of growth factors that do not have a burst release on implantation and that are evenly distributed through the materials as desired without their inactivation (as a result of the printing process) will also be critical, particularly if exogenous cells are not added to recruit and instruct endogenous cells in the regeneration of bone or even joint tissues. If such a construct can be generated that results in robust bone formation, preferably via endochondral ossification without the need to add cells, a much shorter route to the clinic would be envisioned. This is because of reduced costs associated with the good manufacturing practice (GMP)-grade manufacture of large volumes of cells and the regulatory hurdles required to bring any cell-based therapy to the clinic. The other issue of scale, already somewhat alluded to, is the issue of patient numbers. Any approach that should replace autologous bone transplantation should be applicable in hospitals across the world and not only in some very specialized medical institutions that, for example, are equipped with GMP facilities to allow the mass production of clinical-grade cells or materials. Once again, this would most easily be achieved via the generation of an off-the-shelf product that can be stored long term and therefore most likely cell-free. However, if a cell-free approach cannot be developed that surpasses results achieved using chondrogenically primed cells, production will be ultimately streamlined over time to bring costs within reach of developing a product available to the public.

Another approach to ideally scale up the generation of bone via endochondral ossification would be to use a growth factor/biomaterial construct that would be as effective at generating bone as BMP2 without the requirement for massive doses or the potential risks associated with it (ectopic bone formation and tumor development) [114]. One such factor could be growth and differentiation factor 5. This protein is member of a subgroup of BMPs that are most well known for their role in joint formation and chondrogenesis but have also recently been implicated in hypertrophic differentiation and endochondral ossification [115]. Other proteins, for example, connective tissue growth factor (also known as CCN2) and high mobility group box 1, have been shown to be important for the process of endochondral ossification [116–119]. These are a small number of examples of proteins that have been identified as being important in one or more of the processes of endochondral ossification (cell recruitment, vascularization, osteogenesis, etc.) that are being investigated for their ability to generate de novo bone or repair bone defects as a replacement for BMP2 at more physiological doses [114]. Coupled to newly developed biomaterials that can better control the release kinetics of such proteins, there is promise for the development of a scalable treatment for bone defect repair via endochondral ossification.

CONCLUSIONS

In this chapter we discuss the possibility of exploiting the developmental process of endochondral ossification for the purpose of repairing large bone defects. The approach has several potential advantages compared with other TERM-based options; the ability to instruct and control the body to not only generate new bone but also to vascularize it and generate a marrow cavity, the apparent ability of chondrogenically primed cells to survive an initial hypoxic insult and nutrient deprivation until the host vasculature invades and perhaps even the possibility to evade an immune response from the host. Despite the promise and advantages of such an approach, it has yet to be translated to the clinic or even a large animal model. For this to occur, it will be necessary to overcome certain obstacles. Firstly, physical scale will need to be addressed. At present, it is not clear how long cells would survive before being vascularized. Secondly, to avoid the use of cells the ideal growth factor and biomaterial combination will need to be identified. Third and finally, any solution that will induce large bone defect repair by inducing endochondral ossification needs to be easy to produce and handle, as well as cost effective and superior to the current gold standard approach of autologous bone transplantation. By further increasing our understanding of the process of endochondral ossification during bone formation, we will be better equipped to fulfill these needs and create more effective and sustainable TERM-based solutions to large bone defect repair.

LIST OF ACRONYMS AND ABBREVIATIONS

COLX Collagen type X
MSC Mesenchymal stem cell/marrow stromal cell
TERM Tissue engineering and regenerative medicine

REFERENCES

[1] Aciduman A, Belen D. The earliest document regarding the history of cranioplasty from the Ottoman era. Surg Neurol September 2007;68(3):349−52. discussion 52−53. PMID:17719987.

[2] Hayden RE, Mullin DP, Patel AK. Reconstruction of the segmental mandibular defect: current state of the art. Curr Opin Otolaryngol Head Neck Surg August 2012;20(4):231−6. PMID:22894990.

[3] Chang YM, Tsai CY, Wei FC. One-stage, double-barrel fibula osteoseptocutaneous flap and immediate dental implants for functional and aesthetic reconstruction of segmental mandibular defects. Plast Reconstr Surg July 2008;122(1):143−5. PMID:18594398.

[4] Bahr W, Stoll P, Wachter R. Use of the "double barrel" free vascularized fibula in mandibular reconstruction. J Oral Maxillofac Surg January 1998;56(1):38−44. PMID:9437980.

[5] Gadre PK, Ramanojam S, Patankar A, Gadre KS. Nonvascularized bone grafting for mandibular reconstruction: myth or reality? J Craniofac Surg September 2011;22(5):1727−35. PMID:21959421.

[6] Duskova M, Leamerova E, Sosna B, Gojis O. Guided tissue regeneration, barrier membranes and reconstruction of the cleft maxillary alveolus. J Craniofac Surg November 2006;17(6):1153−60. PMID:17119421.

[7] Janssen NG, Weijs WL, Koole R, Rosenberg AJ, Meijer GJ. Tissue engineering strategies for alveolar cleft reconstruction: a systematic review of the literature. Clin Oral Investig February 22, 2013. PMID:23430342.

[8] Shapiro F. Bone development and its relation to fracture repair. The role of mesenchymal osteoblasts and surface osteoblasts. Eur Cell Mater 2008;15:53−76. PMID:18382990. Epub 2008/04/03. eng.

[9] Yang Y. Skeletal morphogenesis during embryonic development. Crit Rev Eukaryot Gene Expr 2009;19(3):197−218. PMID:19883365. Epub 2009/11/04. eng.

[10] Pittenger MF, Mackay AM, Beck SC, Jaiswal RK, Douglas R, Mosca JD, et al. Multilineage potential of adult human mesenchymal stem cells. Science April 2, 1999;284(5411):143−7. PMID:10102814.

[11] Meijer GJ, de Bruijn JD, Koole R, van Blitterswijk CA. Cell based bone tissue engineering in jaw defects. Biomaterials July 2008;29(21):3053−61. PMID:18433864.

[12] Chatterjea A, Meijer G, van Blitterswijk C, de Boer J. Clinical application of human mesenchymal stromal cells for bone tissue engineering. Stem Cells Int 2010;2010:215625. PMID:21113294. PMCID:2989379. Epub 2010/11/30. eng.

[13] Mackie EJ, Ahmed YA, Tatarczuch L, Chen KS, Mirams M. Endochondral ossification: how cartilage is converted into bone in the developing skeleton. Int J Biochem Cell Biol 2008;40(1):46−62. PMID:17659995.

[14] Thompson EM, Matsiko A, Farrell E, Kelly DJ, O'Brien FJ. Recapitulating endochondral ossification: a promising route to in vivo bone regeneration. J Tissue Eng Regen Med August 2015;9(8):889−902. PMID:24916192.

[15] Gawlitta D, Farrell E, Malda J, Creemers LB, Alblas J, Dhert WJ. Modulating endochondral ossification of multipotent stromal cells for bone regeneration. Tissue Eng Part B Rev August 2010;16(4):385−95. PMID:20131956. Epub 2010/02/06. eng.

[16] Medici D, Olsen BR. The role of endothelial-mesenchymal transition in heterotopic ossification. J Bone Miner Res August 2012;27(8):1619–22. PMID:22806925. PMCID:3432417.

[17] Yeung Tsang K, Wa Tsang S, Chan D, Cheah KS. The chondrocytic journey in endochondral bone growth and skeletal dysplasia. Birth Defects Res C Embryo Today March 2014;102(1):52–73. PMID:24677723. Epub 2014/03/29. eng.

[18] Cervantes-Diaz F, Contreras P, Marcellini S. Evolutionary origin of endochondral ossification: the transdifferentiation hypothesis. Dev Genes Evol March 2017;227(2):121–7. PMID:27909803. Epub 2016/12/03. eng.

[19] Long F, Ornitz DM. Development of the endochondral skeleton. Cold Spring Harb Perspect Biol January 2013;5(1):a008334. PMID:23284041.

[20] Gentili C, Cancedda R. Cartilage and bone extracellular matrix. Curr Pharm Des 2009;15(12):1334–48. PMID:19355972.

[21] Heinegard D. Fell-Muir lecture: proteoglycans and more—from molecules to biology. Int J Exp Pathol December 2009;90(6):575–86. PMID:19958398. PMCID:PMC2803248.

[22] Johansson N, Saarialho-Kere U, Airola K, Herva R, Nissinen L, Westermarck J, et al. Collagenase-3 (MMP-13) is expressed by hypertrophic chondrocytes, periosteal cells, and osteoblasts during human fetal bone development. Dev Dyn March 1997;208(3):387–97. PMID:9056642. Epub 1997/03/01. eng.

[23] van der Eerden BC, Karperien M, Wit JM. Systemic and local regulation of the growth plate. Endocr Rev December 2003;24(6):782–801. PMID:14671005. Epub 2003/12/13. eng.

[24] Karsenty G, Wagner EF. Reaching a genetic and molecular understanding of skeletal development. Dev Cell April 2002;2(4):389–406. PMID:11970890. Epub 2002/04/24. eng.

[25] Kronenberg HM. The role of the perichondrium in fetal bone development. Ann NY Acad Sci November 2007;1116:59–64. PMID:18083921. Epub 2007/12/18. eng.

[26] Nakashima K, Zhou X, Kunkel G, Zhang Z, Deng JM, Behringer RR, et al. The novel zinc finger-containing transcription factor osterix is required for osteoblast differentiation and bone formation. Cell January 11, 2002;108(1):17–29. PMID:11792318. Epub 2002/01/17. eng.

[27] Taniguchi N, Yoshida K, Ito T, Tsuda M, Mishima Y, Furumatsu T, et al. Stage-specific secretion of HMGB1 in cartilage regulates endochondral ossification. Mol Cell Biol August 2007;27(16):5650–63. PMID:17548469.

[28] Caplan AI. Bone development and repair. Bioessays April 1987;6(4):171–5. PMID:3593327. Epub 1987/04/01. eng.

[29] Marie PJ. Transcription factors controlling osteoblastogenesis. Arch Biochem Biophys May 15, 2008;473(2):98–105. PMID:18331818.

[30] Shapiro IM, Adams CS, Freeman T, Srinivas V. Fate of the hypertrophic chondrocyte: microenvironmental perspectives on apoptosis and survival in the epiphyseal growth plate. Birth Defects Res C Embryo Today December 2005;75(4):330–9. PMID:16425255.

[31] Warnke PH, Springer IN, Wiltfang J, Acil Y, Eufinger H, Wehmoller M, et al. Growth and transplantation of a custom vascularised bone graft in a man. Lancet 2004;364(9436):766. PMID:167.

[32] Chan CK, Chen CC, Luppen CA, Kim JB, DeBoer AT, Wei K, et al. Endochondral ossification is required for haematopoietic stem-cell niche formation. Nature January 22, 2009;457(7228):490–4. PMID:19078959. PMCID:2648141. Epub 2008/12/17. eng.

[33] Farrell E, van der Jagt OP, Koevoet W, Kops N, van Manen CJ, Hellingman CA, et al. Chondrogenic priming of human bone marrow stromal cells: a better route to bone repair? Tissue Eng Part C Methods June 2009;15(2):285–95. PMID:19505182.

[34] Jukes JM, Both SK, Leusink A, Sterk LM, van Blitterswijk CA, de Boer J. Endochondral bone tissue engineering using embryonic stem cells. Proc Natl Acad Sci USA May 13, 2008;105(19):6840−5. PMID:18467492.

[35] Janicki P, Kasten P, Kleinschmidt K, Luginbuehl R, Richter W. Chondrogenic pre-induction of human mesenchymal stem cells on beta-TCP: enhanced bone quality by endochondral heterotopic bone formation. Acta Biomater August 2010;6(8):3292−301. PMID:20123138. Epub 2010/02/04. eng.

[36] Eric S, Jennifer F, Helga L, Tobias G, Martin J, Wiltrud R. Mesenchymal stem cell differentiation in an experimental cartilage defect: restriction of hypertrophy to bone-close neocartilage. Stem Cell Dev 2009;18.

[37] Dickhut A, Pelttari K, Janicki P, Wagner W, Eckstein V, Egermann M, et al. Calcification or dedifferentiation: requirement to lock mesenchymal stem cells in a desired differentiation stage. J Cell Physiol April 2009;219(1):219−26. PMID:19107842.

[38] Karoliina P, Eric S, Wiltrud R. The use of mesenchymal stem cells for chondrogenesis. Injury 2008;39.

[39] Hennig T, Lorenz H, Thiel A, Goetzke K, Dickhut A, Geiger F, et al. Reduced chondrogenic potential of adipose tissue derived stromal cells correlates with an altered TGFbeta receptor and BMP profile and is overcome by BMP-6. J Cell Physiol June 2007;211(3):682−91. PMID:17238135.

[40] Pelttari K, Winter A, Steck E, Goetzke K, Hennig T, Ochs BG, et al. Premature induction of hypertrophy during in vitro chondrogenesis of human mesenchymal stem cells correlates with calcification and vascular invasion after ectopic transplantation in SCID mice. Arthritis Rheum October 2006;54(10):3254−66. PMID:17009260.

[41] Hellingman CA, Verwiel ET, Slagt I, Koevoet W, Poublon RM, Nolst-Trenite GJ, et al. Differences in cartilage-forming capacity of expanded human chondrocytes from ear and nose and their gene expression profiles. Cell Transplant 2011;20(6):925−40. PMID:21054934. Epub 2010/11/09. eng.

[42] Pleumeekers MM, Nimeskern L, Koevoet WL, Kops N, Poublon RM, Stok KS, et al. The in vitro and in vivo capacity of culture-expanded human cells from several sources encapsulated in alginate to form cartilage. Eur Cell Mater April 06, 2014;27:264−80. discussion 78−80. PMID:24706178.

[43] Scotti C, Tonnarelli B, Papadimitropoulos A, Scherberich A, Schaeren S, Schauerte A, et al. Recapitulation of endochondral bone formation using human adult mesenchymal stem cells as a paradigm for developmental engineering. Proc Natl Acad Sci USA April 20, 2010;107(16):7251−6. PMID:20406908. PMCID:2867676. Epub 2010/04/22. eng.

[44] van der Stok J, Koolen MK, Jahr H, Kops N, Waarsing JH, Weinans H, et al. Chondrogenically differentiated mesenchymal stromal cell pellets stimulate endochondral bone regeneration in critical-sized bone defects. Eur Cell Mater 2014;27:137−48. discussion 48. PMID:24554271.

[45] Bahney CS, Hu DP, Taylor AJ, Ferro F, Britz HM, Hallgrimsson B, et al. Stem cell-derived endochondral cartilage stimulates bone healing by tissue transformation. J Bone Miner Res 2014;29(5):1269−82. PMID:24259230. Epub 2013/11/22. eng.

[46] Harada N, Watanabe Y, Sato K, Abe S, Yamanaka K, Sakai Y, et al. Bone regeneration in a massive rat femur defect through endochondral ossification achieved with chondrogenically differentiated MSCs in a degradable scaffold. Biomaterials September 2014;35(27):7800−10. PMID:24952976. Epub 2014/06/24. eng.

[47] Thompson EM, Matsiko A, Kelly DJ, Gleeson JP, O'Brien FJ. An endochondral ossification-based approach to bone repair: chondrogenically primed mesenchymal stem cell-laden scaffolds support greater repair of critical-sized cranial defects than osteogenically stimulated constructs in vivo. Tissue Eng Part A March 2016;22(5–6):556–67. PMID:26896424.

[48] Luvalle P, Daniels K, Hay ED, Olsen BR. Type X collagen is transcriptionally activated and specifically localized during sternal cartilage maturation. Matrix November 1, 1992;12(5):404–13.

[49] Shen G. The role of type X collagen in facilitating and regulating endochondral ossification of articular cartilage. Orthod Craniofac Res 2005;8(1):11–7.

[50] Schmid TM, Linsenmayer TF. Immunohistochemical localization of short chain cartilage collagen (type X) in avian tissues. J Cell Biol 1985;100(2):598–605.

[51] He Y, Siebuhr AS, Brandt-Hansen NU, Wang J, Su D, Zheng Q, et al. Type X collagen levels are elevated in serum from human osteoarthritis patients and associated with biomarkers of cartilage degradation and inflammation. BMC Muscoskel Disord 2014;15:309. PMID:PMC4179849.

[52] Aigner T, Reichenberger E, Bertling W, Kirsch T, Stöss H, Von der Mark K. Type X collagen expression in osteoarthritic and rheumatoid articular cartilage. Virchows Arch B 1993;63(1):205.

[53] Wasylenko MJ, Wedge JH, Houston CS. Metaphyseal chondrodysplasia, Schmid type. A defect of ultrastructural metabolism: case report. J Bone Joint Surg Am 1980;62(4):660–3.

[54] Chan D, Jacenko O. Phenotypic and biochemical consequences of collagen X mutations in mice and humans. Matrix Biol July 1, 1998;17(3):169–84.

[55] Kirsch T, Harrison G, Golub EE, Nah H-D. The roles of annexins and types II and X collagen in matrix vesicle-mediated mineralization of growth plate cartilage. J Biol Chem 2000;275(45):35577–83.

[56] Jacenko O, Roberts DW, Campbell MR, McManus PM, Gress CJ, Tao Z. Linking hematopoiesis to endochondral skeletogenesis through analysis of mice transgenic for collagen X. Am J Pathol 2002;160(6):2019–34. PMID:PMC1850848.

[57] Farrell E, Both SK, Odorfer KI, Koevoet W, Kops N, O'Brien FJ, et al. In-vivo generation of bone via endochondral ossification by in-vitro chondrogenic priming of adult human and rat mesenchymal stem cells. BMC Musculoskelet Disord 2011;12:31. PMID:21281488. PMCID:3045394. Epub 2011/02/02. eng.

[58] Tasso R, Augello A, Boccardo S, Salvi S, Carida M, Postiglione F, et al. Recruitment of a host's osteoprogenitor cells using exogenous mesenchymal stem cells seeded on porous ceramic. Tissue Eng Part A August 2009;15(8):2203–12. PMID:19265473.

[59] Tasso R, Fais F, Reverberi D, Tortelli F, Cancedda R. The recruitment of two consecutive and different waves of host stem/progenitor cells during the development of tissue-engineered bone in a murine model. Biomaterials March 2010;31(8):2121–9. PMID:20004968. Epub 2009/12/17. eng.

[60] Tortelli F, Tasso R, Loiacono F, Cancedda R. The development of tissue-engineered bone of different origin through endochondral and intramembranous ossification following the implantation of mesenchymal stem cells and osteoblasts in a murine model. Biomaterials January 2010;31(2):242–9. PMID:19796807.

[61] Scotti C, Piccinini E, Takizawa H, Todorov A, Bourgine P, Papadimitropoulos A, et al. Engineering of a functional bone organ through endochondral ossification. Proc Natl Acad Sci USA March 5, 2013;110(10):3997–4002. PMID:23401508. PMCID:3593845.

[62] Yang L, Tsang KY, Tang HC, Chan D, Cheah KS. Hypertrophic chondrocytes can become osteoblasts and osteocytes in endochondral bone formation. Proc Natl Acad Sci USA August 19, 2014;111(33):12097−102. PMID:25092332. PMCID:4143064.

[63] Ono N, Ono W, Nagasawa T, Kronenberg HM. A subset of chondrogenic cells provides early mesenchymal progenitors in growing bones. Nat Cell Biol December 2014;16(12):1157−67. PMID:25419849. PMCID:4250334.

[64] Zhou X, von der Mark K, Henry S, Norton W, Adams H, de Crombrugghe B. Chondrocytes transdifferentiate into osteoblasts in endochondral bone during development, postnatal growth and fracture healing in mice. PLoS Genet December 2014;10(12):e1004820. PMID:25474590. PMCID:4256265.

[65] Mountziaris PM, Mikos AG. Modulation of the inflammatory response for enhanced bone tissue regeneration. Tissue Eng Part B Rev June 2008;14(2):179−86. PMID:18544015. PMCID:2962857. Epub 2008/06/12. eng.

[66] Kolar P, Schmidt-Bleek K, Schell H, Gaber T, Toben D, Schmidmaier G, et al. The early fracture hematoma and its potential role in fracture healing. Tissue Eng Part B Rev August 2010;16(4):427−34. PMID:20196645.

[67] Pape HC, Marcucio R, Humphrey C, Colnot C, Knobe M, Harvey EJ. Trauma-induced inflammation and fracture healing. J Orthop Trauma September 2010;24(9):522−5. PMID:20736786. Epub 2010/08/26. eng.

[68] Gerstenfeld LC, Cullinane DM, Barnes GL, Graves DT, Einhorn TA. Fracture healing as a post-natal developmental process: molecular, spatial, and temporal aspects of its regulation. J Cell Biochem April 01, 2003;88(5):873−84. PMID:12616527. Epub 2003/03/05. eng.

[69] Kovach TK, Dighe AS, Lobo PI, Cui Q. Interactions between MSCs and immune cells: implications for bone healing. J Immunol Res 2015;2015:752510. PMID:26000315. PMCID:4427002. Epub 2015/05/23. eng.

[70] Dimitriou R, Jones E, McGonagle D, Giannoudis PV. Bone regeneration: current concepts and future directions. BMC Med 2011;9:66. PMID:21627784. PMCID:3123714. Epub 2011/06/02. eng.

[71] Loi F, Cordova LA, Pajarinen J, Lin TH, Yao Z, Goodman SB. Inflammation, fracture and bone repair. Bone May 2016;86:119−30. PMID:26946132. PMCID:4833637. Epub 2016/03/08. eng.

[72] Mizuno K, Mineo K, Tachibana T, Sumi M, Matsubara T, Hirohata K. The osteogenetic potential of fracture haematoma. Subperiosteal and intramuscular transplantation of the haematoma. J Bone Joint Surg Br September 1990;72(5):822−9. PMID:2211764. Epub 1990/09/01. eng.

[73] Brighton CT. The biology of fracture repair. Instr Course Lect 1984;33:60−82. PMID:6546128.

[74] Einhorn TA, Gerstenfeld LC. Fracture healing: mechanisms and interventions. Nat Rev Rheumatol January 2015;11(1):45−54. PMID:25266456. PMCID:PMC4464690.

[75] Kuntzman A, Tortora GJ. The process of fracture repair. Anatomy and physiology for manual therapies. Wiley; 2010.

[76] Cho TJ, Gerstenfeld LC, Einhorn TA. Differential temporal expression of members of the transforming growth factor beta superfamily during murine fracture healing. J Bone Miner Res March 2002;17(3):513−20. PMID:11874242.

[77] Konnecke I, Serra A, El Khassawna T, Schlundt C, Schell H, Hauser A, et al. T and B cells participate in bone repair by infiltrating the fracture callus in a two-wave fashion. Bone July 2014;64:155−65. PMID:24721700. Epub 2014/04/12. eng.

[78] Toben D, Schroeder I, El Khassawna T, Mehta M, Hoffmann JE, Frisch JT, et al. Fracture healing is accelerated in the absence of the adaptive immune system. J Bone Miner Res January 2011;26(1):113−24. PMID:20641004. Epub 2010/07/20. eng.

[79] Reinke S, Geissler S, Taylor WR, Schmidt-Bleek K, Juelke K, Schwachmeyer V, et al. Terminally differentiated CD8(+) T cells negatively affect bone regeneration in humans. Sci Transl Med March 20, 2013;5(177):177ra36. PMID:23515078. Epub 2013/03/22. eng.

[80] Nam D, Mau E, Wang Y, Wright D, Silkstone D, Whetstone H, et al. T-lymphocytes enable osteoblast maturation via IL-17F during the early phase of fracture repair. PLoS One 2012;7(6):e40044. PMID:22768215. PMCID:3386936. Epub 2012/07/07. eng.

[81] Mueller SM, Glowacki J. Age-related decline in the osteogenic potential of human bone marrow cells cultured in three-dimensional collagen sponges. J Cell Biochem 2001;82(4):583−90. PMID:11500936. Epub 2001/08/14. eng.

[82] Aggarwal S, Pittenger MF. Human mesenchymal stem cells modulate allogeneic immune cell responses. Blood February 15, 2005;105(4):1815−22. PMID:15494428.

[83] Le Blanc K, Ringden O. Immunomodulation by mesenchymal stem cells and clinical experience. J Intern Med November 2007;262(5):509−25. PMID:17949362.

[84] Nauta AJ, Fibbe WE. Immunomodulatory properties of mesenchymal stromal cells. Blood November 15, 2007;110(10):3499−506. PMID:17664353.

[85] Djouad F, Charbonnier LM, Bouffi C, Louis-Plence P, Bony C, Apparailly F, et al. Mesenchymal stem cells inhibit the differentiation of dendritic cells through an interleukin-6-dependent mechanism. Stem Cell August 2007;25(8):2025−32. PMID:17510220. Epub 2007/05/19. eng.

[86] Jiang XX, Zhang Y, Liu B, Zhang SX, Wu Y, Yu XD, et al. Human mesenchymal stem cells inhibit differentiation and function of monocyte-derived dendritic cells. Blood May 15, 2005;105(10):4120−6. PMID:15692068.

[87] Liu WH, Liu JJ, Wu J, Zhang LL, Liu F, Yin L, et al. Novel mechanism of inhibition of dendritic cells maturation by mesenchymal stem cells via interleukin-10 and the JAK1/STAT3 signaling pathway. PLoS One 2013;8(1):e55487. PMID:23383203. PMCID:PMC3559548.

[88] Zhang W, Ge W, Li C, You S, Liao L, Han Q, et al. Effects of mesenchymal stem cells on differentiation, maturation, and function of human monocyte-derived dendritic cells. Stem Cells Dev June 2004;13(3):263−71. PMID:15186722. Epub 2004/06/10. eng.

[89] Zhao ZG, Xu W, Sun L, You Y, Li F, Li QB, et al. Immunomodulatory function of regulatory dendritic cells induced by mesenchymal stem cells. Immunol Invest 2012;41(2):183−98. PMID:21936678.

[90] Di Ianni M, Del Papa B, De Ioanni M, Moretti L, Bonifacio E, Cecchini D, et al. Mesenchymal cells recruit and regulate T regulatory cells. Exp Hematol March 2008;36(3):309−18. PMID:18279718. Epub 2008/02/19. eng.

[91] Di Nicola M, Carlo-Stella C, Magni M, Milanesi M, Longoni PD, Matteucci P, et al. Human bone marrow stromal cells suppress T-lymphocyte proliferation induced by cellular or nonspecific mitogenic stimuli. Blood May 15, 2002;99(10):3838−43. PMID:11986244. Epub 2002/05/03. eng.

[92] Asari S, Itakura S, Ferreri K, Liu CP, Kuroda Y, Kandeel F, et al. Mesenchymal stem cells suppress B-cell terminal differentiation. Exp Hematol May 2009;37(5):604−15. PMID:19375651. PMCID:2747661. Epub 2009/04/21. eng.

[93] Corcione A, Benvenuto F, Ferretti E, Giunti D, Cappiello V, Cazzanti F, et al. Human mesenchymal stem cells modulate B-cell functions. Blood January 01, 2006;107(1):367−72. PMID:16141348. Epub 2005/09/06. eng.

[94] Bartholomew A, Sturgeon C, Siatskas M, Ferrer K, McIntosh K, Patil S, et al. Mesenchymal stem cells suppress lymphocyte proliferation in vitro and prolong skin graft survival in vivo. Exp Hematol January 2002;30(1):42−8. PMID:11823036. Epub 2002/02/02. eng.

[95] Sotiropoulou PA, Perez SA, Gritzapis AD, Baxevanis CN, Papamichail M. Interactions between human mesenchymal stem cells and natural killer cells. Stem Cell January 2006;24(1):74−85. PMID:16099998. Epub 2005/08/16. eng.

[96] Spaggiari GM, Abdelrazik H, Becchetti F, Moretta L. MSCs inhibit monocyte-derived DC maturation and function by selectively interfering with the generation of immature DCs: central role of MSC-derived prostaglandin E2. Blood June 25, 2009;113(26):6576−83. PMID:19398717. Epub 2009/04/29. eng.

[97] Spaggiari GM, Capobianco A, Abdelrazik H, Becchetti F, Mingari MC, Moretta L. Mesenchymal stem cells inhibit natural killer-cell proliferation, cytotoxicity, and cytokine production: role of indoleamine 2,3-dioxygenase and prostaglandin E2. Blood February 1, 2008;111(3):1327−33. PMID:17951526. Epub 2007/10/24. eng.

[98] Dighe AS, Yang S, Madhu V, Balian G, Cui Q. Interferon gamma and T cells inhibit osteogenesis induced by allogeneic mesenchymal stromal cells. J Orthop Res February 2013;31(2):227−34. PMID:22886855. PMCID:PMC3510319.

[99] Liu Y, Wang L, Kikuiri T, Akiyama K, Chen C, Xu X, et al. Mesenchymal stem cell-based tissue regeneration is governed by recipient T lymphocytes via IFN-gamma and TNF-alpha. Nat Med December 2011;17(12):1594−601. PMID:22101767. PMCID:3233650. Epub 2011/11/22. eng.

[100] Omar OM, Graneli C, Ekstrom K, Karlsson C, Johansson A, Lausmaa J, et al. The stimulation of an osteogenic response by classical monocyte activation. Biomaterials November 2011;32(32):8190−204. PMID:21835463. Epub 2011/08/13. eng.

[101] Hoogduijn MJ, Popp F, Verbeek R, Masoodi M, Nicolaou A, Baan C, et al. The immunomodulatory properties of mesenchymal stem cells and their use for immunotherapy. Int Immunopharmacol December 2010;10(12):1496−500. PMID:20619384. Epub 2010/07/14. eng.

[102] Gao F, Chiu SM, Motan DA, Zhang Z, Chen L, Ji HL, et al. Mesenchymal stem cells and immunomodulation: current status and future prospects. Cell Death Dis January 21, 2016;7:e2062. PMID:26794657. PMCID:4816164. Epub 2016/01/23. eng.

[103] English K. Mechanisms of mesenchymal stromal cell immunomodulation. Immunol Cell Biol January 2013;91(1):19−26. PMID:23090487. Epub 2012/10/24. eng.

[104] Zheng ZH, Li XY, Ding J, Jia JF, Zhu P. Allogeneic mesenchymal stem cell and mesenchymal stem cell-differentiated chondrocyte suppress the responses of type II collagen-reactive T cells in rheumatoid arthritis. Rheumatology January 2008;47(1):22−30. PMID:18077486. Epub 2007/12/14. eng.

[105] Adkisson HD, Martin JA, Amendola RL, Milliman C, Mauch KA, Katwal AB, et al. The potential of human allogeneic juvenile chondrocytes for restoration of articular cartilage. Am J Sports Med July 2010;38(7):1324−33. PMID:20423988. PMCID:3774103. Epub 2010/04/29. eng.

[106] Le Blanc K, Tammik C, Rosendahl K, Zetterberg E, Ringden O. HLA expression and immunologic properties of differentiated and undifferentiated mesenchymal stem cells. Exp Hematol October 2003;31(10):890−6. PMID:14550804. Epub 2003/10/11. eng.

[107] Kiernan CH, Hoogduijn MJ, Franquesa M, Wolvius EB, Brama PA, Farrell E. Allogeneic chondrogenically differentiated human mesenchymal stromal cells do not induce immunogenic responses from T lymphocytes in vitro. Cytotherapy August 2016;18(8):957−69. PMID:27288309.

[108] Mukonoweshuro B, Brown CJ, Fisher J, Ingham E. Immunogenicity of undifferentiated and differentiated allogeneic mouse mesenchymal stem cells. J Tissue Eng 2014;5. 2041731414534255. PMID:24812582. PMCID:4014080. Epub 2014/05/09. eng.

[109] Ryan AE, Lohan P, O'Flynn L, Treacy O, Chen X, Coleman C, et al. Chondrogenic differentiation increases antidonor immune response to allogeneic mesenchymal stem cell transplantation. Mol Ther March 2014;22(3):655−67. PMID:24184966. PMCID:3944342. Epub 2013/11/05. eng.

[110] Bourgine PE, Scotti C, Pigeot S, Tchang LA, Todorov A, Martin I. Osteoinductivity of engineered cartilaginous templates devitalized by inducible apoptosis. Proc Natl Acad Sci USA December 9, 2014;111(49):17426−31. PMID:25422415. PMCID:4267331. Epub 2014/11/26. eng.

[111] Cunniffe GM, Vinardell T, Murphy JM, Thompson EM, Matsiko A, O'Brien FJ, et al. Porous decellularized tissue engineered hypertrophic cartilage as a scaffold for large bone defect healing. Acta Biomater May 31, 2015. PMID:26038199. Epub 2015/06/04. eng.

[112] Mesallati T, Sheehy EJ, Vinardell T, Buckley CT, Kelly DJ. Tissue engineering scaled-up, anatomically shaped osteochondral constructs for joint resurfacing. Eur Cell Mater September 28, 2015;30:163−85. discussion 85−86. PMID:26412388.

[113] Daly AC, Cunniffe GM, Sathy BN, Jeon O, Alsberg E, Kelly DJ. 3D bioprinting of developmentally inspired templates for whole bone organ engineering. Adv Healthc Mater September 2016;5(18):2353−62. PMID:27281607.

[114] Fahmy-Garcia S, van Driel M, Witte-Buoma J, Walles H, van Leeuwen JP, van Osch G, et al. Nell-1, HMGB1 and CCN2 enhance migration and vasculogenesis, but not osteogenic differentiation compared to BMP2. Tissue Eng Part A May 02, 2017. PMID:28463604.

[115] Coleman CM, Vaughan EE, Browe DC, Mooney E, Howard L, Barry F. Growth differentiation factor-5 enhances in vitro mesenchymal stromal cell chondrogenesis and hypertrophy. Stem Cells Dev July 1, 2013;22(13):1968−76. PMID:23388029. PMCID:3685316.

[116] Taniguchi N, Yoshida K, Ito T, Tsuda M, Mishima Y, Furumatsu T, et al. Stage-specific secretion of HMGB1 in cartilage regulates endochondral ossification. Mol Cell Biol 2007;27.

[117] Khattab HM, Aoyama E, Kubota S, Takigawa M. Physical interaction of CCN2 with diverse growth factors involved in chondrocyte differentiation during endochondral ossification. J Cell Commun Signal September 2015;9(3):247−54. PMID:25895141. PMCID:4580687.

[118] Takigawa M. CCN2: a master regulator of the genesis of bone and cartilage. J Cell Commun Signal August 2013;7(3):191−201. PMID:23794334. PMCID:3709051.

[119] Kubota S, Takigawa M. The role of CCN2 in cartilage and bone development. J Cell Commun Signal August 2011;5(3):209−17. PMID:21484188. PMCID:3145877.

Chapter 7

Challenges in Cell-Based Therapies for Intervertebral Disc Regeneration: Lessons Learned From Embryonic Development and Pathophysiology

Pauline Colombier[1], Makarand V. Risbud[2]
[1]University of California, San Francisco (UCSF), San Francisco, CA, United States; [2]Thomas Jefferson University, Philadelphia, PA, United States

INTRODUCTION

Despite ongoing research and recent clinical advancements, degeneration of the intervertebral discs (IVDs) and associated back pain remain ubiquitous and expensive conditions to treat. It is estimated that about 80% of the population will suffer from back pain at some time during their lifetime, and much of this pain is directly attributable to degeneration of the IVDs commonly referred as disc disease [1,2]. Underscoring this point, a recently concluded 20-year study showed that low back pain (LBP) is ranked the first in number of years lived with disability, whereas neck pain is ranked fourth. In addition, LBP was the fourth highest in disability-adjusted life-year ranking—a measure of missed "healthy" years of life [3]. Costs from spinal diseases continue to rise; in 2005 in the United States alone, the costs associated with treating spine-related pathologies were estimated at about $85.9 billion. To the dismay of medical community, from 1997 to 2005, a Medical Expenditure Panel Survey found no significant improvements in multiple parameters surveyed, including self-assessed functional disability, work limitations, health status, and social functioning [4]. Furthermore, spinal fusion, one of the commonly practiced surgeries to relieve back pain stemming from degenerative discs, has been demonstrated to alter biomechanics of the adjacent discs raising concerns

Developmental Biology and Musculoskeletal Tissue Engineering. https://doi.org/10.1016/B978-0-12-811467-4.00007-3
149

about the procedure [5]. It is amply clear from these studies that there is an urgent need for novel, effective strategies for the treatment of IVD degeneration and associated back pain.

The nucleus pulposus (NP) of the healthy adult disc is relatively cell-sparse and rich in proteoglycan and collagen matrix, providing the tissue with its ability to imbibe water and thus, mechanical function of distributing loads applied to the spine [6]. Resident NP cells that arise from the embryonic notochord are responsible for maintenance of this critical extracellular matrix (ECM) through synthesis and secretion of proteoglycans, mainly aggrecan, versican (VCAN) as well as several collagens [7,8]. An early characteristic of disc degeneration is a massive cell depletion in the central, gelatinous NP followed by the disruption of the ECM [9,10]. It is also well recognized that the sterile inflammation of the disc is one of the major contributors of disc degeneration. Inflammatory processes, mainly initiated by cytokines, TNF-α and IL-1β, are thought to be key cytokines in promoting disc degeneration and LBP [11]. These cytokines elevate expression of catabolic enzymes, such as a disintegrin and metalloproteinase with thrombospondin motifs (ADAMTSs) and matrix metalloproteinases (MMPs), and inhibit expression of ECM proteins aggrecan and collagen II [12−22]. Séguin et al. showed that TNF-α-induced increase in MMP2 activity was associated with upregulation of membrane type MT1-MMP [23]. Further studies have shown that in NP cells, TNF-α mediates activation of ADAMTS4 through mitogen-activated protein kinase and nuclear factor-kappa B (NFκB) pathways [24]. In an organ culture model, treatment with TNF-α promoted degenerative changes, including suppression of multiple collagen types, aggrecan, and fibromodulin, along with increased expression of MMPs and nerve growth factor (NGF) [25]. Noteworthy, the relative contribution of TNF-α and IL-1β is still debated.

A positive relationship between TNF-α and Wnt/β-catenin signaling pathways and MMP13 expression has been shown [26,27]. TNF-α and IL-1β cytokines also suppress expression of proanabolic, matricellular protein connective tissue growth factor (CTGF/CCN2). CCN2 is critical for ECM synthesis by disc cells. Animal models support the role of these endogenous cytokines in the degenerative process of the IVD. Mice lacking functional IL-1Ra, an endogenous antagonist for IL-1R, had higher histological grade of degeneration, and IVD cells exhibited diminished proliferative capacity [28]. Recent studies from Risbud and colleagues have shown that the majority of effects of TNF-α and IL-1β in NP cells are through the NFκB signaling pathway [11]. In addition to affecting matrix gene expression, cytokines also induce NP cell expression of many cytokines and chemokines that can progress to the inflammatory state by recruiting and activating immune cells. Thus, the niche of the degenerating disc is characterized by elevated levels of inflammatory cytokines, an altered matrix and cellular landscape, which are important considerations for any therapeutic approach aimed at restoring disc function. Fig. 7.1 describes roles of cytokines in degenerating cascade.

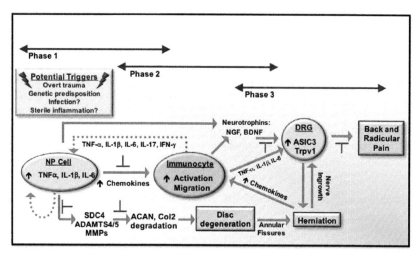

FIGURE 7.1 Schematic showing the major phases of intervertebral disc degeneration and discogenic back pain. Potential triggering events include overt trauma to the disc, altered loading, genetic predisposition, and spinal infection. These triggers result in nucleus pulposus (NP) and annulus fibrosus (AF) cell-mediated synthesis of cytokines such as TNF-α, IL-1β, and IL-6. These cytokines have at least three major effects: increased matrix breakdown through production of SDC4, a disintegrin and metalloproteinase with thrombospondin motifs 4/5 (ADAMTS4/5), and matrix metalloproteinases (MMPs), activation and chemotaxis of immune cells to the disc, induction of neurotrophins (NGF, BDNF) by disc cells and immunocytes. NGF and BDNF induce production of pain-associated channels ASIC3 and Trpv1 in the DRG. Presence of these factors in the inflammatory microenvironment of the degenerated disc is thought to promote DRG sensitization and pain. Blocking of cytokine production in phases 1 and 2 may have most the profound positive effect on disease progression and back pain. Red blocking symbols indicate possible areas for clinical intervention. *Reproduced from Johnson ZI, Schoepflin ZR, Choi H, Shapiro IM, Risbud MV. Disc in flames: roles of TNF-α and IL-1β in intervertebral disc degeneration. Eur Cell Mater September 21, 2015;30:104—16; discussion 116—7.*

Because the activity of NP cells underlies functioning of the disc, and as their capacity to support tissue homeostasis declines with degeneration and aging, several studies have investigated cell-based therapy for treating disc degeneration. Use of endogenous disc progenitor cells or transplantation of mature disc cells or mesenchymal stem cells (MSCs) has been explored. Direct transplantation of NP cells or chondrocytes, from both autologous and allogeneic sources, has been proven to decrease degenerative phenotype in animal models [29—32], and some clinical studies indicate the promise of this approach [33,34]. Noteworthy, identification of an endogenous progenitor cell population within the disc has further expanded the possibilities of cell-based therapies [35]. Finally, MSCs derived from bone marrow, fat, or other tissue sources have been studied as a potential source of regeneration for the NP and have shown promising results [36]. Although these studies provide hope that stem cell therapy can be used in the near future to maintain disc health, several obstacles still remain. One critical constituent of a regenerative therapy is that the new tissue should be able to replace or support function of diseased tissue. In case of IVD, to recapitulate the healthy NP and AF cell phenotype, it must first be clearly defined and that the role of signaling pathways and molecules that are critical in the formation and maintenance of these tissues requires a better understanding [37]. This chapter describes the development of the IVD during embryogenesis and focuses on the key phenotypic characteristics of NP cells that should be recapitulated for a successful cell replacement therapy. The chapter also provides a broad overview of stem/progenitor cell—based regenerative therapies for disc degeneration.

Embryonic Development of the Intervertebral Disc

Embryonic development of the spine centers around the notochord and somites. The notochord, also called axial mesoderm, undergoes morphogenetic events to give rise to NP [38,39], whereas the medioventral part of the somites differentiates and forms vertebrae and annulus fibrosus (AF) [40]. Both notochord and somites are a part of the mesoderm germ layer, but the notochord does not derive from the primitive streak. The precursors of this last structure are located in the embryonic node, which is a distinct population located at the anterior end of the primitive streak [41,42].

Embryogenesis of the Nucleus Pulposus

Elegant lineage-tracing studies using *sonic hedgehog* (*Shh*, mainly expressed by notochord and neural tube) and *notochord homeobox* (*Noto*, only expressed in the notochordal lineage), reporters demonstrated that populating cells of the NP derived from the notochord [38,39]. This structure plays a central role during embryogenesis by notably acting as signaling center responsible for patterning and regionalizing the neural tube and the somites, as well as behaving as the axial skeleton anlage [43]. The specification of the notochordal lineage has been studied in several animal models; but in this chapter, mechanisms mainly defined in mouse will be described.

The Embryonic Node

Fate mapping studies demonstrated that the first precursors of the notochord are found in the early gastrula organizer at E6.5 (EGO) [42]. This specific group of cells is located at the anterior tip of the early primitive streak and expresses the transcription factors *Forkhead box A2* (*FoxA2*), *Goosecoid* (*Gsc*), and *Noto* [44—46]. Some of these precursors migrate toward the anterior part of the embryo and contribute to the anterior head process. The remaining precursors integrate the midgastrula organizer (MGO) at E7.0. Cells within the MGO also migrate anteriorly, and both EGO and MGO-derived cells form the anterior axial mesoderm (anterior head process and prechordal plate).

A specific MGO subpopulation forms the embryonic node. This horseshoe-shaped structure located at the anterior tip of the primitive streak is visible from E7.5, and populating cells are characterized by the expression of *FoxA2*, *Brachyury* (*T*), *Noto*, *Chordin*, *Noggin*, *Shh*, *Forkhead box J1* (*FoxJ1*), and by the loss of *Gsc* expression [47,48]. The embryonic node is a key signaling structure responsible notably for the left—right asymmetry establishment and the formation of posterior axial mesoderm/notochord [42,44,49].

Morphogenesis of the Notochord

Nascent notochord is observed as soon as the end of gastrulation (E8.0) [50]. First, node cells migrate posteriorly and form the notochordal plate through convergent-extension movements [49,51]. At this stage, the notochordal plate is adjacent to the floor plate of the neural tube and integrated into the endoderm layer. Around E10.5, the notochord is detached itself from the endoderm but remains adjacent to the floor plate of the neural tube. Finally, at E11.5 the notochord is formed as a rodlike structure, isolated from the neural tube and the endoderm, and surrounded by mesenchymal cells [50].

Morphogenetic events leading to the formation of the notochord in mouse are different from the one described in avian. In mouse, the node does not regress toward the posterior end of the embryo, as demonstrated by the stable distance between the node and the allantois during notochord formation and elongation [49,52]. Yamanaka and colleagues elegantly demonstrated that node cells migrate toward the posterior end of the embryo and notochord elongation is ensured by convergent-extension movements [49]. Noto transcription factor plays a crucial role in notochord elongation as showed by the abnormal short tail in Tc mutants (spontaneous mutation in the DNA-binding domain of Noto) [44]. This hypothesis has been confirmed by using knockin mutants where GFP is expressed under the control of the Noto promoter (Noto$^{GFP/+}$). Authors reported the migratory behavior of node cells and confirmed the role of Noto in maintaining notochordal cell identity (mechanism previously demonstrated in zebrafish). Thus, Noto is a crucial notochord lineage-specific transcription factor involved in notochord formation and elongation as well as in notochordal cell identity maintenance [49,53—55].

The notochord is also characterized by the presence of a surrounding rigid sheath called perinotochordal sheath. This acellular structure is composed of a basal lamina, collagen, and proteoglycans and appears around E10.0. The perinotochordal sheath progressively extends from the rostral to the caudal part of the embryo [56,57]. Whereas the origin of the sheath is not fully understood, the expression of *Col2a1* and *Acan* detected within the notochord suggests that this sheath may be produced by notochordal cells themselves [58—61]. In addition, Danforth's short tail mutants (Sd mutants) display a discontinuous notochord and an abnormally thin perinotochordal sheath leading to vertebrae and NP defects [57,62]. It has also been demonstrated that notochordal cells of *SRY box 5* and *6* knockout embryos (*sox5* and *Sox6*, two key transcription factors driving the expression of *Col2a1* and *Acan* in chondrogenic cells) undergo a massive apoptotic cell death [61]. All together, these observations suggest an important role of notochordal cells in sheath formation and reciprocally, perinotochordal sheath ensures cell survival and NP formation.

A last major event occurring during notochord morphogenesis is the appearance of vacuoles within the cytoplasm of notochordal cells around E12.5. Molecular mechanisms involved in the formation of these intracellular vacuoles remain unclear. Nevertheless, two studies performed in zebrafish claimed that these vacuoles may be formed from the endoplasmic reticulum or the Golgi apparatus. Whereas a role in transportation of perinotochordal sheath components (glycosaminoglycans, collagen) has been suggested, the intravacuolar content remains uncharacterized [63,64]. A potential role of these vacuoles in osmolarity response of notochordal cells has been suggested in dogs [65]. Nevertheless, the role of osmolarity during notochord and NP morphogenesis remains poorly described in mice or human. The presence and the role of these intracellular vacuoles remains a basic question without a clear answer.

Notochord to Nucleus Pulposus Transition

The formation of the NP occurs between E12.5 and E15.5. All events affecting the notochord within this window are responsible for the notochord to NP transition. First, this transition is characterized by changes in the morphological appearance of the notochord. Notochordal cells accumulate where the future IVD will be formed, while a constriction of the notochord is observed within the future vertebrae, resulting in the formation of a moniliform-shaped notochord [50]. During this notochord to NP transition, the perinotochordal sheath remains visible and the volume of the intracellular vacuoles increases to reach around 80% of the cytoplasm at birth [64,65]. Although understanding of molecular mechanisms driving this transition remains incomplete, several studies showed that the expression pattern of the notochord changes during this transition. The expression of *FoxA2* and *Noto* decreases from E12.5, and the role of the transcription factors Sox5, Sox6, and Sox9 becomes crucial as shown by the disappearance of notochordal cells at the intervertebral levels observed in $Sox5^{-/-};Sox6^{-/-}$, and $Sox9^{-/-}$ mouse embryos [44,66–68]. During this transition, the secretory activity of notochordal cells increases and the NP-specific ECM becomes established. Notochordal cells mainly secrete aggrecan, laminin, and fibronectin components under the control of CTGF/CCN2 [58,69–71] and probably Sox5, Sox6, and Sox9 transcription factors regarding their importance in controlling the secretory activity of chondrogenic cells [72,73]. Nevertheless, the role of these transcription factors has not been reported during notochord and NP embryogenesis.

To conclude, several morphogenesis events occur during notochord development resulting in NP formation. Mistakenly described as a type of cartilage, the NP composition is actually distinct form articular cartilage (AC) at birth. Indeed, the NP is populated by notochordal cells, which secrete an immature ECM. A long maturation process occurs during childhood, where a morphologically distinct second cell-type resembling chondrocytes appears. This second cell type, so-called nucleopulpocytes, is considered to be the mature and resident cell type of the NP in postnatal life. Mechanisms governing the appearance of this second cell type within the NP remain unreported. Nevertheless, two main hypotheses may explain the appearance of nucleopulpocytes: (1) a direct differentiation of notochordal cells in situ or (2) a recruitment of nucleopulpocytes from surrounding tissues (AF or cartilage endplates).

During childhood, a molecular dialog exists between notochordal cells and nucleopulpocytes mainly mediated by CTGF/CCN2, transforming growth factor beta (TGF-β), and Shh. This molecular dialog is required to establish the NP-specific ECM and thus to confer to the NP its swelling and shock-absorbing properties.

Embryogenesis of Vertebrae and Annulus Fibrosus

During embryogenesis, the notochord plays a central role in the formation of the entire spine. In addition to its transition into NP, the notochord patterns the paraxial mesoderm and is responsible for the differentiation of the medioventral part of the somites into sclerotome. Sclerotomal cells migrate around the notochord and the neural tube and start to differentiate to give rise to either AF or vertebrae [40] (Fig. 7.2).

Differentiation of Sclerotome

The first step of the formation of vertebrae and AF is the differentiation of the somites under the control of notochord-secreted morphogens. The notochordal cells mainly secrete Shh, Chordin, and Noggin morphogens (BMP signaling pathway antagonists). The involvement of Shh signaling pathway in the differentiation of somites has been shown by inactivating *Gli2* and *Gli3* genes, leading to blockade of Shh signaling transduction. Abrogation of Shh signaling transduction leads to a decrease in *Paired box 1* (*Pax1*), *Pax9*, and *Sox9* gene expression in somites, and a dramatic impairment of somite differentiation [74–76]. The role of Shh and Noggin in somite differentiation has also been shown in vitro and both morphogens act synergistically to drive the expression of *Pax1* [77,78]. This result suggests that activating Shh signaling and blocking BMP signaling in somites is required for their differentiation into the sclerotome.

Transcription factors mesenchyme homeobox 1/2 (Meox1 and Meox2) expressed by somites also play a crucial role in establishing the sclerotome [79]. Indeed, an absence of mesenchymal sclerotome surrounding the notochord and the neural tube is observed in double knockout $Meox1^{-/-};Meox2^{-/-}$ embryos. The observed sclerotome is not segmented, which directly leads to defects in IVD and vertebrae formation [80].

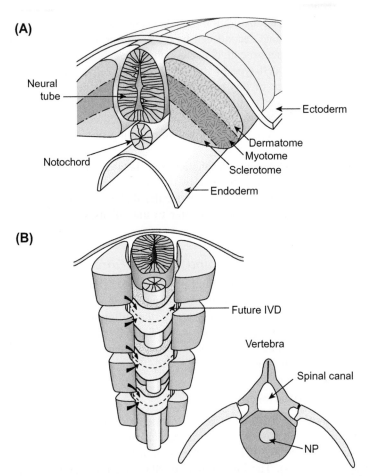

(A)

Neural tube

Ectoderm

Notochord

Dermatome
Myotome
Sclerotome

Endoderm

(B)

Future IVD

Vertebra

Spinal canal

NP

FIGURE 7.2 Axial skeleton formation during embryogenesis. (A) During segmentation the paraxial mesoderm forms pairs of somites along the neural tube (light orange) and notochord (blue). Each somite is composed of a dermatome (light purple), myotome (light brown), and sclerotome (dark yellow). The ectoderm lies above and the endoderm below. (B) Sclerotomal cells migrate from adjacent somites above and below each future vertebra. Dermatomal cells stream beneath the ectoderm to form the dermis, whereas the myotomal cells form muscle. Insert B shows the architecture of the vertebrae with the spinal canal, spinal processes and the nucleus pulposus (NP) in blue. *Adapted from Risbud MV, Shapiro IM. Notochordal cells in the adult intervertebral disc: new perspective on an old question. Crit Rev Eukaryot Gene Expr 2011;21(1):29–41; NIH Public Access. (B) Courtesy of Dr. Richard Dryden.*

Thus, proper expression of *Pax1*, *Pax9*, *Sox9*, *Meox1*, and *Meox2* under the control of Shh and BMP signaling pathways ensures somite differentiation into sclerotome. Those factors are also involved in sclerotomal cell differentiation into chondrogenic cells.

Resegmentation Phenomenon

During migration and condensation of sclerotomal cells around the neural tube and notochord, alternative highly condensed and low-condensed metameres are formed. A spatial rearrangement occurs at this time characterized by the fusion of the caudal part of the anterior metamere with the rostral part of the following metamere. After this second segmentation (resegmentation), low-condensed segments will give rise to the vertebrae, whereas highly condensed segments will give rise to the AF [81,82].

Vertebral Endochondral Ossification and Cartilage Endplate Formation

Vertebrae are formed through an endochondral ossification process, which starts around E11.5 in mice. Mesenchymal cells within low-condensed segments differentiate into chondrocyte progenitors (chondroblasts) followed by a chondrocyte differentiation. After these first steps of differentiation, chondrocytes undergo a terminal hypertrophic differentiation process and start secreting calcified ECM and angiogenic factors (e.g., vascular growth factor) to recruit blood vessels. This process is common between the formation of vertebrae and long bones [83].

The role of the perinotochordal sheath in chondrogenic differentiation of mesenchymal cells within future vertebrae has been investigated in vitro. Somites and notochords isolated from mouse embryos were cocultured (with intact or enzymatically digested perinotochordal sheath) and authors showed that somite cells produce glycosaminoglycans only in the presence of an intact perinotochordal sheath. This result illustrates the positive influence of the perinotochordal sheath on chondrogenic differentiation of somite cells. Nevertheless, molecular mechanisms regulating the chondrogenic differentiation of mesenchymal cells through their interaction (direct or indirect) with perinotochordal sheath remain largely unknown [84].

In addition, knockout mouse embryos for *Col2a1*, *Pax1*, *Pax9*, *Bapx1*, and *Shh* display an incomplete endochondral ossification process. Defects in hypertrophic differentiation of mesenchymal cells, reduction in vascularization, and mineralization have been observed [61,85–87]. This endochondral ossification process ensures mineralization of the vertebral body from the center (primary ossification center) toward both ends of the vertebrae. Similar to long bones, cartilaginous growth plates called vertebral endplates, are present at both rostral and caudal ends of each vertebrae. Nevertheless, the vertebral growth plates are adjacent to hyaline cartilage layers and not comprised between two bony compartments. Vertebral endplates are thus composed of a subchondral bone and a thin layer of hyaline cartilage. This cartilaginous layer is composed of only one cell type; chondrocytes and a type II collagen–enriched ECM, as the image of AC. Cartilage endplates contain the remaining blood vessels ensuring diffusion of oxygen and nutrients to the IVD as well as anchoring of AF fibers.

Signaling Pathways Involved in Annulus Fibrosus Formation

Through an unknown program, the endochondral ossification process does not occur within the highly condensed segments surrounding the notochord. Around E15.0, mesenchymal cells differentiate and rearrange to form the AF (data obtained from rat) [88]. The absence of proper AF in $Pax1^{-/-}$ and $Pax9^{-/-}$ mutant embryos suggests that in addition to their role in somite differentiation and vertebral endochondral ossification, both genes are also involved in the formation of the AF [89]. The expression of $Pax1$ is regulated by Shh signaling suggesting a role of this notochord-secreted morphogen in AF formation. In addition, several studies attested the major role of TGF-β signaling in the formation of the AF. Conditional inactivation of TGF-β $receptor$ II gene under the control of $Col2a1$ promoter leads to the formation of a mineralized tissue around the NP. Thus, by inactivating this TGF-β signaling pathway, mesenchymal cells undergo endochondral ossification process independently of their level of condensation [90,91]. These data, observed in genetically modified mouse embryos, have been completed by a large-scale transcriptome analysis study demonstrating the preventive role of TGF-β signaling in chondrogenic differentiation of mesenchymal cells in the future AF [70]. Altogether, these studies showed that Shh and TGF-β signaling pathways are involved in the formation of AF.

Annulus Fibrosus Fiber Formation

The AF is characterized by a chondrogenic inner part and a fibrous outer part. The inner AF is composed of chondrogenic cells secreting a type II collagen—enriched ECM, whereas the outer AF is composed of fibroblastic cells and type I collagen fibers. During inner AF formation, sclerotomal cells extend and differentiate while they secrete type II collagen fibers perpendicularly to the spine axis [88]. These cells also secrete proteoglycans (fibromodulin, aggrecan, and VCAN) [92]. Meanwhile, outer AF cells rearrange their intracellular actin fiber network and display intercellular tight junctions resulting in the formation of cell sheaths parallel to the spinal axis [93]. Outer AF cells secrete specific ECM components as fibronectin and type I/III collagen creating fibers oriented at 55 degrees to the spine axis [88]. No proteoglycans are synthetized by outer AF cells [92].

To conclude, the formation of IVDs during embryonic development is complex and involves tightly regulated molecular mechanisms. While our knowledge is far from being complete, it appears that notochordal cells play a key role in NP and AF formation/maturation.

The highly specified cell types within IVD may guide scientists to find most relevant therapeutic strategies to treat IVD degeneration.

Cell-Based Therapeutic Strategies

As previously mentioned in this chapter, an early feature of disc degeneration is the loss of cells in the NP compartment [94—96]. Because cells are important for disc functioning, supplying damaged NP before dramatic degenerative events occur might be a relevant therapeutic strategy for treating LBP. In addition, tissue engineering strategies for AF repair are under development. These strategies mainly consist in the use of bioengineered glues to seal the AF in case of herniation, prolapse, or NP cell supplementation. Although the development of these approaches is crucial for a complete IVD regeneration, they do not necessarily aim at regenerating the AF [97]. Thus, this chapter will focus on cell-based strategies for NP regeneration.

An important issue to consider in NP cell transplantation is the availability of nutrients and factors present within the disc microenvironment during degeneration because they will significantly influence survival and activity of transplanted cells [98]. Nutrient deprivation is an important factor in disc degeneration, and as such nutrient supply and demand is tightly regulated in healthy IVD. Addition of growth factors to supplement transplanted cells can thus further exacerbate already low nutrient level in degenerative disc by stimulating cell metabolism [99]. Considering this, cell transplantation may be better suited as an early interventional therapy for disc degeneration when routes of nutritional delivery are still open and chronic inflammation is not set in.

In addition to these considerations, for any cell therapy to be successful, it is essential that the criteria of the target cell phenotype are clearly defined. Several investigators have attempted to prime MSCs into NP-like cells. However, the state of MSC differentiation into NP-like cells was assessed using markers that are either not NP-specific or using an incomplete list of NP markers [100–107]. Minogue et al. reported results of a microarray study comparing bovine NP, AF, and AC cells [108]. The authors identified 34 genes specific to NP and 49 specific to either type of IVD-derived cell. Out of those chosen for validation in human NP cells, 11 genes (*SNAP25, K8, K18, K19, CDH2, IBSP, VCAN, TNMD, BASP1, FOXF1,* and *FBLN1*) were confirmed. The same group identified markers to distinguish NP-like differentiation in human bone marrow–derived MSCs (BM-MSCs) and adipose stromal cells (ASCs), by using five NP marker genes: *PAX1, FOXF1, HBB, CAXII,* and *OVOS2* to validate differentiation to an NP-like phenotype [109]. Differentiated MSCs showed significant increase in expression levels of classical markers *COL2A1* and *ACAN* plus *PAX1* and *FOXF1*. Because in this study as opposed to differentiated ASCs, BM-MSCs expressed negative NP markers *FBLN1* and *IBSP*, it was concluded that ASCs may represent a superior cell type for differentiation to an NP-like phenotype. Nevertheless, the differentiation of MSCs in three dimensions in a type I collagen–based hydrogel may be difficult to translate to the clinic. More recently, the generation of NP cells from ASCs has been confirmed and further investigated [110]. Authors showed that the commitment of the human ASCs was robust and highly specific, as attested by the expression of NP-related genes characteristic of young healthy human NP cells. In addition, the engineered NP-like cells secreted an abundant aggrecan and type II collagen—rich ECM comparable with that of native NP. Furthermore, they demonstrated that these in vitro engineered cells survived and maintained their specialized phenotype and secretory activity after in vivo transplantation in *nude* mice subcutis (Fig. 7.3). Although this study offers valuable insights into the generation of NP-like cells from ASCs, a clear and complete definition and understanding of NP cell phenotype is required to use the concept of cell "markers" for differentiation and before considering this differentiated stem cell—based strategy in clinics. Moreover, cell priming techniques using microenvironmental conditions that simulate disc cells will increase the likelihood of success of such approaches.

Definition of Nucleus Pulposus Cell-Specific Phenotype

Adult NP cells (mostly nucleopulpocytes), which are the primary cell type that are considered for cell replacement therapy have a unique phenotypic signature. Risbud and colleagues have extensively discussed their phenotype in a recent consensus article [37]. Markers that are only relevant to NP cells' metabolic and functional state are described here.

Markers Specific to Cell Function

Two distinguishing features of the NP niche are (1) avascularity, which imposes a hypoxic state, and (2) elevated concentration of hydrophilic proteoglycans and thus high sulfated glycosaminoglycans (sGAG) content, which raises osmotic pressure outside of cells. These considerations make it important that transplanted cells are able to function in these challenging conditions. The ability of resident nucleopulpocytes to survive, proliferate, and function within hypoxic and hyperosmolar niche has been largely due to two transcription factors: hypoxia-inducible factor 1 alpha (HIF-1α) and tonicity-responsive enhancer binding protein (TonEBP/NFAT5).

The HIF family of transcription factors is responsible for activating an adaptive cellular response in cells exposed to hypoxia [111]. Careful work by Risbud and coworkers has shown that NP cells constitutively express HIF-1α and HIF-2α (oxygen labile subunits) under both normoxia and hypoxia [112,113]. Consistent with their observation, alternative pathways that are oxygen-independent participate in controlling HIF-1α degradation and activity [114—116]. Further studies with a mouse model of selective HIF-1α loss in the NP demonstrated the necessity of this transcription factor for NP cell survival [117,118]. Specific conditional deletion of HIF-1α in notochordal cells promoted morphologic changes at E15.5 and their disappearance by 1-month of age [118]. This was attributed to requirement of HIF-1α in driving glycolytic metabolic and synthetic activities of NP cells in vivo, as was shown by previous in vitro studies [112,119]. In addition, in NP cells, HIF-1α has been

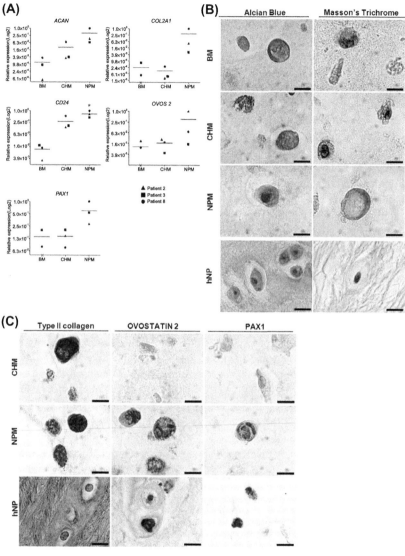

FIGURE 7.3   In vivo biological behavior of human adipose stromal cell−derived nucleopulpocytes. Human adipose stromal cell (ASC) pellets from three patients were cultivated in the presence of basic medium (BM), chondrogenic medium (CHM), and nucleopulpogenic medium (NPM) for 21 days before their association with the si-HPMC hydrogel and their transplantation in *nude* mice subcutis for 6 weeks. (A) The expression of *ACAN*, *COL2A1*, *CD24*, *OVOS2*, and *PAX1* was assessed before transplantation by RT-qPCR. Black crosses on the x-axis represent a Cq value > 38 cycles. (B) Explanted samples and a human nucleus pulposus (hNP) were histologically prepared and sectioned for alcian blue and Masson's trichrome staining. (C) Explanted samples and a hNP were histologically prepared and sectioned for type II collagen, OVOSTATIN 2 and PAX1 immunostainings. Scale bar: 10 μm. *Reproduced from Colombier P, Clouet J, Boyer C, Ruel M, Bonin G, Lesoeur J, et al. TGF-β1 and GDF5 act synergistically to drive the differentiation of human adipose stromal cells toward nucleus pulposus-like cells. Stem Cell March 2016;34(3):653−67.*

shown to regulate the expression of key matrix-related genes including *Acan* [119,120], *1,3-glucuronyltransferase 1* (*GlcAT*-I) [121], and *galectin 3* [122]. More recently, *carbonic anhydrase XII* (*CAXII*) has been proposed as a marker of healthy human NP tissue [123]. This is not unexpected, as carbonic anhydrases play an important role in controlling pH balance, critical in hypoxic tissues [124]. The expression of CAXII is hypoxia inducible in tumor cells, although an HIF-responsive element (HRE) has not yet been described [125]. Therefore, it is possible that other members of the CA family can serve as functional NP markers while a strong expression of *CAIII* transcripts in early notochordal and NP has been shown [126].

In the IVD, the high hydrostatic pressure and the water content gives the tissue its ability to resist compressive forces applied to the spine [6]. Importantly, diurnal loading pattern causes dynamic fluctuations in extracellular osmolarity of the disc (ranging from 430 to 496 mOsm) [127−129]. Transcription factor TonEBP/NFAT5 allows NP cells to sustain normal activities under these dynamic osmotic fluctuations [130,131]. Normally, cells in a hypertonic environment activate transmembrane electrolyte transporters to equilibrate inner- and extracellular solute concentrations resulting in increased intracellular osmotic pressure. If left unchecked, increased ionic concentration can lead to DNA damage and cell death [132]. Under hypertonic conditions, TonEBP promotes expression of genes essential in accumulation of organic nonionic osmolytes that prevent uncontrolled influx of charged ions into the cell [133,134]. *TonEBP* expression is required for NP cell survival in hypertonic medium, and it controls expression of *taurine transporter* (*TauT*), *sodium myoinositol transporter* (*SMIT*), and *betaine-γ-amino butyric acid transporter* (*BGT1*) [131]. TonEBP has also been shown to control expression of *aggrecan* [131,135], *B3GAT3*, an enzyme needed for chondroitin sulfate chain synthesis [136], and the *water channel protein aquaporin 2* (*AQP2*) in NP cells [137]. Although not a classic "cell marker," the robust expression and responsiveness of TonEBP in the NP is certainly required for survival under physiological conditions of the disc.

In addition to these transcription factors, specific proteoglycans and collagens are synthetized by NP cells. The most abundant proteoglycan in the NP is aggrecan, with lesser amounts of versican, biglycan, and decorin [7]. Collagens make up a sizeable component of the NP matrix, mainly collagen II [8]. Importantly, the ratios of aggrecan and collagen II in NP cells differentiate it from chondrocytes. This ratio has been measured as the sGAG to hydroxyproline ratio and is close to 27:1 in the young adult NP compared with 2:1 in healthy hyaline cartilage [138]. With the importance of cell−matrix interaction in mind, Chen et al. investigated expression levels of laminin chains, integrins, and other matrix-binding proteins in NP tissues, finding that compared with surrounding AF, NP tissue had raised levels of laminin α5 chain and α3, α6, and β4, CD239, and CD151 integrin subunits [139]. In a follow-up study, the same group showed that integrins α3, α5, and β1 were important for NP cells attachment to laminin LM-111 and LM-511 [140].

As all these markers are linked to NP cell biology and function, differentiation of stem cells into NP lineage strategies absolutely requires the analysis of these functional NP marker expressions and validate the engineered cells as proper NP-like cells.

Cell Sources and Relevance for Cell-Based Therapies: Nucleus Pulposus Cells, Juvenile Chondrocytes, and Mesenchymal Stem Cells Clinical Reports and the Potential of Bonafide Disc Progenitors

A substantial number of clinical studies have been performed within the last 10 years with a primary objective to evaluate the safety and the tolerance of intradiscal injections of NP regenerative cells. NP cells, juvenile chondrocytes, and MSCs represent the main cell types considered as potential regenerative cell sources for treating LBP.

Improvement of Intervertebral Disc Functions After Nucleus Pulposus Cell Injections

As a primary source of regenerating cells, NP cells from herniated IVD have been considered. The Eurodisc study is based on the use of NP cells from herniation and their preimplantation in the same IVD. More than 100 patients have been enrolled in this clinical study and after a 2-year follow-up, patients showed an increase in IVD MRI signal, stabilization in discal height and a decrease in pain [141]. This clinical study is still ongoing with an MRI follow-up. A second multicenter and randomized clinical phase I/II study, based on a similar therapeutic strategy, is still ongoing (120 patients enrolled, NOVO-CART Disc plus). In both cases, NP cells were amplified in vitro before their injection. However, it is well known that NP cells lose their specific phenotype and dedifferentiate into fibroblasts in vitro [142,143]. To develop a more appropriate therapeutic strategy, injection of reactivated NP cells has been considered. The in vitro coculture of NP cells with MSCs may prevent their dedifferentiation into fibroblasts and improve their viability after injection (preclinical studies) [144]. Thus, NP cells from the fused IVD of nine patients were cocultured in direct contact with autologous bone marrow—derived MSCs and were transplanted into degenerating IVD adjacent to the fused level 7 days after surgery. No deleterious effect has been reported after 3-year follow-up, and a mild improvement has been reported in only one case. Although this clinical trial demonstrated the safety of this procedure, no real efficiency has been noticeable [145].

Improvement of Intervertebral Disc Functions After Articular Chondrocyte Injections

The comparable secretory activity of adult NP cells and articular chondrocytes encouraged scientists and physicians to study the regenerative potential of articular chondrocytes to restore IVD functions. First, injections of juvenile articular chondrocytes have been performed in 15 patients suffering from discogenic LBP (NuQu phase investigational new drug). A decrease in pain has been reported, and an improvement of MRI signal has been observed in 10 patients. Restoration of discal height has not been reported, and three patients underwent a total disc replacement 1 year after intradiscal injections of chondrocytes [146]. A second study, randomized and double-blinded, is still ongoing.

Mesenchymal Stem Cell–Based Clinical Studies

The use of MSCs as a potential regenerative cell source to treat discogenic LBP has been largely investigated. MSCs are multipotent cells easily accessible from bone marrow and adipose tissue. Although MSCs can also be isolated from umbilical cord, skin, synovial fluid and membrane, they are low represented (around 0.002% of the total cell population) and their potential use for IVD regeneration has not been extensively investigated in preclinical studies yet [147,148].

Three clinical studies have been performed consisting in intradiscal injections of BM-MSCs, and five are still ongoing [144,149,150]. A reduction of pain and an improvement of the IVD MRI signal have been observed in most of these clinical cases. Nevertheless, the restoration of discal height is not reported. In addition to the above, no information on the quality of repaired/regenerated tissue is available. Regarding the potential of BM-MSCs to differentiate toward osteochondrogenic cells and hypertrophic chondrocytes, ASCs may represent a more appropriate cell source for IVD regeneration [151]. Three clinical trials are currently investigating the safety and efficiency of ASC injections within degenerated IVDs.

Although the injection of undifferentiated MSCs appears a promising strategy to reduce pain and improve IVD hydration observed by MRI signal, the capacity of MSCs to survive, differentiate into proper cell types and stay within the IVD is questionable. Since a few studies reported the possibility to differentiate MSCs toward adult NP-like cells, no proof of concept has been performed on their regenerative abilities in preclinical studies. Nevertheless, these missing data might fill the gap and bring new insights into IVD regeneration mechanisms, offering an innovative and biologically inspired therapeutic strategy (Fig. 7.4).

Identification of Disc Progenitor Cells and Their Utility for Intervertebral Disc Regeneration

While the cycling of disc cells is believed to be low, limited regeneration has been observed in the disc, especially in the outer and possibly in the inner AF [152–154]. The idea of disc cell self-renewal is underscored by a few studies that showed presence of progenitor cells within the disc compartment in

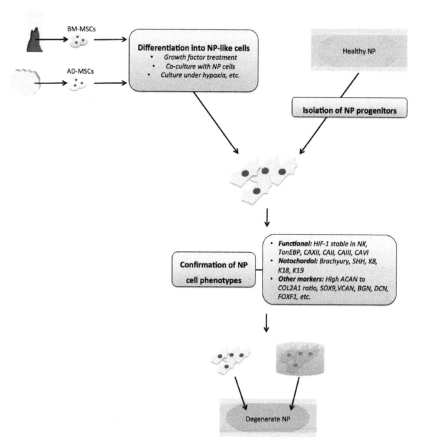

FIGURE 7.4 **Schematic diagram showing cell-based therapy strategy for disc regeneration.**
Mesenchymal stem cells (MSCs) can be derived from either bone marrow (BM-MSCs) or adipose
(AD-MSCs), before undergoing differentiation into nucleus pulposus (NP)–like cells through
various treatments. Alternately, NP progenitor cells may be identified with markers and isolated
directly from healthy disc tissue. Confirmation of NP cell phenotype as defined by functional,
tissue-of-origin, and other markers unique to NP cells before transplantation is necessary for
successful disc regeneration. *Reproduced from Choi H, Johnson ZI, Risbud MV. Understanding
nucleus pulposus cell phenotype: a prerequisite for stem cell–based therapies to treat intervertebral disc degeneration. Curr Stem Cell Res Ther 2015;10(4):307–16.*

healthy and diseased state. Risbud and colleagues identified cells expressing
CD105, CD166, CD63, CD49a, CD90, CD73, p75 low-affinity NGF receptor,
and CD133/1 in diseased human disc tissues, all of which are considered BM-
MSCs markers. These cells differentiated into osteogenic, adipogenic, and
chondrogenic lineages under appropriate conditions [155]. Feng et al. later
made a similar observation with AF cells obtained from nondegenerative
human discs [156]. In addition to previously reported MSC markers, these

cells expressed neuronal stem cell markers *nestin* and neuron-specific *enolase* and evidenced multilineage differentiation. Brisby et al. observed OCT3/4, CD105, CD90, STRO-1, and NOTCH1 expressions in degenerative human disc tissues [157]. Blanco et al. showed that when compared with the BM-MSCs, NP-derived MSCs had comparable morphology, immunophenotype, and differentiation capacities, with the exception that NP-derived MSCs lacked the ability to differentiate into adipocytes [158]. Supporting the notion of dividing cells in IVD, Henriksson et al. showed that NP and AF compartments in rabbit discs, contained a small population of 5-bromo-2′-deoxyuridine (BrdU)−positive cells [159]. Likewise, in the region of the AF-bordering ligament and the perichondrium region, a high number of BrdU-positive cells were seen during early growth that substantially decreased. In a later study, these authors confirmed the presence of BrdU-positive cells in the previously reported stem cell niche near the epiphyseal plate. Although the number of these BrdU-positive cells is reduced in older animals, they appeared prominently in a new niche in the mature AF [160].

Studies by Sakai et al. and Yasen et al. underscored the change in the progenitor cells with age and disease. Age-dependent reduction in the Tie2+ NP progenitor cell population in human and mouse [161] as well as reduction of progenitor cell marker expression in rabbit discs [162] suggest there may be challenges in using these endogenous progenitors for disc repair in a classical cell therapy protocols. However, these studies suggest that these cells can be used to replace old and damaged disc cells, if appropriate stimuli are identified to jump start their activities in vivo.

In addition to above described studies, a few studies have investigated possible contribution of Notch signaling pathway in maintenance of disc progenitor cells. The Notch pathway has been shown to be essential for progenitor cell activity, proliferation, and fate determination in various tissues, and thus, its related molecules are used as stem/progenitor cell markers. In IVD, *Notch1* expression pattern overlaps that of BrdU-positive areas [159] and correlates with the disc progenitor cell niche [163]. Noteworthy, the Notch signaling pathway is activated by hypoxia in disc cells. Hiyama and colleagues confirmed expression of Notch receptors and ligands in rat NP and AF tissues and demonstrated that hypoxia increased expression of *Notch1*, *Notch4*, and *Jagged2* transcripts in both tissues and *Jagged1* expression in AF [164]. Inhibition of Notch signaling significantly reduced NP cell proliferation. Interestingly, the expression of *Hes1*, a Notch target and an important regulator of stem cell maintenance, was also induced by hypoxia in AF cells. Although not yet confirmed in IVD progenitor cells, an earlier study by Gustafsson et al. showed functional cross talk between HIF-1α and Notch-ICD leads to inhibition of myogenic and neural differentiation [165]. Interestingly, in NP, Notch signaling activity is responsive to the inflammatory cytokine TNF and may thus act as a compensatory mechanism to maintain endogenous cell population [164,166].

These studies indicate that Notch signaling molecules can be used as additional disc progenitor markers and targeting activity of this pathway may have therapeutic potential to slow down or reverse IVD degeneration.

The Use of Pluripotent Stem Cells for Nucleus Pulposus Regeneration: A Future Biologically Inspired Treatment for Disc Degeneration?

Embryonic stem cells (ES cells) are derived from the inner cell mass of blastocysts (E3.5 in mouse and E6.5 in human). These cells are able to self-renew and can give rise to every cell type of the organism, making them a potent regenerating cell source to generate a large number of specified cells. Nevertheless, the use of human ES cells in regenerative medicine suffers from two main obstacles: their ethical use and the incompatibility with an autologous therapeutic strategy. Another source of pluripotent stem cells is now available; human induced pluripotent stem cells (iPS cells) generated by adult somatic cell reprogramming [167]. Pioneering work by Yamanaka demonstrated that by overexpressing four transcription factors (*kruppel-like factor 4, POU class 5 homeobox 1, SRY box 2,* and *avian myelocytomatosis viral oncogene homolog*) in murine and human fibroblasts, these somatic cells reprogram their genome and are able to give rise to mesoderm-, endoderm- and ectoderm-derived tissues in vivo (teratoma formation) [168,169]. Owing to ethical reasons, the generation of chimeras by injecting human iPS cells in a mouse oocyte is forbidden. Nevertheless, a recent study, which consisted in injecting human iPS cells in the anterior region of the primitive streak of a mouse gastrula, showed that injected iPS cells contributed to endoderm-, mesoderm-, and ectoderm-derived structures [170]. Altogether, these data demonstrate the ability of human iPS cells to generate cells derived from the three germ layers and thus attest their pluripotency.

Differentiation of Pluripotent Stem Cells Toward Notochord/Nucleus Pulposus Lineage

The development of differentiation protocols for pluripotent stem cells (ES or iPS cells) is based on reproducing major events that occur during embryogenesis. As previously described in this chapter, the embryonic node contains precursors of the notochord/NP lineage. Unfortunately, molecular mechanisms leading to the formation of the node/notochord lineage and their expression profiles are unknown in humans. Thus, the in vitro generation of node/notochordal cells is mainly based on data obtained from mouse embryos and human pluripotent stem cell differentiation abilities.

During mouse embryogenesis, the formation of the node is mainly regulated by the Wnt and nodal signaling pathways [171,172]. In vitro, these signaling pathways can be respectively activated by the use of CHIR99021 and Activin molecules. On the activation of both signaling pathways, human pluripotent stem cells (ES or iPS cells) differentiate toward mesendoderm progenitors, characterized by the expression of *FoxA2*, *Brachyury*, *Gsc*, but the expression of *Shh* or *Noto* remains unreported [173,174]. These data suggest that human ES—cell derived and iPS cell—derived mesendoderm progenitors are different from the node population (compared with mouse node expression profile). These progenitors are able to differentiate toward mesoderm and endoderm cells, but their ability to differentiate toward notochord is not elucidated yet.

However, the differentiation of mouse ES cells toward the notochordal lineage has yet to be elucidated. A first differentiation protocol based on the inhibition of Wnt, retinoic acid, BMP, and the activation of FGF signaling pathways has been reported [175]. This protocol allows the formation of 10% of cells expressing *Noto*, *FoxA2*, *Brachyury*, *Shh*, *Noggin*, *Chordin*, and *FoxJ1* (node/notochord signature). However, these cells also express *Aristaless-related homeodomain transcription factor* (*Arx*), a marker of the floor plate of the neural tube, suggesting that the generated population is heterogeneous. In parallel, a study investigated the possibility to generate notochordal cells from murines iPS cells. In this last study, the differentiation protocol is based on the formation of embryoïd bodies (spontaneous formation of the three germ layers in vitro) and the selection of CD24 expressing cells [176]. These CD24-positive cells express *FoxA2*, *Brachyury*, *Noto*, *Shh*, and *Noggin*. After 3 weeks of culture of CD24-positive-selected cells in the presence of notochordal-conditioned culture medium, authors demonstrated that these cells produce an aggrecan and type II collagen—enriched ECM. Regarding these encouraging data, this differentiation protocol has been tested on human iPS cells [177]. Although the generated cells produced a proteoglycan and type II collagen—enriched ECM at the end of the protocol, the expression of *NOTO*, *FOXA2*, and *SHH* has not been reported in the CD24-positive cell population.

Thus, notochordal differentiation of human ES or iPS cells remains to be not clearly demonstrated.

CONCLUSIONS

IVD degeneration is ubiquitous and chronic. Although there have been several advances to better understand disc cell function and pathophysiology, complete recovery of disc function and alleviation of back/neck pain still remains a major health problem. Regenerative therapies, both those that harness the ability of native progenitor cells in the NP and AF compartments and those that rely on transplanted cells may lead the way in biological approach to disease treatment. However, the success of these approaches relies on recapitulation of tissue phenotype of healthy NP and AF cells. For this purpose, the understanding of the embryonic origin of IVD and morphogenetic events leading to the formation of this complex structure is crucial to generate proper functional NP and AF cells. In addition, from a standpoint of cell introduction to the NP, it is crucial that the cells are primed to express factors allowing their survival, such as HIF-1α and TonEBP. Finally, consideration of both functional markers and cell surface markers, and others for cell sorting, will be critical in future cell-based therapies for treating disc degeneration.

REFERENCES

[1] Deyo RA, Weinstein JN. Low back pain. N Engl J Med February 1, 2001;344(5):363−70.

[2] Freemont AJ. The cellular pathobiology of the degenerate intervertebral disc and discogenic back pain. Rheumatology January 2009;48(1):5−10.

[3] Murray CJL, Atkinson C, Chou D, Dellavalle R, Danaei G, Fahimi A, et al. The state of US health, 1990−2010: burden of diseases, injuries, and risk factors. J Am Med Assoc August 14, 2013;310(6):591−608.

[4] Martin BI, Deyo RA, Mirza SK, Turner JA, Comstock BA, Hollingworth W, et al. Expenditures and health status among adults with back and neck problems. J Am Med Assoc February 13, 2008;299(6):656−64.

[5] Phillips FM, Reuben J, Wetzel FT. Intervertebral disc degeneration adjacent to a lumbar fusion. An experimental rabbit model. J Bone Joint Surg Br March 2002;84(2):289−94.

[6] Kraemer J, Kolditz D, Gowin R. Water and electrolyte content of human intervertebral discs under variable load. Spine January 1985;10(1):69−71.

[7] Cs-Szabo G, Ragasa-San Juan D, Turumella V, Masuda K, Thonar EJ, An HS. Changes in mRNA and protein levels of proteoglycans of the anulus fibrosus and nucleus pulposus during intervertebral disc degeneration. Spine October 15, 2002;27(20):2212−9.

[8] Le Maitre CL, Pockert A, Buttle DJ, Freemont AJ, Hoyland JA. Matrix synthesis and degradation in human intervertebral disc degeneration. Biochem Soc Trans August 2007;35(Pt. 4):652−5. Portland Press Limited.

[9] Antoniou J, Steffen T, Nelson F, Winterbottom N, Hollander AP, Poole RA, et al. The human lumbar intervertebral disc: evidence for changes in the biosynthesis and denaturation of the extracellular matrix with growth, maturation, ageing, and degeneration. J Clin Invest August 15, 1996;98(4):996−1003. American Society for Clinical Investigation.

[10] Buckwalter JA. Aging and degeneration of the human intervertebral disc. Spine June 1, 1995;20(11):1307−14.

[11] Risbud MV, Shapiro IM. Role of cytokines in intervertebral disc degeneration: pain and disc content. Nat Rev Rheumatol January 2014;10(1):44−56. Nature Research.

[12] Bachmeier BE, Nerlich A, Mittermaier N, Weiler C, Lumenta C, Wuertz K, et al. Matrix metalloproteinase expression levels suggest distinct enzyme roles during lumbar disc herniation and degeneration. Eur Spine J November 2009;18(11):1573−86.

[13] Doita M, Kanatani T, Ozaki T, Matsui N, Kurosaka M, Yoshiya S. Influence of macrophage infiltration of herniated disc tissue on the production of matrix metalloproteinases leading to disc resorption. Spine July 15, 2001;26(14):1522−7.

[14] Jimbo K, Park JS, Yokosuka K, Sato K, Nagata K. Positive feedback loop of interleukin-1beta upregulating production of inflammatory mediators in human intervertebral disc cells in vitro. J Neurosurg Spine May 2005;2(5):589−95. Journal of Neurosurgery Publishing Group.

[15] Le Maitre CL, Hoyland JA, Freemont AJ. Interleukin-1 receptor antagonist delivered directly and by gene therapy inhibits matrix degradation in the intact degenerate human intervertebral disc: an in situ zymographic and gene therapy study. Arthritis Res Ther 2007;9(4):R83. BioMed Central.

[16] Le Maitre CL, Freemont AJ, Hoyland JA. The role of interleukin-1 in the pathogenesis of human intervertebral disc degeneration. Arthritis Res Ther 2005;7(4):R732—45. BioMed Central.

[17] Murata Y, Onda A, Rydevik B, Takahashi I, Takahashi K, Olmarker K. Changes in pain behavior and histologic changes caused by application of tumor necrosis factor-alpha to the dorsal root ganglion in rats. Spine March 1, 2006;31(5):530—5.

[18] Pockert AJ, Richardson SM, Le Maitre CL, Lyon M, Deakin JA, Buttle DJ, et al. Modified expression of the ADAMTS enzymes and tissue inhibitor of metalloproteinases 3 during human intervertebral disc degeneration. Arthritis Rheum February 2009;60(2):482—91. Wiley Subscription Services, Inc., A Wiley Company.

[19] Séguin CA, Pilliar RM, Roughley PJ, Kandel RA. Tumor necrosis factor-alpha modulates matrix production and catabolism in nucleus pulposus tissue. Spine September 1, 2005;30(17):1940—8.

[20] Shen B, Melrose J, Ghosh P, Taylor F. Induction of matrix metalloproteinase-2 and -3 activity in ovine nucleus pulposus cells grown in three-dimensional agarose gel culture by interleukin-1beta: a potential pathway of disc degeneration. Eur Spine J February 2003;12(1):66—75.

[21] Wang J, Markova D, Anderson DG, Zheng Z, Shapiro IM, Risbud MV. TNF-α and IL-1β promote a disintegrin-like and metalloprotease with thrombospondin type I motif-5-mediated aggrecan degradation through syndecan-4 in intervertebral disc. J Biol Chem November 18, 2011;286(46):39738—49.

[22] Wang X, Wang H, Yang H, Li J, Cai Q, Shapiro IM, et al. Tumor necrosis factor-α- and interleukin-1β-dependent matrix metalloproteinase-3 expression in nucleus pulposus cells requires cooperative signaling via syndecan 4 and mitogen-activated protein kinase-NF-κB axis: implications in inflammatory disc disease. Am J Pathol September 2014;184(9):2560—72.

[23] Séguin CA, Pilliar RM, Madri JA, Kandel RA. TNF-alpha induces MMP2 gelatinase activity and MT1-MMP expression in an in vitro model of nucleus pulposus tissue degeneration. Spine February 15, 2008;33(4):356—65.

[24] Tian Y, Yuan W, Fujita N, Wang J, Wang H, Shapiro IM, et al. Inflammatory cytokines associated with degenerative disc disease control aggrecanase-1 (ADAMTS-4) expression in nucleus pulposus cells through MAPK and NF-κB. Am J Pathol June 2013;182(6):2310—21.

[25] Ponnappan RK, Markova DZ, Antonio PJD, Murray HB, Vaccaro AR, Shapiro IM, et al. An organ culture system to model early degenerative changes of the intervertebral disc. Arthritis Res Ther 2011;13(5):R171. BioMed Central.

[26] Ye S, Wang J, Yang S, Xu W, Xie M, Han K, et al. Specific inhibitory protein Dkk-1 blocking Wnt/β-catenin signaling pathway improve protectives effect on the extracellular matrix. J Huazhong Univ Sci Technol — Med Sci October 2011;31(5):657—62.

[27] Hiyama A, Yokoyama K, Nukaga T, Sakai D, Mochida J. A complex interaction between Wnt signaling and TNF-α in nucleus pulposus cells. Arthritis Res Ther November 14, 2013;15(6):R189. BioMed Central.

[28] Phillips KLE, Jordan-Mahy N, Nicklin MJH, Le Maitre CL. Interleukin-1 receptor antagonist deficient mice provide insights into pathogenesis of human intervertebral disc degeneration. Ann Rheum Dis November 2013;72(11):1860—7.

[29] Acosta FL, Metz L, Adkisson HD, Liu J, Carruthers-Liebenberg E, Milliman C, et al. Porcine intervertebral disc repair using allogeneic juvenile articular chondrocytes or mesenchymal stem cells. Tissue Eng Part A December 2011;17(23—24):3045—55. Mary Ann Liebert, Inc. 140 Huguenot Street, 3rd Floor New Rochelle, NY 10801 USA.

[30] Gorensek M, Jaksimović C, Kregar-Velikonja N, Gorensek M, Knezevic M, Jeras M, et al. Nucleus pulposus repair with cultured autologous elastic cartilage derived chondrocytes. Cell Mol Biol Lett 2004;9(2):363—73.

[31] Nishimura K, Mochida J. Percutaneous reinsertion of the nucleus pulposus. An experimental study. Spine July 15, 1998;23(14):1531—8. discussion 1539.

[32] Nomura T, Mochida J, Okuma M, Nishimura K, Sakabe K. Nucleus pulposus allograft retards intervertebral disc degeneration. Clin Orthop Relat Res August 2001;(389):94—101.

[33] Pennicooke B, Moriguchi Y, Hussain I, Bonssar L, Härtl R. Biological treatment approaches for degenerative disc disease: a review of clinical trials and future directions. Cureus November 22, 2016;8(11):e892.

[34] Sakai D, Andersson GBJ. Stem cell therapy for intervertebral disc regeneration: obstacles and solutions. Nat Rev Rheumatol April 2015;11(4):243—56. Nature Research.

[35] Risbud MV, Shapiro IM. Notochordal cells in the adult intervertebral disc: new perspective on an old question. Crit Rev Eukaryot Gene Expr 2011;21(1):29—41. NIH Public Access.

[36] Sakai D, Mochida J, Shapiro IM, Risbud MV. Use of stem cells for regeneration of the intervertebral disc. In: Shapiro IM, Risbud MV, editors. The intervertebral disc: molecular and structural studies of the disc in health and disease. Springer: Vienna; 2014. p. 373—83.

[37] Risbud MV, Schoepflin ZR, Mwale F, Kandel RA, Grad S, Iatridis JC, et al. Defining the phenotype of young healthy nucleus pulposus cells: recommendations of the Spine Research Interest Group at the 2014 annual ORS meeting. J Orthop Res March 2015;33(3):283—93.

[38] Choi K-S, Cohn MJ, Harfe BD. Identification of nucleus pulposus precursor cells and notochordal remnants in the mouse: implications for disk degeneration and chordoma formation. Dev Dyn December 2008;237(12):3953—8. Wiley-Liss, Inc.

[39] McCann MR, Tamplin OJ, Rossant J, Séguin CA. Tracing notochord-derived cells using a Noto-cre mouse: implications for intervertebral disc development. Dis Model Mech January 2012;5(1):73—82.

[40] Christ B, Wilting J. From somites to vertebral column. Ann Anat February 1992;174(1):23—32.

[41] Beddington RS. Induction of a second neural axis by the mouse node. Development March 1994;120(3):613—20.

[42] Kinder SJ, Tsang TE, Wakamiya M, Sasaki H, Behringer RR, Nagy A, et al. The organizer of the mouse gastrula is composed of a dynamic population of progenitor cells for the axial mesoderm. Development September 2001;128(18):3623—34.

[43] Corallo D, Trapani V, Bonaldo P. The notochord: structure and functions. Cell Mol Life Sci August 2015;72(16):2989—3008. Springer Basel.

[44] Abdelkhalek HB, Beckers A, Schuster-Gossler K, Pavlova MN, Burkhardt H, Lickert H, et al. The mouse homeobox gene Not is required for caudal notochord development and affected by the truncate mutation. Genes Dev July 15, 2004;18(14):1725—36. Cold Spring Harbor Lab.

[45] Lawson KA, Meneses JJ, Pedersen RA. Clonal analysis of epiblast fate during germ layer formation in the mouse embryo. Development November 1991;113(3):891—911.

[46] Tam PP, Behringer RR. Mouse gastrulation: the formation of a mammalian body plan. Mech Dev November 1997;68(1—2):3—25.

[47] Bachiller D, Klingensmith J, Kemp C, Belo JA, Anderson RM, May SR, et al. The organizer factors chordin and noggin are required for mouse forebrain development. Nature February 10, 2000;403(6770):658—61. Nature Publishing Group.

[48] Blum M, Andre P, Muders K, Schweickert A, Fischer A, Bitzer E, et al. Ciliation and gene expression distinguish between node and posterior notochord in the mammalian embryo. Differentiation February 2007;75(2):133—46.

[49] Yamanaka Y, Tamplin OJ, Beckers A, Gossler A, Rossant J. Live imaging and genetic analysis of mouse notochord formation reveals regional morphogenetic mechanisms. Dev Cell December 2007;13(6):884—96.

[50] Jurand A. Some aspects of the development of the notochord in mouse embryos. J Embryol Exp Morphol August 1974;32(1):1—33.

[51] Sulik K, Dehart DB, Iangaki T, Carson JL, Vrablic T, Gesteland K, et al. Morphogenesis of the murine node and notochordal plate. Dev Dyn November 1994;201(3):260—78. Wiley Subscription Services, Inc., A Wiley Company.

[52] Catala M, Teillet MA, De Robertis EM, Le Douarin ML. A spinal cord fate map in the avian embryo: while regressing, Hensen's node lays down the notochord and floor plate thus joining the spinal cord lateral walls. Development September 1996;122(9):2599—610.

[53] Amacher SL, Kimmel CB. Promoting notochord fate and repressing muscle development in zebrafish axial mesoderm. Development April 1998;125(8):1397—406.

[54] Beckers A, Alten L, Viebahn C, Andre P, Gossler A. The mouse homeobox gene Noto regulates node morphogenesis, notochordal ciliogenesis, and left right patterning. Proc Natl Acad Sci USA October 2, 2007;104(40):15765—70.

[55] Melby AE, Kimelman D, Kimmel CB. Spatial regulation of floating head expression in the developing notochord. Dev Dyn June 1997;209(2):156—65. Wiley-Liss, Inc.

[56] Choi K-S, Harfe BD. Hedgehog signaling is required for formation of the notochord sheath and patterning of nuclei pulposi within the intervertebral discs. Proc Natl Acad Sci USA June 7, 2011;108(23):9484—9.

[57] Paavola LG, Wilson DB, Center EM. Histochemistry of the developing notochord, perichordal sheath and vertebrae in Danforth's short-tail (sd) and normal C57BL/6 mice. J Embryol Exp Morphol February 1980;55:227—45.

[58] Cheah KS, Lau ET, Au PK, Tam PP. Expression of the mouse alpha 1(II) collagen gene is not restricted to cartilage during development. Development April 1991;111(4):945—53.

[59] Lefebvre V, Li P, de Crombrugghe B. A new long form of Sox5 (L-Sox5), Sox6 and Sox9 are coexpressed in chondrogenesis and cooperatively activate the type II collagen gene. EMBO J October 1, 1998;17(19):5718—33. EMBO Press.

[60] Ng LJ, Wheatley S, Muscat GE, Conway-Campbell J, Bowles J, Wright E, et al. SOX9 binds DNA, activates transcription, and coexpresses with type II collagen during chondrogenesis in the mouse. Dev Biol March 1, 1997;183(1):108—21.

[61] Smits P, Lefebvre V. Sox5 and Sox6 are required for notochord extracellular matrix sheath formation, notochord cell survival and development of the nucleus pulposus of intervertebral discs. Development March 2003;130(6):1135—48.

[62] Center EM, Spigelman SS, Wilson DB. Perinotochordal sheath of heterozygous and homozygous Danforth's short-tail mice. J Hered July 1982;73(4):299—300.

[63] Ellis K, Bagwell J, Bagnat M. Notochord vacuoles are lysosome-related organelles that function in axis and spine morphogenesis. J Cell Biol March 4, 2013;200(5):667—79.

[64] Ellis K, Hoffman BD, Bagnat M. The vacuole within: how cellular organization dictates notochord function. BioArchitecture May 2013;3(3):64—8.

[65] Hunter CJ, Bianchi S, Cheng P, Muldrew K. Osmoregulatory function of large vacuoles found in notochordal cells of the intervertebral disc running title: an osmoregulatory vacuole. Mol Cell BioMech December 2007;4(4):227—37. PMC Canada manuscript submission.

[66] Barrionuevo F, Taketo MM, Scherer G, Kispert A. Sox9 is required for notochord maintenance in mice. Dev Biol July 1, 2006;295(1):128—40.

[67] Hildebrand A, Romarís M, Rasmussen LM, Heinegård D, Twardzik DR, Border WA, et al. Interaction of the small interstitial proteoglycans biglycan, decorin and fibromodulin with transforming growth factor beta. Biochem J September 1, 1994;302(Pt 2):527—34. Portland Press Ltd.

[68] Maier JA, Lo Y, Harfe BD. Foxa1 and Foxa2 are required for formation of the intervertebral discsZhang X, editor. PLoS ONE 2013;8(1):e55528.

[69] Bedore J, Sha W, McCann MR, Liu S, Leask A, Séguin CA. Impaired intervertebral disc development and premature disc degeneration in mice with notochord-specific deletion of CCN2. Arthritis Rheum October 2013;65(10):2634−44.

[70] Sohn P, Cox M, Chen D, Serra R. Molecular profiling of the developing mouse axial skeleton: a role for Tgfbr2 in the development of the intervertebral disc. BMC Dev Biol March 9, 2010;10(1):29. BioMed Central.

[71] Uetzmann L, Burtscher I, Lickert H. A mouse line expressing Foxa2-driven Cre recombinase in node, notochord, floorplate, and endoderm. Genesis October 2008;46(10):515−22. Wiley Subscription Services, Inc., A Wiley Company.

[72] Cameron TL, Belluoccio D, Farlie PG, Brachvogel B, Bateman JF. Global comparative transcriptome analysis of cartilage formation in vivo. BMC Dev Biol March 10, 2009;9(1):20. BioMed Central.

[73] Zhu Y, McAlinden A, Sandell LJ. Type IIA procollagen in development of the human intervertebral disc: regulated expression of the NH(2)-propeptide by enzymic processing reveals a unique developmental pathway. Dev Dyn April 2001;220(4):350−62. John Wiley & Sons, Inc.

[74] Buttitta L, Mo R, Hui C-C, Fan C-M. Interplays of Gli2 and Gli3 and their requirement in mediating Shh-dependent sclerotome induction. Development December 2003;130(25):6233−43. The Company of Biologists Ltd.

[75] Murtaugh LC, Chyung JH, Lassar AB. Sonic hedgehog promotes somitic chondrogenesis by altering the cellular response to BMP signaling. Genes Dev January 15, 1999;13(2):225−37. Cold Spring Harbor Laboratory Press.

[76] Parker MH, Seale P, Rudnicki MA. Looking back to the embryo: defining transcriptional networks in adult myogenesis. Nat Rev Genet July 2003;4(7):497−507. Nature Publishing Group.

[77] McMahon JA, Takada S, Zimmerman LB, Fan CM, Harland RM, McMahon AP. Noggin-mediated antagonism of BMP signaling is required for growth and patterning of the neural tube and somite. Genes Dev May 15, 1998;12(10):1438−52. Cold Spring Harbor Laboratory Press.

[78] Stafford DA, Brunet LJ, Khokha MK, Economides AN, Harland RM. Cooperative activity of noggin and gremlin 1 in axial skeleton development. Development March 2011;138(5):1005−14.

[79] Zeng L, Kempf H, Murtaugh LC, Sato ME, Lassar AB. Shh establishes an Nkx3.2/Sox9 autoregulatory loop that is maintained by BMP signals to induce somitic chondrogenesis. Genes Dev August 1, 2002;16(15):1990−2005.

[80] Jukkola T, Trokovic R, Maj P, Lamberg A, Mankoo B, Pachnis V, et al. Meox1Cre: a mouse line expressing Cre recombinase in somitic mesoderm. Genesis November 2005;43(3):148−53. Wiley Subscription Services, Inc., A Wiley Company.

[81] Kaplan KM, Spivak JM, Bendo JA. Embryology of the spine and associated congenital abnormalities. Spine J September 2005;5(5):564−76.

[82] Peacock A. Observations on the prenatal development of the intervertebral disc in man. J Anat July 1951;85(3):260−74. Wiley-Blackwell.

[83] Alman BA. The role of hedgehog signalling in skeletal health and disease. Nat Rev Rheumatol September 2015;11(9):552−60. Nature Research.

[84] Kosher RA, Lash JW. Notochordal stimulation of in vitro somite chondrogenesis before and after enzymatic removal of perinotochordal materials. Dev Biol February 1975;42(2):362−78.

[85] Aszódi A, Pfeifer A, Wendel M, Hiripi L, Fässler R. Mouse models for extracellular matrix diseases. J Mol Med March 1998;76(3−4):238−52.

[86] Furumoto TA, Miura N, Akasaka T, Mizutani-Koseki Y, Sudo H, Fukuda K, et al. Notochord-dependent expression of MFH1 and PAX1 cooperates to maintain the proliferation of sclerotome cells during the vertebral column development. Dev Biol June 1, 1999;210(1):15−29.

[87] Tribioli C, Lufkin T. The murine Bapx1 homeobox gene plays a critical role in embryonic development of the axial skeleton and spleen. Development December 1999;126(24):5699−711.

[88] Hayes AJ, Isaacs MD, Hughes C, Caterson B, Ralphs JR. Collagen fibrillogenesis in the development of the annulus fibrosus of the intervertebral disc. Eur Cell Mater October 11, 2011;22:226−41.

[89] Peters H, Wilm B, Sakai N, Imai K, Maas R, Balling R. Pax1 and Pax9 synergistically regulate vertebral column development. Development December 1999;126(23):5399−408.

[90] Baffi MO, Moran MA, Serra R. Tgfbr2 regulates the maintenance of boundaries in the axial skeleton. Dev Biol August 15, 2006;296(2):363−74.

[91] Baffi MO, Slattery E, Sohn P, Moses HL, Chytil A, Serra R. Conditional deletion of the TGF-beta type II receptor in Col2a expressing cells results in defects in the axial skeleton without alterations in chondrocyte differentiation or embryonic development of long bones. Dev Biol December 1, 2004;276(1):124−42.

[92] Hayes AJ, Benjamin M, Ralphs JR. Extracellular matrix in development of the intervertebral disc. Matrix Biol April 2001;20(2):107−21.

[93] Hayes AJ, Benjamin M, Ralphs JR. Role of actin stress fibres in the development of the intervertebral disc: cytoskeletal control of extracellular matrix assembly. Dev Dyn July 1999;215(3):179−89. John Wiley & Sons, Inc.

[94] Ding F, Shao Z-W, Xiong L-M. Cell death in intervertebral disc degeneration. Apoptosis July 2013;18(7):777−85. Springer US.

[95] Zhao C-Q, Jiang L-S, Dai L-Y. Programmed cell death in intervertebral disc degeneration. Apoptosis December 2006;11(12):2079−88.

[96] Colombier P, Clouet J, Hamel O, Lescaudron L, Guicheux J. The lumbar intervertebral disc: from embryonic development to degeneration. Joint Bone Spine March 2014;81(2):125−9.

[97] Sloan SR, Lintz M, Hussain I, Härtl R, Bonassar LJ. Biologic annulus fibrosus repair: a review of preclinical in vivo investigations. Tissue Eng Part B Rev November 4, 2017. https://doi.org/10.1089/ten.TEB.2017.0351.

[98] Zhu W, Chen J, Cong X, Hu S, Chen X. Hypoxia and serum deprivation-induced apoptosis in mesenchymal stem cells. Stem Cell February 2006;24(2):416−25. John Wiley & Sons, Ltd.

[99] Urban JPG, Smith S, Fairbank JCT. Nutrition of the intervertebral disc. Spine December 1, 2004;29(23):2700−9.

[100] Feng G, Jin X, Hu J, Ma H, Gupte MJ, Liu H, et al. Effects of hypoxias and scaffold architecture on rabbit mesenchymal stem cell differentiation towards a nucleus pulposus-like phenotype. Biomaterials November 2011;32(32):8182−9.

[101] Gantenbein-Ritter B, Benneker LM, Alini M, Grad S. Differential response of human bone marrow stromal cells to either TGF-β(1) or rhGDF-5. Eur Spine J June 2011;20(6):962−71.

[102] Henriksson HB, Svanvik T, Jonsson M, Hagman M, Horn M, Lindahl A, et al. Transplantation of human mesenchymal stems cells into intervertebral discs in a xenogeneic porcine model. Spine January 15, 2009;34(2):141−8.

[103] Richardson SM, Walker RV, Parker S, Rhodes NP, Hunt JA, Freemont AJ, et al. Intervertebral disc cell-mediated mesenchymal stem cell differentiation. Stem Cell March 2006;24(3):707−16. John Wiley & Sons, Ltd.

[104] Risbud MV, Albert TJ, Guttapalli A, Vresilovic EJ, Hillibrand AS, Vaccaro AR, et al. Differentiation of mesenchymal stem cells towards a nucleus pulposus-like phenotype in vitro: implications for cell-based transplantation therapy. Spine December 1, 2004;29(23):2627—32.

[105] Steck E, Bertram H, Abel R, Chen B, Winter A, Richter W. Induction of intervertebral disc-like cells from adult mesenchymal stem cells. Stem Cell March 2005;23(3):403—11. John Wiley & Sons, Ltd.

[106] Stoyanov JV, Gantenbein-Ritter B, Bertolo A, Aebli N, Baur M, Alini M, et al. Role of hypoxia and growth and differentiation factor-5 on differentiation of human mesenchymal stem cells towards intervertebral nucleus pulposus-like cells. Eur Cell Mater June 20, 2011;21:533—47.

[107] Strassburg S, Richardson SM, Freemont AJ, Hoyland JA. Co-culture induces mesenchymal stem cell differentiation and modulation of the degenerate human nucleus pulposus cell phenotype. Regen Med September 2010;5(5):701—11. Future Medicine Ltd. London, UK.

[108] Minogue BM, Richardson SM, Zeef LA, Freemont AJ, Hoyland JA. Transcriptional profiling of bovine intervertebral disc cells: implications for identification of normal and degenerate human intervertebral disc cell phenotypes. Arthritis Res Ther 2010;12(1):R22. BioMed Central.

[109] Minogue BM, Richardson SM, Zeef LAH, Freemont AJ, Hoyland JA. Characterization of the human nucleus pulposus cell phenotype and evaluation of novel marker gene expression to define adult stem cell differentiation. Arthritis Rheum December 2010;62(12):3695—705. Wiley Subscription Services, Inc., A Wiley Company.

[110] Colombier P, Clouet J, Boyer C, Ruel M, Bonin G, Lesoeur J, et al. TGF-β1 and GDF5 act synergistically to drive the differentiation of human adipose stromal cells toward nucleus pulposus-like cells. Stem Cell March 2016;34(3):653—67.

[111] Palmer BF, Clegg DJ. Oxygen sensing and metabolic homeostasis. Mol Cell Endocrinol November 2014;397(1—2):51—8.

[112] Agrawal A, Gajghate S, Smith H, Anderson DG, Albert TJ, Shapiro IM, et al. Cited2 modulates hypoxia-inducible factor-dependent expression of vascular endothelial growth factor in nucleus pulposus cells of the rat intervertebral disc. Arthritis Rheum December 2008;58(12):3798—808. Wiley Subscription Services, Inc., A Wiley Company.

[113] Risbud MV, Guttapalli A, Stokes DG, Hawkins D, Danielson KG, Schaer TP, et al. Nucleus pulposus cells express HIF-1 alpha under normoxic culture conditions: a metabolic adaptation to the intervertebral disc microenvironment. J Cell Biochem May 1, 2006;98(1):152—9. Wiley Subscription Services, Inc., A Wiley Company.

[114] Fujita N, Chiba K, Shapiro IM, Risbud MV. HIF-1α and HIF-2α degradation is differentially regulated in nucleus pulposus cells of the intervertebral disc. J Bone Miner Res February 2012;27(2):401—12. Wiley Subscription Services, Inc., A Wiley Company.

[115] Fujita N, Markova D, Anderson DG, Chiba K, Toyama Y, Shapiro IM, et al. Expression of prolyl hydroxylases (PHDs) is selectively controlled by HIF-1 and HIF-2 proteins in nucleus pulposus cells of the intervertebral disc: distinct roles of PHD2 and PHD3 proteins in controlling HIF-1α activity in hypoxia. J Biol Chem May 11, 2012;287(20):16975—86.

[116] Gogate SS, Fujita N, Skubutyte R, Shapiro IM, Risbud MV. Tonicity enhancer binding protein (TonEBP) and hypoxia-inducible factor (HIF) coordinate heat shock protein 70 (Hsp70) expression in hypoxic nucleus pulposus cells: role of Hsp70 in HIF-1α degradation. J Bone Miner Res May 2012;27(5):1106—17. Wiley Subscription Services, Inc., A Wiley Company.

[117] Merceron C, Mangiavini L, Merrill BJ, Wilson TL, Robling A, Guicheux J, et al. Loss of HIF-1a in the notochord results in cell death and complete disappearance of the nucleus pulposus. PLoS One 2014;9(10):e110768.

[118] Merceron C, Mangiavini L, Robling A, Wilson TL, Giaccia AJ, Shapiro IM, et al. Loss of HIF-1α in the notochord results in cell death and complete disappearance of the nucleus pulposusKletsas D, editor. PLoS ONE 2014;9(10):e110768.

[119] Agrawal A, Guttapalli A, Narayan S, Albert TJ, Shapiro IM, Risbud MV. Normoxic stabilization of HIF-1alpha drives glycolytic metabolism and regulates aggrecan gene expression in nucleus pulposus cells of the rat intervertebral disk. Am J Physiol Cell Physiol August 2007;293(2):C621−31.

[120] Rajpurohit R, Risbud MV, Ducheyne P, Vresilovic EJ, Shapiro IM. Phenotypic characteristics of the nucleus pulposus: expression of hypoxia inducing factor-1, glucose transporter-1 and MMP-2. Cell Tissue Res June 2002;308(3):401−7.

[121] Gogate SS, Nasser R, Shapiro IM, Risbud MV. Hypoxic regulation of β-1,3-glucuronyltransferase 1 expression in nucleus pulposus cells of the rat intervertebral disc: role of hypoxia-inducible factor proteins. Arthritis Rheum July 2011;63(7):1950−60. Wiley Subscription Services, Inc., A Wiley Company.

[122] Zeng Y, Danielson KG, Albert TJ, Shapiro IM, Risbud MV. HIF-1 alpha is a regulator of galectin-3 expression in the intervertebral disc. J Bone Miner Res December 2007;22(12):1851−61. John Wiley and Sons and The American Society for Bone and Mineral Research (ASBMR).

[123] Power KA, Grad S, Rutges JPHJ, Creemers LB, van Rijen MHP, O'Gaora P, et al. Identification of cell surface-specific markers to target human nucleus pulposus cells: expression of carbonic anhydrase XII varies with age and degeneration. Arthritis Rheum December 2011;63(12):3876−86. Wiley Subscription Services, Inc., A Wiley Company.

[124] Parks SK, Chiche J, Pouyssegur J. pH control mechanisms of tumor survival and growth. J Cell Physiol February 2011;226(2):299−308. Wiley Subscription Services, Inc., A Wiley Company.

[125] Chiche J, Ilc K, Laferrière J, Trottier E, Dayan F, Mazure NM, et al. Hypoxia-inducible carbonic anhydrase IX and XII promote tumor cell growth by counteracting acidosis through the regulation of the intracellular pH. Cancer Res January 1, 2009;69(1):358−68. American Association for Cancer Research.

[126] Lyons GE, Buckingham ME, Tweedie S, Edwards YH. Carbonic anhydrase III, an early mesodermal marker, is expressed in embryonic mouse skeletal muscle and notochord. Development January 1991;111(1):233−44.

[127] van Dijk B, Potier E, Ito K. Culturing bovine nucleus pulposus explants by balancing medium osmolarity. Tissue Eng Part C Methods November 2011;17(11):1089−96. Mary Ann Liebert, Inc. 140 Huguenot Street, 3rd Floor New Rochelle, NY 10801 USA.

[128] Roberts N, Hogg D, Whitehouse GH, Dangerfield P. Quantitative analysis of diurnal variation in volume and water content of lumbar intervertebral discs. Clin Anat 1998;11(1):1−8. Wiley Subscription Services, Inc., A Wiley Company.

[129] Urban JP. The chondrocyte: a cell under pressure. Br J Rheumatol October 1994;33(10):901−8.

[130] Johnson ZI, Shapiro IM, Risbud MV. Extracellular osmolarity regulates matrix homeostasis in the intervertebral disc and articular cartilage: evolving role of TonEBP. Matrix Biol November 2014;40:10−6.

[131] Tsai T-T, Danielson KG, Guttapalli A, Oguz E, Albert TJ, Shapiro IM, et al. TonEBP/OREBP is a regulator of nucleus pulposus cell function and survival in the intervertebral disc. J Biol Chem September 1, 2006;281(35):25416−24.

[132] Cheung CY, Ko BC. NFAT5 in cellular adaptation to hypertonic stress − regulations and functional significance. J Mol Signal April 23, 2013;8(1):5. Ubiquity Press.

[133] Garcia-Perez A, Burg MB. Renal medullary organic osmolytes. Physiol Rev October 1991;71(4):1081−115.

[134]   Yancey PH, Clark ME, Hand SC, Bowlus RD, Somero GN. Living with water stress: evolution of osmolyte systems. Science September 24, 1982;217(4566):1214−22.

[135]   Tsai T-T, Guttapalli A, Agrawal A, Albert TJ, Shapiro IM, Risbud MV. MEK/ERK signaling controls osmoregulation of nucleus pulposus cells of the intervertebral disc by transactivation of TonEBP/OREBP. J Bone Miner Res July 2007;22(7):965−74. John Wiley and Sons and The American Society for Bone and Mineral Research (ASBMR).

[136]   Hiyama A, Gajghate S, Sakai D, Mochida J, Shapiro IM, Risbud MV. Activation of TonEBP by calcium controls {beta}1,3-glucuronosyltransferase-I expression, a key regulator of glycosaminoglycan synthesis in cells of the intervertebral disc. J Biol Chem April 10, 2009;284(15):9824−34. American Society for Biochemistry and Molecular Biology.

[137]   Gajghate S, Hiyama A, Shah M, Sakai D, Anderson DG, Shapiro IM, et al. Osmolarity and intracellular calcium regulate aquaporin2 expression through TonEBP in nucleus pulposus cells of the intervertebral disc. J Bone Miner Res June 2009;24(6):992−1001. John Wiley and Sons and The American Society for Bone and Mineral Research (ASBMR).

[138]   Mwale F, Roughley P, Antoniou J. Distinction between the extracellular matrix of the nucleus pulposus and hyaline cartilage: a requisite for tissue engineering of intervertebral disc. Eur Cell Mater December 15, 2004;8:58−63. discussion 63−4.

[139]   Bridgen DT, Gilchrist CL, Richardson WJ, Isaacs RE, Brown CR, Yang KL, et al. Integrin-mediated interactions with extracellular matrix proteins for nucleus pulposus cells of the human intervertebral disc. J Orthop Res October 2013;31(10):1661−7.

[140]   Chen J, Jing L, Gilchrist CL, Richardson WJ, Fitch RD, Setton LA. Expression of laminin isoforms, receptors, and binding proteins unique to nucleus pulposus cells of immature intervertebral disc. Connect Tissue Res 2009;50(5):294−306. NIH Public Access.

[141]   Meisel HJ, Siodla V, Ganey T, Minkus Y, Hutton WC, Alasevic OJ. Clinical experience in cell-based therapeutics: disc chondrocyte transplantation A treatment for degenerated or damaged intervertebral disc. Biomol Eng February 2007;24(1):5−21.

[142]   Hu M-H, Hung L-W, Yang S-H, Sun Y-H, Shih TT-F, Lin F-H. Lovastatin promotes redifferentiation of human nucleus pulposus cells during expansion in monolayer culture. Artif Organs April 2011;35(4):411−6. Blackwell Publishing Inc.

[143]   Vinatier C, Magne D, Weiss P, Trojani C, Rochet N, Carle GF, et al. A silanized hydroxypropyl methylcellulose hydrogel for the three-dimensional culture of chondrocytes. Biomaterials November 2005;26(33):6643−51.

[144]   Yamamoto Y, Mochida J, Sakai D, Nakai T, Nishimura K, Kawada H, et al. Upregulation of the viability of nucleus pulposus cells by bone marrow-derived stromal cells: significance of direct cell-to-cell contact in coculture system. Spine July 15, 2004;29(14):1508−14.

[145]   Mochida J, Sakai D, Nakamura Y, Watanabe T, Yamamoto Y, Kato S. Intervertebral disc repair with activated nucleus pulposus cell transplantation: a three-year, prospective clinical study of its safety. Eur Cell Mater March 20, 2015;29:202−12. discussion 212.

[146]   Coric D, Pettine K, Sumich A, Boltes MO. Prospective study of disc repair with allogeneic chondrocytes presented at the 2012 Joint Spine Section Meeting. J Neurosurg Spine January 2013;18(1):85−95.

[147]   Jones EA, English A, Henshaw K, Kinsey SE, Markham AF, Emery P, et al. Enumeration and phenotypic characterization of synovial fluid multipotential mesenchymal progenitor cells in inflammatory and degenerative arthritis. Arthritis Rheum March 2004;50(3):817−27. Wiley Subscription Services, Inc., A Wiley Company.

[148]   Kern S, Eichler H, Stoeve J, Klüter H, Bieback K. Comparative analysis of mesenchymal stem cells from bone marrow, umbilical cord blood, or adipose tissue. Stem Cell May 2006;24(5):1294−301. John Wiley & Sons, Ltd.

[149] Elabd C, Centeno CJ, Schultz JR, Lutz G, Ichim T, Silva FJ. Intra-discal injection of autologous, hypoxic cultured bone marrow-derived mesenchymal stem cells in five patients with chronic lower back pain: a long-term safety and feasibility study. J Transl Med September 1, 2016;14(1):253. BioMed Central.

[150] Orozco L, Soler R, Morera C, Alberca M, Sánchez A, García-Sancho J. Intervertebral disc repair by autologous mesenchymal bone marrow cells: a pilot study. Transplantation October 15, 2011;92(7):822−8.

[151] Li C-Y, Wu X-Y, Tong J-B, Yang X-X, Zhao J-L, Zheng Q-F, et al. Comparative analysis of human mesenchymal stem cells from bone marrow and adipose tissue under xeno-free conditions for cell therapy. Stem Cell Res Ther April 13, 2015;6(1):55. BioMed Central.

[152] Korecki CL, Costi JJ, Iatridis JC. Needle puncture injury affects intervertebral disc mechanics and biology in an organ culture model. Spine February 1, 2008;33(3):235−41.

[153] Melrose J, Smith SM, Fuller ES, Young AA, Roughley PJ, Dart A, et al. Biglycan and fibromodulin fragmentation correlates with temporal and spatial annular remodelling in experimentally injured ovine intervertebral discs. Eur Spine J December 2007;16(12):2193−205.

[154] Smith JW, Walmsley R. Experimental incision of the intervertebral disc. J Bone Joint Surg Br November 1951;33-B(4):612−25.

[155] Risbud MV, Guttapalli A, Tsai T-T, Lee JY, Danielson KG, Vaccaro AR, et al. Evidence for skeletal progenitor cells in the degenerate human intervertebral disc. Spine November 1, 2007;32(23):2537−44.

[156] Feng G, Yang X, Shang H, Marks IW, Shen FH, Katz A, et al. Multipotential differentiation of human anulus fibrosus cells: an in vitro study. J Bone Joint Surg Am March 2010;92(3):675−85.

[157] Brisby H, Papadimitriou N, Brantsing C, Bergh P, Lindahl A, Barreto Henriksson H. The presence of local mesenchymal progenitor cells in human degenerated intervertebral discs and possibilities to influence these in vitro: a descriptive study in humans. Stem Cells Dev March 1, 2013;22(5):804−14. Mary Ann Liebert, Inc. 140 Huguenot Street, 3rd Floor New Rochelle, NY 10801 USA.

[158] Blanco JF, Graciani IF, Sanchez-Guijo FM, Muntión S, Hernandez-Campo P, Santamaria C, et al. Isolation and characterization of mesenchymal stromal cells from human degenerated nucleus pulposus: comparison with bone marrow mesenchymal stromal cells from the same subjects. Spine December 15, 2010;35(26):2259−65.

[159] Henriksson H, Thornemo M, Karlsson C, Hägg O, Junevik K, Lindahl A, et al. Identification of cell proliferation zones, progenitor cells and a potential stem cell niche in the intervertebral disc region: a study in four species. Spine October 1, 2009;34(21):2278−87.

[160] Henriksson HB, Svala E, Skioldebrand E, Lindahl A, Brisby H. Support of concept that migrating progenitor cells from stem cell niches contribute to normal regeneration of the adult mammal intervertebral disc: a descriptive study in the New Zealand white rabbit. Spine April 20, 2012;37(9):722−32.

[161] Sakai D, Nakamura Y, Nakai T, Mishima T, Kato S, Grad S, et al. Exhaustion of nucleus pulposus progenitor cells with ageing and degeneration of the intervertebral disc. Nat Commun 2012;3:1264. Nature Publishing Group.

[162] Yasen M, Fei Q, Hutton WC, Zhang J, Dong J, Jiang X, et al. Changes of number of cells expressing proliferation and progenitor cell markers with age in rabbit intervertebral discs. Acta Biochim Biophys Sin May 2013;45(5):368−76.

[163] Shu C, Hughes C, Smith SM, Smith MM, Hayes A, Caterson B, et al. The ovine newborn and human foetal intervertebral disc contain perlecan and aggrecan variably substituted with native 7D4 CS sulphation motif: spatiotemporal immunolocalisation and co-distribution with Notch-1 in the human foetal disc. Glycoconj J October 2013;30(7):717−25. Springer US.

[164] Hiyama A, Skubutyte R, Markova D, Anderson DG, Yadla S, Sakai D, et al. Hypoxia activates the notch signaling pathway in cells of the intervertebral disc: implications in degenerative disc disease. Arthritis Rheum May 2011;63(5):1355−64.

[165] Gustafsson MV, Zheng X, Pereira T, Gradin K, Jin S, Lundkvist J, et al. Hypoxia requires notch signaling to maintain the undifferentiated cell state. Dev Cell November 2005;9(5):617−28.

[166] Wang H, Tian Y, Wang J, Phillips KLE, Binch ALA, Dunn S, et al. Inflammatory cytokines induce NOTCH signaling in nucleus pulposus cells: implications in intervertebral disc degeneration. J Biol Chem June 7, 2013;288(23):16761−74.

[167] Bellin M, Marchetto MC, Gage FH, Mummery CL. Induced pluripotent stem cells: the new patient? Nat Rev Mol Cell Biol November 2012;13(11):713−26. Nature Publishing Group.

[168] Takahashi K, Yamanaka S. Induction of pluripotent stem cells from mouse embryonic and adult fibroblast cultures by defined factors. Cell August 25, 2006;126(4):663−76.

[169] Takahashi K, Tanabe K, Ohnuki M, Narita M, Ichisaka T, Tomoda K, et al. Induction of pluripotent stem cells from adult human fibroblasts by defined factors. Cell November 30, 2007;131(5):861−72.

[170] Mascetti VL, Pedersen RA. Human-mouse chimerism validates human stem cell pluripotency. Cell Stem Cell January 7, 2016;18(1):67−72.

[171] Merrill BJ, Pasolli HA, Polak L, Rendl M, García-García MJ, Anderson KV, et al. Tcf3: a transcriptional regulator of axis induction in the early embryo. Development January 2004;131(2):263−74.

[172] Vincent SD, Dunn NR, Hayashi S, Norris DP, Robertson EJ. Cell fate decisions within the mouse organizer are governed by graded Nodal signals. Genes Dev July 1, 2003;17(13):1646−62. Cold Spring Harbor Lab.

[173] Loh KM, Ang LT, Zhang J, Kumar V, Ang J, Auyeong JQ, et al. Efficient endoderm induction from human pluripotent stem cells by logically directing signals controlling lineage bifurcations. Cell Stem Cell February 6, 2014;14(2):237−52.

[174] Vallier L, Touboul T, Brown S, Cho C, Bilican B, Alexander M, et al. Signaling pathways controlling pluripotency and early cell fate decisions of human induced pluripotent stem cells. Stem Cell November 2009;27(11):2655−66. Wiley Subscription Services, Inc., A Wiley Company.

[175] Winzi MK, Hyttel P, Dale JK, Serup P. Isolation and characterization of node/notochord-like cells from mouse embryonic stem cells. Stem Cells Dev November 2011;20(11):1817−27. Mary Ann Liebert, Inc. 140 Huguenot Street, 3rd Floor New Rochelle, NY 10801 USA.

[176] Chen J, Lee EJ, Jing L, Christoforou N, Leong KW, Setton LA. Differentiation of mouse induced pluripotent stem cells (iPSCs) into nucleus pulposus-like cells in vitroAgarwal S, editor. PLoS ONE 2013;8(9):e75548.

[177] Liu Y, Rahaman MN, Bal BS. Modulating notochordal differentiation of human induced pluripotent stem cells using natural nucleus pulposus tissue matrixGuicheux J, editor. PLoS One 2014;9(7):e100885.

[178] Johnson ZI, Schoepflin ZR, Choi H, Shapiro IM, Risbud MV. Disc in flames: roles of TNF-α and IL-1β in intervertebral disc degeneration. Eur Cell Mater September 21, 2015;30:104−16. discussion 116−7.

[179] Choi H, Johnson ZI, Risbud MV. Understanding nucleus pulposus cell phenotype: a prerequisite for stem cell based therapies to treat intervertebral disc degeneration. Curr Stem Cell Res Ther 2015;10(4):307−16.

Chapter 8

Developmental Biology in Tendon Tissue Engineering

Mor Grinstein[1,2], Jenna L. Galloway[1,2]

[1]*Massachusetts General Hospital, Harvard Medical School, Boston, MA, United States;* [2]*Harvard Stem Cell Institute, Cambridge, MA, United States*

INTRODUCTION

In the United States, approximately 30% of doctor visits for musculoskeletal pain are related to tendon disorders, and more than 15 billion dollars are spent each year on tendon injuries [1,2]. Surgery is a common treatment to repair injured tendons; however, surgical repair often can be unsuccessful, resulting in persistent pain and problems with mobility [3]. When a tendon is injured, the organized hierarchical collagen structure is damaged. The healing response is generally characterized by three main phases: inflammation, cell proliferation, and remodeling. Although these processes are well described morphologically, the molecular mechanisms that control them are not well understood. Furthermore, the healed tendon rarely regains its original biomechanical properties, making regenerative medicine and tissue engineering approaches a promising future treatment strategy for tendon injury. To pursue such stem cell—based approaches, it is essential to better understand the molecular mechanisms that govern tendon development and maturation, and the pathological events that interfere with proper tendon healing. This chapter will describe our current understanding of tendon structure, development, and injury response, and how knowledge in these areas can provide the foundation for developing regenerative medicine—based strategies to treat tendon injuries.

OVERVIEW OF TENDON BIOLOGY

Tendon Structure and Function

By connecting muscles to bones, tendons serve as force transmitters that enable all the movements of our body. They can experience loads up to 9 kN, which corresponds to more than 10 times the body weight of the average person. Their ability to withstand and function under such forces is because of

Developmental Biology and Musculoskeletal Tissue Engineering. https://doi.org/10.1016/B978-0-12-811467-4.00008-5
181

their unique composition and structure [4]. Tendons are predominantly made of type I collagen arranged into a triple helical rod-shaped molecule that associates with other collagen molecules. Together they form a fibril array that forms the highly organized tendon structure. Bundles of fibrils form larger primary fiber bundles called fascicles; groups that associate form tertiary fiber bundles. These fibers are surrounded by a connective tissue called the endotenon that holds the blood vessels, nerves, and lymphocytes [5]. The fiber bundles and endotenon are covered by the epitenon, a layer of connective tissue surrounding the tendon. Tendons such as the Achilles do not have a true synovial sheath, but rather a layer called the paratenon, which is attached to the outside of the epitenon. These tissues have an important role filling the space between tendons and the immovable fascial compartments, secreting lubricant, and facilitating gliding to reduce friction during tendon movement (Fig. 8.1).

Although type I collagen is the major matrix component accounting for 65%–80% of the dry mass of tendons, there are multiple molecular elements that also contribute to its structure and function. Other collagens such as II, III, V, VI, IX, X, XI, XII, and XIV are present, and each is thought to have an active role in tendon development and function [6]. Collagen II, IX, and X are expressed at tendon bone insertion sites and are important in the organization of the fibrocartilage transition to bone [7]. Collagen III is expressed in the embryo during tendon differentiation and is involved in the assembly of the collagen fibrils [8]. At later stages of development, it is expressed in regions of smaller-diameter collagen fibrils, suggesting it has a role in

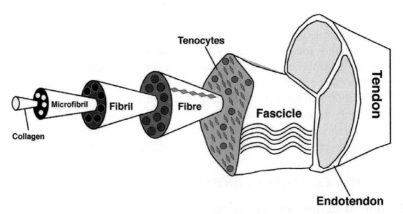

FIGURE 8.1 Structure and anatomy of the tendon. Tendons consist of dense regular extracellular matrix (ECM) encased in connective tissue sheaths. Most of the ECM is comprised of collagen I, which is important for the tensile strength of the tendon. Collagen fibrils are heterotrimeric collagen molecules, which associate together to form fibers. The fascicles hold bundles of fibers that surround tenoblasts and tenocytes. These fascicles are encased in a sheath called the endotenon.

restricting the lateral growth of collagen I fibrils [9]. Although not as abundant as type I and III, collagen V and XI are fibril-forming collagens. They are involved in nucleating fibril formation as decreasing their amounts in culture results in fewer fibrils assembled with larger diameters [10]. Collagen V can be found in differing amounts in distinct tendon and ligaments in the body, suggesting that there are tissue-specific differences in the distributions of this collagen type. Loss of collagen V in the tendons results in altered tendon and ligament mechanical properties and joint hypermobility as observed in the human condition, Ehlers−Danlos syndrome [3]. Collagen V and XI also have redundant functions as loss of one allele of collagen V in the collagen XI null, has an extremely severe phenotype with significantly reduced fibril number and structure [11]. The microfibrillar matrix collagen VI has been primarily characterized for its role in human myopathies. It is found in the tendon pericellular matrix [12], and its loss affects the tendon structure [13] However, it is unclear if the entirety of the tendon phenotype is due to its direct requirement in the tendon or an indirect result of the myopathy. Collagen XII and XIV are fibril associate collagens with interrupted triple helices that are believed to have a function in assembling and integrating the developing matrix [8]. Collagen XII can bind decorin (Dcn) and fibromodulin (Fbmd) as well as type I collagen fibrils [14]. Although collagen XIV−deficient mice have normal adult tendons, the tendons at early postnatal stages are abnormal, suggesting an early role for this collagen in development [15].

The extracellular matrix (ECM) also contains proteoglycans/glycoproteins, glycosaminoglycans (GAGs), and several other small molecules [4]. The hydrophilic nature of proteoglycans provides a viscous environment, allowing the diffusion of water-soluble molecules, cell migration, and the sliding of collagen fibers relative to each other with less friction [4]. Among them, the leucine-rich proteoglycans such as Fbmd, Dcn, and lumican have a role in fibrillogenesis, which in turn affects tendon mechanics [16]. Interestingly, double knockouts of Fbmd and biglycan have ectopic tendon ossification and premature osteoarthritis (OA) [17].

In addition to the rich ECM, there are cells residing within the tendon called tenoblasts and tenocytes. Tenoblasts are immature tendon cells with a more rounded morphology and numerous cytoplasmic organelles, reflecting their high metabolic activity. As they mature, the tenoblasts become elongated and are thought to transform into tenocytes, which have a lower nucleus to cytoplasm ratio and decreased metabolic activity [4]. After injury, tenoblast-like cells proliferate in the epitenon and were thought to be instrumental for the repair process [18]. However, we currently lack molecular markers that distinguish tenoblasts from tenocytes, making it difficult to understand how each may contribute to the healing capacity of the adult tendon.

Tendon Development in the Embryo

Tendons arise in all areas of the vertebrate body where muscle connects bone, in the cranial, axial, and appendicular skeleton. Tendons in each of these regions are derived from different embryonic origins: cranial tendons arise from the neural crest, axial tendons from the paraxial mesoderm, and the limb tendons derive from the lateral plate mesoderm (LPM). Although the development of the muscle and skeletal tissues are reasonably well understood, the processes regulating the development of the tendons are less clear. The discovery of *Scleraxis* (*Scx*), a basic helix-loop-helix transcription factor, as the earliest marker of tendon progenitors enabled the study of the initial stages of tendon development and allowed for more mechanistic dissection of the pathways controlling tendon formation [19]. It has become apparent from many previous studies that there are similarities and differences in the molecular regulation of the tendon program in each region of the developing embryo.

The vertebrate axial musculoskeletal system is derived from the somites, which are dorsally located segmental blocks of mesoderm. In response to signals from the surrounding tissues, the somites differentiate into distinct compartments that give rise to different tissue lineages: dermis/muscle (dermomyotome), cartilage/bone (sclerotome), and tendon (syndetome). *Scx*-expressing progenitor cells of the axial trunk tendons first appear between the myotome and sclerotome, the muscle and skeletal somite compartments, respectively. Indeed, this region of *Scx* expression defines a distinct somitic compartment, the syndetome, which gives rise to the axial tendons. Interestingly [20], the syndetome appears at a slightly later stage from a region originating from the early sclerotome: *Scx* expression is limited to the dorsolateral regions of the sclerotome, whereas the expression of *Pax1*, the molecular marker for sclerotome, is more ventromedial. The tendon somitic region is regulated by the interaction of the somitic muscle and the cartilage precursor regions. The removal of dermomyotome just before myotome formation causes a loss of *Scx* expression [20]. Moreover, in the *MyoD* and *Myf5* double-mutant mice, which lack differentiated muscle, the expression of *Scx* was eliminated [21]. Therefore, signals from the muscle are critical for tendon formation. These signals were discovered to include members of the fibroblast growth factor (FGF) family, FGF8 and FGF4 as overexpression of FGF8 robustly upregulates *Scx* and FGF inhibition causes loss of *Scx* expression in the somite [20–22]. Further evidence suggests that FGFs stimulate the expression of *Scx* through the MAPK/mitogen-activated protein kinase (originally called ERK or extracellular signal-regulated kinase) pathway and the activation of E-twenty-six (Ets) transcription factors in tendon progenitors [21]. However, these studies were performed in avian systems, and recent work in the mouse indicates that FGF signaling has inhibitory effects on tendon formation, suggesting that the regulatory mechanisms governing tendon induction may differ between species. The skeletal lineage, in contrast to the muscle, appears to have an inhibitory role in tendon development in the axial skeleton. Pax1, a transcription factor that marks the forming sclerotome, must be downregulated in the region that will later become the tendon-forming syndetome. Combined loss of *Sry-related HMG Box* (*Sox*) 5 and 6, transcription factors important for cartilage differentiation, result in ectopic expression of *Scx* in axial cartilage-forming regions, underscoring a reciprocal relationship between tendon and skeletal fates [23].

In the limb, tendon progenitors form in regions of the subectodermal limb bud mesenchyme that is derived from the LPM. These cells organize into tendon primordia that align between the forming muscles and cartilage [19,24]. In contrast to the axial tendon progenitors, limb tendons do not require the presence of myogenic cells for their induction, but rather signals from the muscle and movement are important for tendon maintenance, differentiation, and individuation [24,25]. Ectoderm appears to secrete tendon-inducing signals as its removal results in a loss of *Scx* expression [19]. Furthermore, limiting Bone Morphogenetic Protein (BMP) signaling can expand tendon gene expression, likely underscoring the regulation of tendon and cartilage fates from a common progenitor pool [19]. Regulation of the tendon program also differs based on the location along the anterior—posterior axis of the limb. Proximal tendons such as those in the forearm do not require *Sox9* for their formation but do require muscle for their differentiation and individuation. In contrast, distal tendons in the hand do not require muscle but do require *Sox9* and cartilage condensation for their initiation [24]. Together, these studies suggest that tendon development is controlled differently in distinct anatomic locations in the embryo or at least the source of the signal(s) responsible for their induction and/or differentiation is unique in each region.

One of the major pathways to emerge as having a pivotal role in tendon development is Transforming Growth Factor beta (TGF-β). TGF-β is a potent inducer of *Scx* expression and is required for tendon development in all embryonic locations throughout vertebrates [25—27]. In addition, the canonical TGF-β intracellular pathway components, SMAD2 and SMAD3, are necessary for Scx expression in mouse limbs [28]. Activation of this pathway can promote tendon fates in mouse embryonic fibroblasts and mesenchymal cell lines [26]. Addition of the ligand in high-density chick micromass leads to an upregulation of *Scx* and *Tenomodulin* (*Tnmd*) via the Smad 2/3 pathway [29]. Notably, addition of TGF-β ligands in cell culture activates Scx but inhibits *Tnmd* expression, suggesting that additional factors are necessary for recapitulating tendon developmental processes in 2D culture [30]. In addition, TGF-β signaling appears to be required at later gestational stages for the continued development of tendons [26] as TGF-β2 and TGF-β3 double knockouts and the TGF-β type 2 receptor knockout express *Scx* at early embryonic stages. This would suggest that signals other than TGF-β are capable of inducing the tendon program in its absence. Currently, our knowledge of the molecular events that lie upstream of *Scx* is limited (Fig. 8.2).

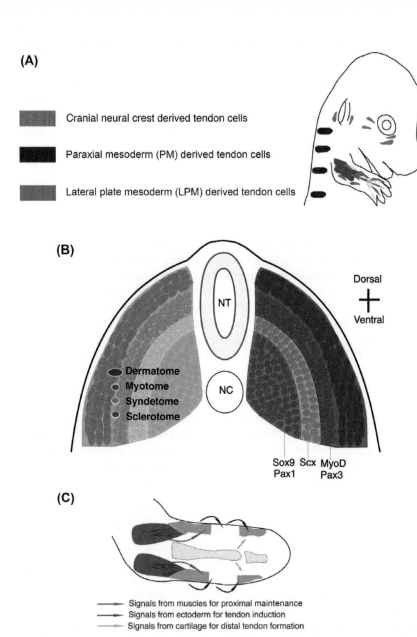

FIGURE 8.2 Tendon development in the mouse embryo. (A) The early tendon cells express scleraxis (Scx) and form connections between the muscle and bones. They originate from three different embryonic sources. The cranial neural crest will give rise to the cranial tendons that populate the head and neck regions. The paraxial mesoderm will give rise to axial tendons in the vertebral column and ribs, and the lateral plate mesoderm, which will give rise to the limb tendons. (B) A transverse section through the embryo shows the central neural tube (NT) and notochord (NC) and the lateral somites. The axial musculoskeletal system is derived from specific somitic compartments. In response to various signals, the somites differentiate into the dermatome, which develops into dermis, the myotome, which expresses *MyoD* and *Pax3* and gives rise to muscle, the sclerotome, which expresses *Sox9* and *Pax1* and forms cartilage and bone, and the syndetome, which generates *Scx*-expressing tendons. (C) The cross talk between tendons and surrounding tissues in limb development.

The Extracellular Matrix in Tendon Differentiation and Maturation

How aggregates of embryonic cells transform a tissue into an adult structure primarily composed of ECM is a dynamic phenomenon and is not well understood. In the embryo, cells secrete collagen fibrils that grow in number and longitudinally, but modestly in diameter. The collagen fibrils appear to be positioned in an actin-dependent manner by membrane extensions from tendon fibroblast cells, termed fibripositors [31]. After birth, the collagen fibrils continue to grow in diameter and length, eventually taking on a bimodal distribution of collagen fibril diameters ranging between 35 nm and 400 nm. The enlargement of the collagen fibril diameter at this later stage is thought to drive the overall growth of the tissue to its mature form as the number of bundles per cell and fibrils per bundle remain constant [32]. Using 3D electron microscopy of mouse tail tendons, it was found that the cell−cell contacts remain intact and cell volume does not change during the matrix expansion. Instead, the cell surface area increases as the cell shape changes from rounded to stellate in appearance. These observations lend weight to the idea that the rounded tenoblasts form the later stellate appearing tenocytes during the growth of the matrix.

Tnmd is a type II transmembrane glycoprotein that is one of the most specific markers for tendon cells [25,33]. Tnmd is first detected during tendon differentiation stages [34] and is dependent on Scx for expression. *Tnmd* transcripts are lost in *Scx* null mice [35] and are upregulated in mesenchymal cells virally transduced with *Tnmd*. In addition, Tnmd has a role in cell proliferation and matrix formation as its loss results in thicker collagen fibrils. This phenotype has similarities to aged tendon matrix, which is interesting given that tendon-derived stem/progenitor cells from *Tnmd*-deficient mice display proliferation defects and earlier signs of senescence.

Additional matrix proteins such as Dcn, Biglycan, Fbmd, and Lumican, which belong to the small leucine-rich proteoglycan (SLRP) family, have important roles in tendon matrix organization and collagen fibril growth (reviewed in Ref. [16]). The expression of several SLRPs peak at postnatal stages: lumican at P9 and Fbmd between P14-30 [36]. Lumican-deficient tendons are normal at adult stages but have some abnormal fibrils at early stages. In the absence of Fbmd, there are increased numbers of smaller diameter collagen fibrils and reduced stiffness, suggesting Fbmd is necessary for fibril growth. Loss of Dcn also results in altered fibril structure and viscoelastic properties, suggesting changes to the interfibril organization. It is also thought that the longitudinal growth of the collagen molecules is, in part, dependent on other matrix proteins such as proteoglycans [37]. The mechanical properties of the tendon change with a positive correlation between collagen fibril diameter and tendon stiffness. Interestingly, part of the increase in tendon stiffness is also due to cross-linking mediated by lysyl oxidase [38].

The timing and amount of expression of such SLRPs and other matrix molecules suggests that key transcription factors directly regulate this process to control the expression of ECM proteins at specific stages. Although also having an earlier role in the formation of certain tendon types, the loss of *Scx* disrupts tenocyte differentiation and leads to disorganized ECM. Scx directly controls *Col1a1* and *Cola2* transcription, and the expression of *Tnmd*, *Col1a1*, *Cola2*, *Col3a1*, and *Col4a1* is reduced in *Scx* mutant mice [34]. *Scx* loss affects the organization of cells that contribute to the tendon sheath [34].

In addition to *Scx*, the transcription factors, *Mohawk* (*Mkx*) and *Early Growth Response 1* (*Egr1*), also guide the complex process of tendon differentiation and matrix production [39,40]. *Mkx* (also known as *Irxl1*) is a member of the three amino acid loop extension (TALE) class of homeodomain proteins. Loss of results in reduced expression of *Tnmd*, *Fbmd*, and *Dcn* in the neonatal tendons and disrupted postnatal growth of the collagen fibrils, resulting in tendon hypoplasia [41]. Although the defects in *Mkx* null mice have been primarily examined in the tendon lineage, this transcription factor is also expressed in the early somite in the skeletal muscle, cartilage, and bone progenitors [42]. *Mkx* may also antagonize muscle development as studies have shown that it inhibits muscle differentiation in mouse cell culture and delays muscle development in zebrafish embryos [30]. However, *Mkx* mutant mice do not show any obvious skeletal muscle defects. In the adult, loss of *Mkx* causes degeneration of the periodontal ligament, and its expression is significantly lower in human tendinopathy tissue, suggesting *Mkx* also has an important role in maintaining adult tendons [30]. Forced expression of *Mkx* promotes tendon differentiation in the mouse embryonic mesenchymal cell line, C3H10T1/2, and bone marrow mesenchymal cells [43], suggesting this gene could be a target for applications to improve tendon healing.

Similar to *Mkx*, *Egr1* is important for tendon differentiation and repair [41]. *Egr1* mutant mice have defects in collagen fibril organization at fetal and postnatal stages as well as harbor reductions in collagen gene expression, *Col1a1*, *Col1a2*, *Col3a1*, *Col5a1*, *Col12a1*, and *Col14a1* and ECM components, *Tnmd*, *Fbmod*, and *Dcn* [43]. *Scx* expression is reduced in these mice, although *Mkx* is expressed normally. Interestingly, *Egr1*-deficient tendons have impaired mechanical properties and delayed healing after injury [43]. Forced expression of *Egr1* promotes tendon gene expression and improves repair in an injury model. This along with other studies suggests that developmental regulators of the tendon program may be instrumental in devising successful regenerative strategies for tendon repair [44].

Tendon Mechanobiology

Mechanotransduction is the process by which cells respond to external forces with biochemical signals. Physical forces have an important role in tendon tissue development, homeostasis, healing, and degeneration [22]. Although recent studies have begun to shed light on the molecular mechanisms regulating tendon mechanobiology, there remain many questions surrounding how changes in load are sensed, processed, and interpreted by cells. Knowledge of these pathways would impact our understanding of the events underlying to tendon disease and has clear applications in their treatments.

During embryogenesis, movement produced by the muscles is necessary for tendon maintenance and differentiation. Loss of muscles or of innervation to the muscles reduces the expression of tendon markers and disrupts tendon differentiation [25,45]. The TGF-β/Smad2/3 and the FGF/ERK/MAPK pathways act downstream of muscle movement as both can rescue *Scx* expression in the absence of movement [25]. However, only TGF-β, not FGF, can rescue the additional tendon genes, *Tnmd* and *Thbs2*, under immobilization conditions during embryogenesis [46]. Elevated secretions of Col4a1 and Col6a1 are also seen in developing chick tendons under stress [47]. In adult tendons, the TGF-β pathway also appears to be downstream of mechanical forces and regulates *Scx* expression [48]. Loading is important for the proper development of entheses, tendon—bone attachment sites, as well as tuberosities, specialized bone eminences attached to tendons [49]. Loss of loading causes mechanically inferior attachment sites with reduced mineralized bone and delayed fibrocartilage development [50].

Tendons respond to physical activity in an adaptive manner [44]. Forces applied to the tendon result in the transmission of tensile, compressive, and shear stresses and strains on cells [51]. Fluid flow shear stress, which is known to regulate osteocyte mechanotransduction, may also occur [52]. Changes in load, in turn, result in increases in tendon cross-sectional area, collagen synthesis, and the expression of *Scx*, *Tnmd*, and *TGF-β* [50,53,54]. Studies using human tendon tissues show dramatic increases in the levels of matrix metalloproteinase (MMP) MMP2, MMP9, and MMP14 after exercise, suggesting that MMPs are upregulated in response to mechanotransduction. This upregulation depends on the duration of loading. In cultured tendon cells, MMP2 and MMP13 are upregulated after very short cycles; however, MMP1 is downregulated after longer cycles of loading [55]. In vitro studies using mouse tenocytes have shown that low levels of shear stress leads to upregulation of *Col1a1* and *Tnmd*, whereas greater stresses lead to the upregulation of *Runt-related transcription factor* (*Runx*) *Runx2* and *Sox9* [53]. In addition, *Tnmd* has been described as a mechanosensitive gene that is upregulated on mechanical stimulation in vitro and in vivo [56]. Loss of *Tnmd* perturbs running performance because of problems with the matrix that are exacerbated by exercise, suggesting *Tnmd* is necessary for the tendon to properly adapt to changes in the loading environment [57].

TGF-β, which can be induced by changes in load, also regulates the expression of MMPs [58], and MMP2, MMP9, and BMP1 proteases are capable of releasing TGF-β. This creates positive feedback loop for regulating MMP levels and ECM remodeling in homeostasis and injury repair [3]. In contrast, repetitive mechanical loading or overuse can result in a catabolic response that includes increased cellularity, matrix disorganization, and the expression of proinflammatory cytokines and MMPs [50]. Expression of the cartilage genes, *Sox9*, *Aggrecan*, and *Col2a1*, has been reported with overuse [59]. A reduction or absence of load can also result in pathological tendon conditions. Tendon disuse leads to lower expression of collagen I, II, and III, Dcn and higher expression of MMPs along with a decline in the mechanical properties. Therefore, overloading and unloading can have similar detrimental effects on the tendon with some of the hallmarks of tendinopathy.

Although we still do not have a comprehensive understanding on how load is sensed by the tendon cells, recent studies have begun to reveal how changes in loading are regulated on the transcriptional level. Several transcription factors, known to be important for tendon development, have also been implicated in tendon mechanotransduction. *Scx* expression is very sensitive to mechanical force [47], and this is likely regulated through force activation of the TGF-β pathway [43]. Interestingly, Egr1 is a possible candidate that may act upstream to TGF-β and downstream of mechanical forces. Egr1 can bind the TGF-β promoter in adult mouse tendons and can activate TGF-β2 expression in vitro [2]. In addition, *Egr1* and *Egr2* expression has been reported to increase within 15 min of physical activity in an injured rat tendon [2]. Mkx has also been implicated in the transduction of mechanical signals by regulating the expression of TGF-β2 [60]. *Mkx* was shown to be essential for the mechanical upregulation of several tendon genes, including *type I collagen* and *Fbmd*. This effect is thought to be mediated by general transcription factor II-I repeat domain containing protein 1 (Gtf2ird1) as this transcription factor was found to translocate to the nucleus and activate the *Mkx* promoter after mechanical stimulation [61].

In addition to transcription factor activity changes after mechanical stimulation, additional cellular factors have been implicated in tendon mechanobiology. Within the adult tendon tissue, the network of cells are connected by gap junctions, which express *Connexin (Cx) 32* and *43* [62]. The addition of a gap junction uncoupling agent blocked the response of tendon cells to mechanical stimulation, suggesting that intercellular communication is important for regulating the response to mechanical loading [54]. It was later shown that the level of loading applied to tendon cells can affect the opening of the gap junction pores, further supporting an active role for gap junctions in tendon mechanotransduction [54]. In addition to gap junctions, changes in cytoskeletal tension can affect cell behaviors in many cell types. In cultured tenocyte alterations in substrate elasticity result in different cell traction forces, and this effect is mediated through nonmuscle myosin II. The chemical inhibition of myosin II function or matrix deformation altered expression of MMP1, suggesting changes in substrate stiffness lead to ECM remodeling [63]. Therefore, mechanical force can stimulate changes in the cytoskeleton that influences tendon cell gene expression resulting in modulation of the ECM. Another cellular structure with a role in mechanotranduction in other cell types is the primary cilium. In tendons, the primarily cilia are aligned along the long axis of the tendon. On changes in load, the angle and length of the cilia are altered, suggesting they may have an important role in sensing mechanical signals [64]. Currently, the function of the primary cilia in tendons is not well defined, making it an interesting area of future study.

Despite these advances, the pathways sensing mechanical signals and the transcriptional regulation of the tendon cell response are not well defined. Several additional pathways such as the Rac and Rho family of GTPases and Hippo signaling have been implicated in mechanotransduction in other tissue types but have yet to be investigated in the context of adult tendon homeostasis and response to a changing load environment. Further investigation of the pathways responding to mechanical stimulation and how they control tendon cell behaviors and matrix remodeling would be invaluable toward devising targeted therapeutic interventions for tendon injury and disease.

TENDON INJURY AND REPAIR

Tendon injury involves a disruption to the matrix, affecting the form and function of the tissue. This injury can be chronic or acute in nature. Chronic injuries can result from overuse and aging, whereas acute injuries are induced by trauma or rupture. Although some of the surgical treatments for tendon injuries can have high success rates, patients often experience chronic pain, limited mobility, and joint instability, leading to OA [65]. This is true especially for anterior cruciate ligament (ACL) tears, which, even with surgical treatment, are associated with a high incidence of OA. It also has been found that acute injuries can be preceded by a degenerative component similar to chronic tendinopathy [66]. Achilles tendon injuries can result from structural and mechanical deficiencies, which are connected to extrinsic factors such as excessive loading during physical training and age [67]. The specific trigger for tendinopathy remains unclear, but many causes such as hypoxia, apoptosis, inflammation, and MMP imbalance have been implicated. In addition to a degenerative response in chronic tendon injuries, tendons can also undergo calcific tendinopathy, in which ectopic chondrogenesis and ossification occur in overuse [4]. Despite its prevalence, the underlying cause for the calcification is unknown. As the treatments for such conditions are limited to surgery and physical therapy, the use of stem cells in tissue replacement strategies is a promising alternative. Advances in our understanding of pathways and cell populations regulating the repair process would impact the development of new treatments for tendon injury and disease.

Stages of Tendon Healing

After an acute tendon injury, there is a hallmark response to heal the damaged tissue involving three overlapping stages: inflammatory, proliferative, and remodeling.

The inflammatory phase begins immediately and persists through the first few days after injury. Neutrophils, monocytes, and macrophages home to the wound site and remove necrotic cells by phagocytosis [68]. Platelets release growth factors and proinflammatory signals, and a fibrin clot forms a temporary scaffold for the tissue to stabilize the injury [68]. Tendon cells recruited to the site begin to synthesize ECM components [69]. Angiogenic factors initiate the formation of the vascular network.

The proliferative stage is characterized by the presence of tendon fibroblasts, which deposit collagen, mainly collagen III, and other ECM components in the wound site. A blood vessel network is created to provide nutrients and increase cell metabolism, which results in proliferation and ECM deposition. Notably, there is no consensus on when the proliferative stage commences after injury. Some research shows that fibroblast cells migrate to the wound site as early as 2 days after the injury, whereas other work shows fibroblast migration and proliferation starts day 7 after injury. These differences could be explained by the type of the injury performed by the different groups and by the paucity of knowledge regarding the origin of cells that actively participate in tendon healing. Although tendon-derived stem/progenitor cells were identified and isolated in vitro [70], until recently [4], few studies have described genetically marked cell populations and their activities in vivo. Therefore, the molecular mechanisms controlling cell proliferation and migration after tendon injury are not well understood. Nevertheless, the proliferation stage begins relatively soon after injury and can last up to 20 or 30 days [4].

The remodeling phase is characterized by the synthesis of matrix components such as collagen and GAGs, the activity of matrix-degrading enzymes, and a decrease in cellularity. The remodeling phase can be divided into the consolidation stage and the maturation stage. The consolidation stage begins at about 6 weeks and can last up until 10 weeks after injury. During this stage, tenocyte metabolism remains high to secrete ECM components, particularly type I collagen fibers, which become aligned in the direction of stress [4]. During maturation, the fibrous tissue gradually changes to scarlike tendon tissue. Tenocyte metabolism decreases, as does the vascularity of the tendon. This process can take up to one year after the injury (Fig. 8.3).

Tendon Cell Populations and Their Role in Homeostasis and Repair

Understanding the mechanisms underlying tendon cell regulation during tissue homeostasis and healing are critical for designing novel repair strategies. There are two proposed mechanisms for tendon repair that are not mutually exclusive: extrinsic (cells originating from a source outside the tendon) and intrinsic (the repair comes from cells within the tendon) repair. The extrinsic mechanism involves inflammatory cells, vascular cells, and fibroblasts, which

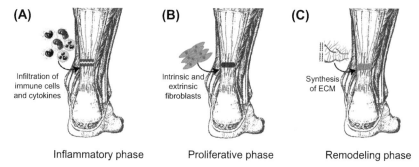

(A)

Infiltration of
immune cells
and cytokines

Inflammatory phase

(B)

Intrinsic and
extrinsic
fibroblasts

Proliferative phase

(C)

Synthesis
of ECM

Remodeling phase

FIGURE 8.3   Stages of tendon repair. Three overlapping stages of acute tendon repair have been described. (A) **The inflammatory phase** begins immediately after injury with the infiltration of blood cells into the wound. Inflammatory cells clear the injury site and release growth factors. (B) **The proliferative stage** in which the cells proliferate and deposit temporary extracellular matrix (ECM) components among them type III collagen. (C) **The remodeling phase** involves the expression of matrix remodeling enzymes and matrix components such as type I collagen. Although this results in a more aligned matrix, the tissue remains scarlike with inferior biomechanical properties.

migrate to the tendon from the surrounding tissues [71,72]. This includes cells in the connective tissue sheath surrounding the tendon such as the epitenon and paratenon. The intrinsic mechanism involves the internal fibroblast cell population from the tendon itself. Previous studies present evidence for each mechanism and suggest that tendon healing is a dynamic process involving multiple cell types.

In support of the intrinsic model, BrdU (5-bromo-2'-deoxyuridine) has been used to identify label-retaining cells in the tendon [73]. In other tissues, label-retaining cells are thought to be stem cells because of their slow-cycling activities [74]. However, the overall turnover rate of adult tendon cells is low [75], making similar studies in the tendon difficult to interpret. Ex vivo isolation and characterization of a subset of tendon cells demonstrated their multipotency and clonogenicity, suggesting that the tendon itself harbors stem/progenitor cells [76]. However, the in vivo function and location of these cells remains unclear. A recent study addressed this by analyzing the characteristics of cells derived from different regions of the tendon and found them to have distinct features [76]. Cells isolated from the peritenon, the outer tendon sheath, and the inner tendon expressed similar stem/progenitor markers such as Cd90, Sca1, and Cd44, but the inner tendon population showed higher expression of *Scx* and *Tnmd* and produced significantly more cell colonies than the peritenon population. This study suggests an enrichment of tendon stem/progenitor behavior in the inner tendon cell population compared with the peritenon cells and supports the theory that progenitor pools in different regions may contribute differently to tendon healing processes [68]. Because their analysis of the cell populations was performed after extraction from the tissue and in culture, the endogenous role of each cell population in tendon healing remains unclear. In contrast to these studies, early experiments suggested an extrinsic origin of the repair population. By using thymidine labeling, which is incorporated into the DNA of actively dividing cells, they followed the location of dividing cells in the tendon after injury. Immediately after injury, thymidine-labeled cells were detected in the paratenon prior the inner tendon region. Over time, the thymidine-labeled cells decreased in the paratenon tissue and increased in the inner tendon region. These observations suggested that cells originating from the paratenon migrated into the inner tendon during healing and supported the extrinsic healing model. Using lineage tracing and transgenic reporter mice, *αSMA+* and *Nestin+* cells in perivascular and peritenon regions were found to contribute to adult tendon healing by expressing *Scx-GFP* and synthesizing new matrix [70]. Strikingly, the origin of the cells contributing to the healing response may underlie the difference between regenerative versus reparative healing. A recent study found that *Scx*-expressing cells actively contributed to tendon regeneration in neonates, while only *αSMA+* cells were found in the adult healed tissue in an Achilles complete transection injury [75]. It is not known how these cell populations relate to the tendon-derived stem/progenitor cells that were characterized previously [77]. However, it is clear that studies aimed as deciphering the molecular mechanisms controlling the proliferation, migration, and differentiation of the cells contributing to tendon repair will be important toward impacting our understanding of tendon healing and devising targeted therapies for their improved repair.

CURRENT AND FUTURE TREATMENT STRATEGIES FOR TENDON REPAIR

Growth Factors and Developmental Regulators as Treatments for Tendon Injury

Although natural tendon healing in some ways recapitulates tendon development [78], the healing process is imperfect and once injured, the tendon properties are never fully restored. Indeed, a deeper understanding of the molecular mechanisms underlying tendon development could advance strategies to improve tendon healing. In line with this strategy, major developmental pathways such as growth differentiation factor (GDF), FGF, and TGF-β, have been targeted as treatments for tendon injuries; however, there have been conflicting reports regarding their effects on the healing process. FGF and TGF-β are expressed during tendon healing and may play an active role in this process. FGF treatment during rotator cuff healing has a positive effect on tendon gene expression and mechanical properties [79]. In chick, FGF expression decreases during the early phase of tendon healing but virally mediated delivery of FGF enhances *Scx* expression and improves the biomechanical properties of the healed tendon [80]. FGFs also cause increased cell proliferation, angiogenesis, and expression of collagen III in rat patellar tendons, yet in other studies, FGF treatment fails to improve mechanical or functional properties of the healed tendons [81]. In addition to an increase in TGF-β, there is also increased expression of its receptors, TGF-βR1 and TGF-βR2, in the injured tendon and sheath [82]. Culturing tendon cells with TGF-β protein increases the production of type I collagen in vitro, indicating that TGF-β may have an important role in reestablishing the proper tendon ECM properties after injury [83]. However, there have been conflicting results regarding TGF-β in improving healing outcomes in vivo [77]. Enforced expression of TGF-β using adenovirus-modified muscle grafts results in accelerated tendon healing with decreased deposition of type III, and increased deposition of type I collagen [82]. In contrast, other studies show that therapeutic TGF-β can increase scar formation [84]. Other members of the TGF-β superfamily also have been shown to have positive effects on tendon healing. Loss of GDF5, which is involved in musculoskeletal development and the formation of joints [85], slows down healing of the Achilles tendon, and its overexpression improves tendon repair by limiting flexor tendon adhesion formation [77]. Adenoviral transfer of GDF7 (BMP12) to injured rat Achilles tendons, results in an increase in the size of the callus, improved remodeling with larger collagen fibers, and improved biomechanical properties compared with the control group, suggesting a positive effect on tendon healing [2].

Additional growth factors such as platelet-derived growth factor (PDGF), connective tissue growth factor (CTGF), vascular endothelial growth factor (VEGF), and insulin-like growth factor (IGF) are among several that are upregulated after tendon injury [86]. Although ectopic application of VEGF and PDGF increase cell proliferation, it does not increase *Tnmd* expression and fails to improve tendon healing. Therefore, it is unclear if the use of these factors increases tendon cell proliferation or selectively targets the cells that participate in angiogenesis [2]. With many of the studies examining growth factors, some results have been promising, but others show negligent effects on tendon healing [43]. This is likely because of the timing, concentration, injury model, and tendon being used. Future studies using temporal and tissue-specific ablation models would provide a powerful approach to identifying their role in the repair process.

In addition to growth factors, transcription factors and ECM components are also upregulated during the healing process. These include *Col1a1*, *Col1a2*, *Col3a1*, *Col12a1*, and *Col14a1*, other matrix associated genes such as *Tnmd* and *Tenascin C*, and the tendon transcription factors, *Scx*, *Egr1*, and *Mkx*. Similar changes in gene expression also occur in human tendinopathy. Some studies have shown a positive effect on healing with misexpression of *Egr1*, *Mkx*, and *Scx* [87]. In addition, several of these factors are able to promote tenogenic differentiation in adult and embryonic stem (ES) cell systems. However, the delivery of transcription factors to the tendon after injury may be challenging in a clinical setting.

Other Factors in Tendon Healing

Platelet-rich plasma (PRP): The ectopic delivery of single factors has helped us to better define their roles in tendon healing. However, the strategy for testing other approaches such as PRP lies in the fact that alpha granules of platelets are enriched in multiple growth factors, including PDGF, TGF-β, IGF-1, and VEGF. However, studies in animal models and clinical trials in humans have not shown that PRP treatment positively affects tendon repair after injury [88]. The negative results could be explained by variability in the PRP itself or the overactivation of multiple intracellular signaling pathways that will have opposite effects on tendon healing. In contrast, when PRP was combined with an ECM carrier, ligaments showed improved healing in vivo. There was increased proliferation and spreading of the tendon cells on the scaffold in vitro. It is thought that the addition of the scaffold with PRP facilitates and targets platelet activation to the wound site. Such an approach led to bridge-enhanced ACL repair in an FDA-approved clinical trial [47].

Mechanical Force in Tendon Repair

As mechanical forces are important for the maturation and differentiation of tendons during development, movement is also crucial for maintaining healthy adult tendon tissues. A loss of force from skeletal muscles in animal models leads to a reduction in tendon size, lower expression of collagen I, II III, *Scx*, *Dcn*, and *Aggrecan*, and higher expression of MMPs, resulting in a degenerative state [2,53,59]. In contrast, increased mechanical loading through treadmill running in vivo or in 2D and 3D human-engineered tendons in vitro increases ECM production and the expression of several tendon markers such as *Scx* and *Col1a1* [50]. However, too much mechanical stimulation can lead to overuse conditions, resulting in hypercellularity, decreased collagen alignment and increased inflammation, vasculature, and expression of *Col2a1*, *Sox9*, and *Aggrecan* [7]. Therefore, the right amount of mechanical loading can have beneficial effects on tendon tissue properties, whereas too much can have a negative effect. For example, studies in rats show that mechanical stimulation can improve tendon healing. With greater loading conditions, expression of IGF-1 increased, which has been found to be important for cell proliferation and remodeling. However, the positive effect of mechanical loading varies in the type of tendons and in the type of tendon injury. In the rotator cuff, immobilization improves healing by preventing reinjury after repair, reducing expression of type III collagen, and increasing expression of *Aggrecan* and type II collagen at the tendon—bone attachment site [89]. Despite many studies examining the effect of load on the tendon, there is not a full understanding of how load is sensed molecularly. Advances in this area could lead to the discovery of molecular targets that could be used as a marker of overload or injured conditions, and more importantly, new pathways to be manipulated for therapeutic purposes.

Mesenchymal Stem Cells and Tendon-Derived Stem/Progenitor Cells

Mesenchymal stem cells (MSCs) are multipotent stem cells that can differentiate into cartilage, bone, and adipocytes. Historically, these cells were identified from the bone marrow as multipotent cell types capable of being clonally expanded in culture and forming hematopoiesis-supporting stroma. These bone marrow–derived MSCs have more recently been characterized and functionally defined in vivo as skeletal stem cells that contribute to the cell types in the bone (stromal cells, adipocytes, osteoblasts) as well as chondrocytes during fracture repair [90]. Others studies broadened the definition of MSCs to include stromal cells from other tissues, including muscle, tendon, and fat [90]. Recent work has pointed toward MSCs having a perivascular origin, with distinct differentiation potentials [91]. Although these cells have shown great potency in vitro for expansion and multilineage differentiation in certain contexts [92], their endogenous roles in tissue repair remain unclear. However, because of relative ease of isolation and procuring such cells, they have gained great attention as sources for potential therapies. Indeed, tendon cells can be induced from bone marrow–derived MSCs by the addition of growth factors, transcription factors, and/or mechanical signals. There remains conflicting evidence for them improving tendon repair with some studies showing positive, neutral, and negative effects of MSCs [75].

Other approaches have used tendon-derived stem/progenitor cells (also referred to as tendon stem/progenitor cells), which have been shown to have multilineage differentiation potential and clonogenicity in culture [93]. Examination of these cells in culture has expanded our molecular understanding of tendon cell biology. Mechanical stimulation of human tendon-derived stem/progenitor cells resulted in the upregulation of several matrix components in addition to an effect on p38 and ERK kinase activity. In addition, tendon-derived stem/progenitor cells from aged patients display early senescence, upregulation of $p16^{INK4A}$, and harbor changes in gene expression. Together this suggests aberrant cell–matrix interactions along with premature senescence underlie tendon aging and degeneration. Tendon-derived stem/progenitor cells have been injected and implanted in different injury models with promising effects, yet many questions remain about the underlying mechanism. Future lineage tracing studies demonstrating the extent of integration and contribution to repair will be important in defining the role of tendon-derived stem cells in tendon healing.

The Use of Induced Pluripotent Stem/Embryonic Stem Cells to Treat Tendon Injuries

Regenerative strategies using pluripotent stem cells are emerging as promising therapies for injured tendons. Not only could pluripotent stem cells provide an unlimited source of cells for tissue replacement strategies, but they would also permit high-throughput screening assays for functionally important pathways in human biology. To date, the stem cells that have primarily been tested in tendon injury studies are adult stem cells, mesenchymal stem cells, which are derived from bone marrow or fat, and tendon-derived stem/progenitor cells. As adult cells, these stem cell types are thought to be more restricted in their differentiation potential after transplantation, and there can be variability in their potential, depending on the age and condition of the individual. Human pluripotent stem cells include ES cells, which are derived from the inner cell mass of a blastocyst-stage embryo, or induced pluripotent stem (iPS) cells, which are fibroblasts that have been transduced with a cocktail of pluripotency factors. These cells have unlimited proliferation capacity and can form any cell type in the body. Because of these characteristics, iPS/ES cells have become popular avenues for tissue engineering applications. However, to use human ES/iPS cells to generate functional tendon tissues for clinical therapies, there are many challenges that must be overcome. Foremost among them is that the transplanted cells must be completely differentiated into mature cells capable of generating a proper tendon tissue. This is essential as undifferentiated iPS/ES cells can form teratomas after transplantation, which would preclude their use for human treatments. Despite this possibility, the use of human iPS/ES cells holds great promise for tendon injury applications. Robust tendon differentiation methods would have great impact for high-throughput drug screening for therapeutic compounds or for tissue replacement strategies.

The first studies to use of human pluripotent stem cells in tendon repair differentiated them into mesenchymal stem cells (hESC-MSCs). When combined with silk scaffolds, they improved tendon-specific ECM production [93]. Following these studies, the group generated human ES cell–derived and iPS cell–derived tendon cells from mesenchymal cell intermediates, which were virally targeted to express *Scx*, and subjected these cells to mechanical stimulation. Together, these factors acted synergistically to drive differentiation to the tendon lineage [93]. Interestingly, the same group found that forced expression of *Scx* in the hESC-MSCs enhanced teno-lineage differentiation and positively affected the healing process, resulting in improved mechanical abilities [94]. Consistent with this idea, studies using mesenchymal stem cell lines overexpressing *Scx* also observed improved healing in Achilles and rotator cuff tendons [95]. These studies suggest that the use of *Scx*-expressing cells is important for improving outcomes of tendon repair. It is interesting to note the parallels between these findings and a recent study that demonstrated that the recruitment of *Scx*-lineage cells to the healing process was a key difference underlying regenerative versus nonregenerative tendon healing [93].

The previous studies were major advances in that they were the first to demonstrate that human iPS and ES cells could form tendon tissue [96]. To use the cells for high-throughput drug screening or human clinical applications, further studies are necessary to develop robust and defined differentiation protocols that avoid the use of serum and viral transduction. More recent studies have used developmental signaling mechanisms to guide the formation of skeletal progenitors. An early protocol formed chondrogenic cells from ES cells [97], but it was Craft and colleagues [98] who developed a robust protocol for the differentiation of human pluripotent stem cells into chondrocytes that formed stable cartilage tissue in vivo. Using the activation and inhibition of known developmental signaling pathways, they were able to form primitive streak, mesoderm, and finally paraxial mesoderm from which they made cartilage tissue. Interestingly, activation of the TGF-β pathway resulted in articular chondrocytes that produced stable cartilage tissue in vivo, whereas activation of BMP-4 generated hypertrophic chondrocytes that enabled endochondral ossification, mirroring the developmental processes. Another study generated multiple mesodermal lineages from human pluripotent stem cells, including heart, muscle, and bone [99]. As tendon lineages are known to derive from similar developmental origins to cartilage and bone [100], these protocols will be instrumental in the development of one for tendon and ligament cell types (Fig. 8.4).

The use of iPS/ES cells as an unlimited source of new tendon tissue could hold great promise for tendon repair. Certainly, their use in drug-screening assays would be a powerful method to identify potential therapeutic molecules that could be delivered systemically or directly to the injury site. However, a few challenges must be resolved before implementation as clinical therapies. The use of iPS/ES cells to generate replacement tissues, although promising, may face regulatory hurdles in terms of demonstrating safety from generating tumor and immune responses. These limitations are expected to be overcome, as researchers using these cells for other organ systems face similar challenges and are working toward solutions, such as generating cells with immune privilege and a "self-destruct genetic switch." Overall, the development of iPS/ES cell−derived tendons has important benefits not only for regenerative medicine−based strategies for treating tendon injury and disease but also as a means to understand human tendon biology.

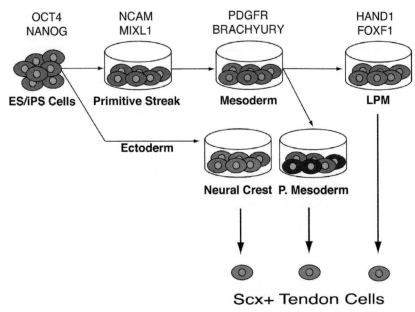

Scx+ Tendon Cells

FIGURE 8.4   Generation of tendon cells from pluripotent stem cells. The robust and efficient generation of tendon cells from embryonic stem/induced pluripotent stem (ES/iPS) cells will require the recapitulation of embryonic stages through the activation and inhibition of specific developmental signaling pathways. The creation of a primitive streak, expressing NCAM, MIXL1, will be followed by its conversion to mesoderm expressing PDGFR and BRACHYURY. This will form the paraxial mesoderm for the axial tendons and the lateral plate mesoderm (LPM) for the generation of limb tendons. The cranial neural crest will be generated from the ectoderm germ layer to form the cranial tendon cells.

CONCLUSION

Tendons are an integral component of our musculoskeletal system with a unique structure, function, and mechanics. Injuries to this vital tissue can result in significant pain and disability. The use of iPS/ES cells has great promise toward the development of better treatment strategies. Whether they are used as drug-screening tools or as a cellular source for tissue engineering approaches, the challenge in effectively working with these cells stems from a limited understanding of the cues governing tendon development, homeostasis, and the repair process. Advances in these areas will provide diagnostic benchmarks for engineering a correctly differentiated tendon. In addition, these pathways could be targeted, enabling the design of more efficient drug-based treatments for tendon injuries. Ultimately, success will require an approach combining expertise from multiple fields to overcome the many challenges that may arise. Together, this interdisciplinary approach holds great promise for the future of tendon injury treatments for patients suffering from imperfect tendon healing.

REFERENCES

[1] Kaux J-F, Forthomme B, Goff CL, Crielaard J-M, Croisier J-L. Current opinions on ten-dinopathy. J Sports Sci Med 2011;10(2):238−53.

[2] Nourissat G, Berenbaum F, Duprez D. Tendon injury: from biology to tendon repair. Nat Rev Rheumatol 2015;11(4):223−33.

[3] Voleti PB, Buckley MR, Soslowsky LJ. Tendon healing: repair and regeneratio. Annu Rev Biomed Eng 2012;14:47−71.

[4] Sharma P, Maffulli N. Biology of tendon injury: healing, modeling and remodeling. J Musculoskelet Neuronal Interact 2006;6(2):181−90.

[5] Benjamin MRJ. The cell and developmental tendons and ligaments. Int Rev Cytol 2000;196:85−130.

[6] Dublet B, van der Rest M. Type XII collagen is expressed in embryonic chick tendons. Isolation of pepsin-derived fragments. J Biol Chem 1987;262(36):17724−7.

[7] Thomopoulos S, Hattersley G, Rosen V, Mertens M, Galatz L, Williams G, et al. The localized expression of extracellular matrix components in healing tendon insertion sites: an in situ hybridization study. J Orthop Res 2002;20(3):454−63.

[8] Zhang G, Young B, Ezura Y, Favata M, Soslowsky L, Chakravarti S, et al. Development of tendon structure and function: regulation of collagen fibrillogenesis. J Musculoskelet Neuronal Interact 2005;5(1):5−21.

[9] Birk D, Mayne R. Localization of collagen types I, III and V during tendon development. Changes in collagen types I and III are correlated with changes in fibril diameter. Euro J Cell Biol 1997;72(4):352−61.

[10] Birk D. Type V collagen: heterotypic type I/V collagen interactions in the regulation of fibril assembly. Micron 2001;32(3):223−37.

[11] Wenstrup RJ, Smith SM, Florer JB, Zhang G, Beason DP, Seegmiller RE, et al. Regulation of collagen fibril nucleation and initial fibril assembly involves coordinate interactions with collagens V and XI in developing tendon. J Biol Chem 2011;286(23):20455−65.

[12] Ritty TM, Roth R, Heuser JE. Tendon cell array isolation reveals a previously unknown fibrillin-2-containing macromolecular assembly. Structure 2003;11(9):1179−88.

[13] Pan T-C, Zhang R-Z, Markova D, Arita M, Zhang Y, Bogdanovich S, et al. COL6A3 protein deficiency in mice leads to muscle and tendon defects similar to human collagen VI congenital muscular dystrophy. J Biol Chem 2013;288(20):14320−31.

[14] Font B, Eichenberger D, Rosenberg L, Van der Rest M. Characterization of the interactions of type XII collagen with two small proteoglycans from fetal bovine tendon, decorin and fibromodulin. Matrix Biol 1996;15(5):341−8.

[15] Ansorge HL, Beredjiklian PK, Soslowsky LJ. CD44 deficiency improves healing tendon mechanics and increases matrix and cytokine expression in a mouse patellar tendon injury model. J Orthop Res 2009;27(10):1386−91.

[16] Ezura Y, Chakravarti S, Oldberg Å, Chervoneva I, Birk DE. Differential expression of lumican and fibromodulin regulate collagen fibrillogenesis in developing mouse tendons. J Cell Biol 2000;151(4):779−88.

[17] Ameye L, Aria D, Jepsen K, Oldberg A, Xu T, Young MF. Abnormal collagen fibrils in tendons of biglycan/fibromodulin-deficient mice lead to gait impairment, ectopic ossifi-cation, and osteoarthritis. Faseb J 2002;16(7):673−80.

[18] Manske P, Gelberman R, Lesker P. Flexor tendon healing. Hand Clinics 1985;1(1):25−34.

[19] Schweitzer R, Chyung JH, Murtaugh LC, Brent AE, Rosen V, Olson EN, et al. Analysis of the tendon cell fate using Scleraxis, a specific marker for tendons and ligaments. Development 2001;128(19):3855−66.

[20] Brent AE, Schweitzer R, Tabin CJ. A somitic compartment of tendon progenitors. Cell 2003;113(2):235−48.

[21] Brent AE, Tabin CJ. FGF acts directly on the somitic tendon progenitors through the Ets transcription factors Pea3 and Erm to regulate scleraxis expression. Development 2004;131(16):3885−96.

[22] Edom-Vovard F, Schuler B, Bonnin M-A, Teillet M-A, Duprez D. Fgf4 positively regulates scleraxis and tenascin expression in chick limb tendons. Dev Biol 2002;247(2):351−66.

[23] Brent AE, Braun T, Tabin CJ. Genetic analysis of interactions between the somitic muscle, cartilage and tendon cell lineages during mouse development. Development 2005;132(3):515−28.

[24] Huang AH, Lu HH, Schweitzer R. Molecular regulation of tendon cell fate during development. J Orthop Res 2015;33(6):800−12.

[25] Havis E, Bonnin M-A, de Lima JE, Charvet B, Milet C, Duprez D. TGFβ and FGF promote tendon progenitor fate and act downstream of muscle contraction to regulate tendon differentiation during chick limb development. Development 2016;143(20):3839−51.

[26] Pryce BA, Watson SS, Murchison ND, Staverosky JA, Dünker N, Schweitzer R. Recruitment and maintenance of tendon progenitors by TGFβ signaling are essential for tendon formation. Development 2009;136(8):1351−61.

[27] Chen JW, Galloway JL. The development of zebrafish tendon and ligament progenitors. Development 2014;141(10):2035−45.

[28] Havis E, Bonnin M-A, Olivera-Martinez I, Nazaret N, Ruggiu M, Weibel J, et al. Transcriptomic analysis of mouse limb tendon cells during development. Development 2014;141(19):3683−96.

[29] Shukunami CTA, Oro M, Hiraki Y. Scleraxis positively regulates the expression of tenomodulin, a differentiation marker of tenocytes. Dev Biol 2006;298:234−47.

[30] Liu H, Zhang C, Zhu S, Lu P, Zhu T, Gong X, Zhang ZHJ, Yin Z, Heng BC. Mohawk promotes the tenogenesis of mesenchymal stem cells through activation of the TGF. Stem Cell 2015;33:443−55.

[31] Canty EG, Lu Y, Meadows RS, Shaw MK, Holmes DF, Kadler KE. Coalignment of plasma membrane channels and protrusions (fibripositors) specifies the parallelism of tendon. J Cell Biol 2004;165(4):553−63.

[32] Kalson NS, Lu Y, Taylor SH, Starborg T, Holmes DF, Kadler KE. A structure-based extracellular matrix expansion mechanism of fibrous tissue growth. Elife 2015;4:e05958.

[33] Dex S, Lin D, Shukunami C, Docheva D. Tenogenic modulating insider factor: ystematic assessment on the functions of tenomodulin gene. Gene 2016;587:1−17.

[34] Murchison ND, Price BA, Conner DA, Keene DR, Olson EN, Tabin CJ, et al. Regulation of tendon differentiation by scleraxis distinguishes force-transmitting tendons from muscle-anchoring tendons. Development 2007;134(14):2697−708.

[35] Chen S, Birk DE. The regulatory roles of small leucine-rich proteoglycans in extracellular matrix assembly. FEBS J 2013;280(10):2120−37.

[36] Graham HK, Holmes DF, Watson RB, Kadler KE. Identification of collagen fibril fusion during vertebrate tendon morphogenesis. The process relies on unipolar fibrils and is regulated by collagen-proteoglycan interaction. J Mole Biol 2000;295(4):891−902.

[37] Marturano JE, Arena JD, Schiller ZA, Georgakoudi I, Kuo CK. Characterization of mechanical and biochemical properties of developing embryonic tendon. Proc Natl Acad Sci USA 2013;110(16):6370−5.

[38] Espira L, Lamoureux L, Jones S, Dixon I, Czubryt M, editors. Scleraxis interaction with DNA and E-box proteins regulate collagen gene expression in cardiac myofibroblasts. Canadian Journal of Cardiology. 2902 S Sheridan Way, Oakville, Ontario L6J 7L6, Canada: Pulsus Group Inc; 2007.

[39] Ito Y, Toriuchi N, Yoshitaka T, Ueno-Kudoh H, Sato T, Yokoyama S, et al. The Mohawk homeobox gene is a critical regulator of tendon differentiation. Proc Natl Acad Sci USA 2010;107(23):10538−42.

[40] Liu W, Watson SS, Lan Y, Keene DR, Ovitt CE, Liu H, et al. The atypical homeodomain transcription factor Mohawk controls tendon morphogenesis. Mole Cell Biol 2010;30(20):4797−807.

[41] Anderson DM, Arredondo J, Hahn K, Valente G, Martin JF, Wilson-Rawls J, et al. Mohawk is a novel homeobox gene expressed in the developing mouse embryo. Dev Dynam 2006;235(3):792−801.

[42] Chuang HN, Hsiao KM, Chang HY, Wu CC, Pan H. The homeobox transcription factor Irxl1 negatively regulates MyoD expression and myoblast differentiation. FEBS J 2014;281(13):2990−3003.

[43] Guerquin M-J, Charvet B, Nourissat G, Havis E, Ronsin O, Bonnin M-A, et al. Transcription factor EGR1 directs tendon differentiation and promotes tendon repair. J Clinic Invest 2013;123(8):3564.

[44] Arnoczky SP, Tian T, Lavagnino M, Gardner K. Ex vivo static tensile loading inhibits MMP-1 expression in rat tail tendon cells through a cytoskeletally based mechanotransduction mechanism. J Orthop Res 2004;22(2):328−33.

[45] Kook S-H, Jang Y-S, Lee J-C. Involvement of JNK-AP-1 and ERK-NF-κB signaling in tension-stimulated expression of type I collagen and MMP-1 in human periodontal ligament fibroblasts. J Appl Physiol 2011;111(6):1575−83.

[46] Schiele NR, Marturano JE, Kuo CK. Mechanical factors in embryonic tendon development: potential cues for stem cell tenogenesis. Curr Opin Biotechnol 2013;24(5):834−40.

[47] Maeda T, Sakabe T, Sunaga A, Sakai K, Rivera AL, Keene DR, et al. Conversion of mechanical force into TGF-β-mediated biochemical signals. Curr Biol 2011;21 (11):933−41.

[48] Thomopoulos S, Kim HM, Rothermich SY, Biederstadt C, Das R, Galatz LM. Decreased muscle loading delays maturation of the tendon enthesis during postnatal development. J Orthop Res 2007;25(9):1154−63.

[49] Schwartz AG, Pasteris JD, Genin GM, Daulton TL, Thomopoulos S. Mineral distributions at the developing tendon enthesis. PLoS One 2012;7(11):e48630.

[50] Archambault J, Hart D, Herzog W. Response of rabbit Achilles tendon to chronic repetitive loading. Connect Tissue Res 2001;42(1):13−23.

[51] Lavagnino M, Arnoczky SP, Kepich E, Caballero O, Haut RC. A finite element model predicts the mechanotransduction response of tendon cells to cyclic tensile loading. Biomechanics Model Mechanobiol 2008;7(5):405−16.

[52] Langberg H, Skovgaard D, Karamouzis M, Bülow J, Kjær M. Metabolism and inflammatory mediators in the peritendinous space measured by microdialysis during intermittent isometric exercise in humans. J Physiol 1999;515(3):919−27.

[53] Mendias CL, Gumucio JP, Bakhurin KI, Lynch EB, Brooks SV. Physiological loading of tendons induces scleraxis expression in epitenon fibroblasts. J Orthop Res 2012;30 (4):606–12.

[54] Maeda E, Sugimoto M, Ohashi T. Cytoskeletal tension modulates MMP-1 gene expression from tenocytes on micropillar substrates. J Biomech 2013;46(5):991–7.

[55] Zhang J, Wang JH. The effects of mechanical loading on tendons-an in vivo and in vitro model study. PLoS One 2013;8(8):e71740.

[56] Dex S, Alberton P, Willkomm L, Sollradl T, Bago S, Milz S, et al. Tenomodulin is required for tendon endurance running and collagen I fibril adaptation to mechanical load. EBio-Medicine 2017;20(Suppl. C):240–54.

[57] Farhat YM, Al-Maliki AA, Easa A, O'Keefe RJ, Schwarz EM, Awad HA. TGF-β1 suppresses plasmin and MMP activity in flexor tendon cells via PAI-1: implications for scarless flexor tendon repair. J Cell Physiol 2015;230(2):318–26.

[58] Subramanian A, Schilling TF. Tendon development and musculoskeletal assembly: emerging roles for the extracellular matrix. Development 2015;142(24):4191–204.

[59] Scott A, Danielson P, Abraham T, Fong G, Sampaio A, Underhill T. Mechanical force modulates scleraxis expression in bioartificial tendons. J Musculoskelet Neuronal Interact 2011;11(2):124–32.

[60] Kayama T, Mori M, Ito Y, Matsushima T, Nakamichi R, Suzuki H, et al. Gtf2ird1-Dependent Mohawk expression regulates mechanosensing properties of the tendon. Mole Cell Biol 2016;36(8):1297–309.

[61] McNeilly C, Banes A, Benjamin M, Ralphs J. Tendon cells in vivo form a three dimensional network of cell processes linked by gap junctions. J Anat 1996;189(Pt 3):593.

[62] Banes AJ, Weinhold P, Yang X, Tsuzaki M, Bynum D, Bottlang M, et al. Gap junctions regulate responses of tendon cells ex vivo to mechanical loading. Clin Orthop Relat Res 1999;367:S356–70.

[63] Gardner K, Arnoczky SP, Lavagnino M. Effect of in vitro stress-deprivation and cyclic loading on the length of tendon cell cilia in situ. J Orthop Res 2011;29(4):582–7.

[64] Alfredson H, Lorentzon R. Chronic tendon pain: no signs of chemical inflammation but high concentrations of the neurotransmitter glutamate. Implications for treatment? Curr Drug Targets 2002;3(1):43–54.

[65] Rees J, Wilson A, Wolman R. Current concepts in the management of tendon disorders. Rheumatology 2006;45(5):508–21.

[66] Selvanetti A, Cipolla M, Puddu G. Overuse tendon injuries: basic science and classification. Operat Tech Sports Med 1997;5(3):110–7.

[67] Fenwick SA, Hazleman BL, Riley GP. The vasculature and its role in the damaged and healing tendon. Arthritis Res Ther 2002;4(4):252.

[68] Lindsay W, Birch J. The fibroblast in flexor tendon healing. Plast Reconstr Surg 1964;34(3):223–32.

[69] Gelberman RH, Vandeberg JS, Manske PR, Akeson WH. The early stages of flexor tendon healing: a morphologic study of the first fourteen days. J Hand Surg 1985;10(6):776–84.

[70] Dyment NA, Hagiwara Y, Matthews BG, Li Y, Kalajzic I, Rowe DW. Lineage tracing of resident tendon progenitor cells during growth and natural healing. PLoS One 2014;9(4):e96113.

[71] Runesson E, Ackermann P, Brisby H, Karlsson J, Eriksson BI. Detection of slow-cycling and stem/progenitor cells in different regions of rat Achilles tendon: response to treadmill exercise. Knee Surg Sports Traumatol Arthrosc 2013;21(7):1694–703.

[72]   Tan Q, Lui PP, Lee YW. In vivo identity of tendon stem cells and the roles of stem cells in tendon healing. Stem Cells Dev 2013.

[73]   Fuchs E, Horsley V. Ferreting out stem cells from their niches. Nat Cell Biol 2011;13(5):513–8.

[74]   Ruchti C, Haller D, Nuber M, Cottier H. Regional differences in renewal rates of fibroblasts in young adult female mice. Cell Tissue Res 1983;232(3):625–36.

[75]   Bi Y, Ehirchiou D, Kilts TM, Inkson CA, Embree MC, Sonoyama W, et al. Identification of tendon stem/progenitor cells and the role of the extracellular matrix in their niche. Nat Med 2007;13(10):1219.

[76]   Mienaltowski MJ, Adams SM, Birk DE. Tendon proper-and peritenon-derived progenitor cells have unique tenogenic properties. Stem Cell Res Ther 2014;5(4):86.

[77]   Gaut L, Duprez D. Tendon development and diseases. Dev Biol 2016;5:5–23.

[78]   Tokunaga T, Ide J, Arimura H, Nakamura T, Uehara Y, Sakamoto H, et al. Local application of gelatin hydrogel sheets impregnated with platelet-derived growth factor BB promotes tendon-to-bone healing after rotator cuff repair in rats. Arthrosc J Arthrosc Relat Surg 2015;31(8):1482–91.

[79]   Tang JB, Cao Y, Zhu B, Xin K-Q, Wang XT, Liu PY. Adeno-associated virus-2-mediated bFGF gene transfer to digital flexor tendons significantly increases healing strength: an in vivo study. JBJS 2008;90(5):1078–89.

[80]   Thomopoulos S, Kim HM, Das R, Silva MJ, Sakiyama-Elbert S, Amiel D, et al. The effects of exogenous basic fibroblast growth factor on intrasynovial flexor tendon healing in a canine model. J Bone Jt Surg Am Vol 2010;92(13):2285.

[81]   Ngo M, Pham H, Longaker MT, Chang J. Differential expression of transforming growth factor-beta receptors in a rabbit zone II flexor tendon wound healing model. Plast Reconstr Surg 2001;108(5):1260–7.

[82]   Klein MB, Yalamanchi N, Pham H, Longaker MT, Chan J. Flexor tendon healing in vitro: effects of TGF-β on tendon cell collagen production. J Hand Surg 2002;27(4):615–20.

[83]   Manning CN, Kim HM, Sakiyama-Elbert S, Galatz LM, Havlioglu N, Thomopoulos S. Sustained delivery of transforming growth factor beta three enhances tendon-to-bone healing in a rat model. J Orthop Res 2011;29(7):1099–105.

[84]   Francis-West P, Abdelfattah A, Chen P, Allen C, Parish J, Ladher R, et al. Mechanisms of GDF-5 action during skeletal development. Development 1999;126(6):1305–15.

[85]   Hasslund S, Dadali T, Ulrich-Vinther M, Søballe K, Schwarz EM, Awad HA. Freeze-dried allograft-mediated gene or protein delivery of growth and differentiation factor 5 reduces reconstructed murine flexor tendon adhesions. J Tissue Eng 2014;5. 2041731414528736.

[86]   Halper J, Kjaer M. Basic components of connective tissues and extracellular matrix: elastin, fibrillin, fibulins, fibrinogen, fibronectin, laminin, tenascins and thrombospondins. In: Progress in heritable soft connective tissue diseases. Springer; 2014. p. 31–47.

[87]   Rodeo SA, Delos D, Williams RJ, Adler RS, Pearle A, Warren RF. The effect of platelet-rich fibrin matrix on rotator cuff tendon healing: a prospective, randomized clinical study. Am J Sports Med 2012;40(6):1234–41.

[88]   Perrone GS, Proffen BL, Kiapour AM, Sieker JT, Fleming BC, Murray MM. Bench-to-Bedside: bridge-enhanced anterior cruciate ligament repair. J Orthop Res 2017.

[89]   Bianco P, Robey PG. Skeletal stem cells. Development 2015;142(6):1023–7.

[90]   Caplan AI, Correa D. The MSC: an injury drugstore. Cell Stem Cell 2011;9(1):11–5.

[91]   Dominici M, Le Blanc K, Mueller I, Slaper-Cortenbach I, Marini F, Krause D, et al. Minimal criteria for defining multipotent mesenchymal stromal cells. The International Society for Cellular Therapy position statement. Cytotherapy 2006;8(4):315–7.

[92] Selek O, Buluc L, Muezzinoğlu B, Ergün RE, Ayhan S, Karaöz E. Mesenchymal stem cell application improves tendon healing via anti-apoptotic effect (Animal study). Acta Orthop Traumatol Turc 2014;48(2):187−95.

[93] Chen J-P, Chen S-H, Lai G-J. Preparation and characterization of biomimetic silk fibroin/chitosan composite nanofibers by electrospinning for osteoblasts culture. Nanoscale Res Lett 2012;7(1):170.

[94] Hsieh C-F, Alberton P, Loffredo-Verde E, Volkmer E, Pietschmann M, Müller P, et al. Scaffold-free Scleraxis-programmed tendon progenitors aid in significantly enhanced repair of full-size Achilles tendon rupture. Nanomedicine 2016;11(9):1153−67.

[95] Howell K, Chien C, Bell R, Laudier D, Tufa SF, Keene DR, et al. Novel model of tendon regeneration reveals distinct cell mechanisms underlying regenerative and fibrotic tendon healing. Sci Rep 2017;7:45238.

[96] Oldershaw RA, Baxter MA, Lowe ET, Bates N, Grady LM, Soncin F, et al. Directed differentiation of human embryonic stem cells toward chondrocytes. Nat Biotechnol 2010;28(11):1187−94.

[97] Craft AM, Rockel JS, Nartiss Y, Kandel RA, Alman BA, Keller GM. Generation of articular chondrocytes from human pluripotent stem cells. Nat Biotechnol 2015;33(6):638−45.

[98] Loh KM, Ang LT, Zhang J, Kumar V, Ang J, Auyeong JQ, et al. Efficient endoderm induction from human pluripotent stem cells by logically directing signals controlling lineage bifurcations. Cell Stem Cell 2014;14(2):237−52.

[99] Soeda T, Deng JM, de Crombrugghe B, Behringer RR, Nakamura T, Akiyama H. Sox9-expressing precursors are the cellular origin of the cruciate ligament of the knee joint and the limb tendons. Genesis 2010;48(11):635−44.

[100] Zhong B, Trobridge GD, Zhang X, Watts KL, Ramakrishnan A, Wohlfahrt M, et al. Efficient generation of nonhuman primate induced pluripotent stem cells. Stem Cell Dev 2010;20(5):795−807.

Chapter 9

Biomimetic Tissue Engineering for Musculoskeletal Tissues

Nailah M. Seale[a,1], Yuze Zeng[a,2], Shyni Varghese[1,2]
[1]University of California, San Diego, La Jolla, CA, United States; [2]Duke University, Durham, NC, United States

INTRODUCTION

Tissue engineering is an interdisciplinary approach that aims to create functional tissue replacements for clinical applications. Choice of biomaterials, growth factors and other signaling molecules, cells, and culture conditions plays a pivotal role in determining the structure and function of engineered tissues [1,2]. Although tissue-specific cells are an ideal choice, their limited availability and self-renewal ability make them difficult to produce the billions of viable cells necessary to engineer functional tissues. Hence, adult and pluripotent stem cells have been considered as alternative cell sources. Although adult stem cells have their advantages, pluripotent stem cells (which include both embryonic stem cells [ESCs] and induced pluripotent stem cells [iPSCs]) can be maintained and expanded ex vivo as undifferentiated cells indefinitely [3,4]. Furthermore, the ability of pluripotent cells such as hiPSCs to be differentiated into all cell types found in the human body holds great potential in studying human development and disease progression, producing more pathophysiologically relevant tissue surrogates and disease models, advancing technological platforms for drug screening and discovery, and engineering biomimetic tissues for clinical transplantation [5,6].

Studies over the years have led to the development of culture conditions and protocols to drive lineage-specific differentiation of stem cells and tissue formation. Initial tissue engineering approaches used biomaterials as a scaffold to provide the three-dimensional structural support while using growth factors to regulate cellular behaviors. However, emerging evidence over the last few decades have shown that the insoluble extracellular matrix (ECM) found in

---

a. Equal contribution.

Developmental Biology and Musculoskeletal Tissue Engineering. https://doi.org/10.1016/B978-0-12-811467-4.00009-7
207

native tissue plays an equally important role in controlling cell fates and function [7−10]. The ECM is a dynamic ensemble of proteins and proteoglycans with multiple functions ranging from providing anchoring sites to the cells to regulating growth factor signaling [2] (Fig. 9.1).

The physicochemical properties of the ECM play an active role in maintaining tissue homeostasis and function and, when perturbed, contribute to disease progression [11]. These understandings have resulted in an evolving area of active research discipline named "biomimetic tissue engineering." Biomimetic tissue engineering is an approach that recapitulates multiple attributes of the native tissue or development process. Toward this, biomaterials, recapitulating tissue-specific biophysical and biochemical cues, have been developed and used as artificial ECMs to direct various cellular functions including stem cell differentiation and to create functional tissues [1]. An ideal biomaterial/scaffold is expected to provide the necessary

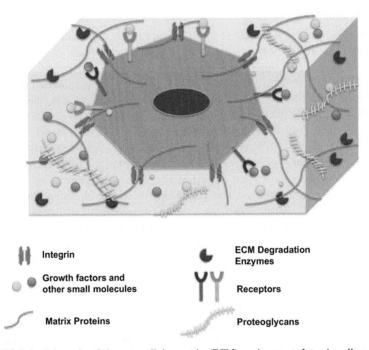

Integrin

Growth factors and other small molecules

Matrix Proteins

ECM Degradation Enzymes

Receptors

Proteoglycans

FIGURE 9.1 Schematic of the extracellular matrix (ECM) environment. Integrins allow cell attachment to specific peptide sequences of matrix proteins found in the ECM environment. Matrix proteins such as collagen and elastin not only facilitate cell attachment but also provide mechanical and structural cues that can influence cell behavior. Proteoglycans in the ECM help to sequester growth factors and other small molecules and bring them to the cell receptors. These growth factors and small molecules can direct cell differentiation, growth, proliferation, and migration. ECM degradation enzymes such as plasmin and metalloproteinases help the cell remodel its environment.

biochemical and physical cues to modulate the tissue-specific cellular functions and maturation of the neotissue with appropriate degradation characteristics without introducing any adverse effects [12−16]. A wealth of biomaterials is already in place, and new ones are being rapidly developed. Whereas most initial efforts focused on the ability of the biomaterials to function as a structural support (i.e., permissive scaffold), later efforts designed biomaterials to emulate tissue-specific niche and to provide an instructive environment to the cells [2,17,18]. In this chapter, we will focus on biomimetic musculoskeletal tissue engineering, specifically on recent advancements at the intersection of developmental biology and biomaterials as it pertains to bone, cartilage, and skeletal muscle tissue engineering.

Engineering Biochemical Environment

When stem cells, exhibiting multipotency or pluripotency, are used as a cell source for tissues, it is very important to provide an environment that promotes their differentiation into the targeted cellular phenotype. During embryogenesis and development, stem cells or progenitor cells take on distinct fates according to the positional information driven by the morphogen gradients (i.e., biochemical environment). Morphogens are chemical molecules, which establish a concentration gradient [13]. The concentration gradient elicits distinct cellular response resulting in pattern formation and eventually contributing to formation of functional organs. In vitro cultures use a range of growth factors, soluble small molecules, and their combinations to direct musculoskeletal differentiation of stem cells. Some examples of this are the use of transforming growth factor beta (TGF-β) for chondrogenesis [19] and bone morphogenic proteins (BMPs) for osteogenic differentiation [20] of stem cells. Most in vitro approaches use medium supplemented with growth factors to achieve desirable cellular functions. Conventional 2D cultures are not adequate to create gradients seen in vivo and lack 3D geometries. ECM in native tissue not only provides a 3D structural support and mediates cell−matrix and cell−cell attachment but also regulates growth factor signaling [21]. In tissue engineering, biomaterials (as scaffolds) have been used as an analogue to recapitulate different functions of native ECM and to support diverse cellular functions such as cell attachment, growth, migration, and differentiation. Hence, several strategies ranging from decellularized or devitalized tissues to major ECM components, to synthetic polymers, individually or in combination, have been extensively used as scaffolds for musculoskeletal tissue engineering [22].

Although both decellularized and devitalized tissues are derived from native tissues, they vary in their biological functions because of the variance in processing—decellularization involves the use of chemical reagents, whereas devitalization is a physical process involving repeated freeze/thawing [19]. One of the advantages of decellularized and devitalized tissues is that the ECM bears most of the tissue-specific biochemical cues and direct cellular functions accordingly. Additionally, if properly stripped of their cellular content, using detergents, proteases or hypertonic mixtures [23], decellularized tissues have the potential to prevent adverse immune responses. These tissue-derived scaffolds, when used in vitro and in vivo, have been shown to support musculoskeletal tissue formation [19,24−26]. For example, Sicari et al. have used decellularized porcine urinary bladder to treat volumetric muscle loss in rodents and humans [24]. While the rodents showed formation of stimulus-responsive skeletal muscle, which was attributed to the ECM-mediated constructive remodeling, three out of the five human recipients were reported to have functional muscle improvement. Grayson et al. have harnessed decellularized bone tissues to create anatomically shaped bone tissue replacements for temporomandibular joint [26]. Yang et al. prepared decellularized human cartilage and molded it into a porous scaffold, which preserved most of the ECM components and supported mesenchymal stem cells (MSCs) differentiation in chondrogenic medium [25]. When implanted subcutaneously, the ECM scaffold loaded with fluorescent-labeled MSCs formed cartilage-like tissue, as confirmed by in vivo fluorescent imaging and immunofluorescence staining. In another study, comparison of tissue regeneration from the devitalized cartilage with that from the decellularized cartilage showed the devitalized cartilage to be more chondroinductive and even outperformed the decellularized groups supplemented with exogenous TGF-β [19]. Although results seem promising, and some of these decellularized interventions are even currently on the market [27], functional outcomes have shown high variability [28]. Variance in preparation protocols and quality of the resulting products is a major hurdle that can have a significant effect on the biochemical and biophysical properties of the decellularized tissue and even make some of these products immunogenic. With better regulation and systematic analyses, these tissue-derived scaffolds have the potential to provide a more reliable intervention for large musculoskeletal defects [29].

A further approach to mimicking the biochemical environment is the use of ECM proteins and polysaccharides. These biopolymers can be used either as themselves or incorporated with other polymers (synthetic or natural). Among the ECM molecules, collagen is the most extensively explored protein in scaffold engineering [30]. For instance, one of the early approaches has used collagen type I, the most abundant ECM protein in adult muscle, to engineer skeletal muscle mimics [31]. While embedding myotubes in the collagen matrix and anchoring it to two fixed points promoted better tissue alignment and longer culture times compared with 2D dish culture with a force generation of $2 \, mN/mm^2$, subsequent studies augmented collagen matrices with other ECM components such as Matrigel to improve the force generation. This approach has been shown to generate forces around $100 \, mN/mm^2$ using rodent cells (on the order of native tissue $230 \, mN/mm^2$ [32,33]), although, similar forces have not been achieved using human cells

[34−36]. Collagen-based scaffolds are also extensively used for cartilage and bone tissue engineering [30].

In addition to ECM proteins, polysaccharides such as hyaluronic acid (HA) are extensively used as scaffolds, especially in cartilage tissue engineering, given the key role that HA molecules play in the formation and function of cartilage tissues. HA is differentially regulated during limb bud formation and mesenchymal condensation, which is a prerequisite for cartilage tissue formation. Unlike proteins, ECM polysaccharides do not possess cell adhesive units and therefore require modifications to support cell attachment. Attachment of cells to such nonadhesive scaffolds can be improved by conjugating them with proteins and peptides. Compared with proteins, short peptides are easy to synthesize, purify, and modify without compromising their biological functions. The Arg−Gly−Asp tripeptide (RGD) and its derivatives, collagen mimetic peptides, and matrix metalloproteinase (MMP)−sensitive peptides are a few examples of peptides that have been incorporated into scaffolds for tissue engineering [37−40]. Studies using HA-based scaffolds for cartilage tissue engineering have showed that they support chondrogenic differentiation of stem cells (e.g., MSCs) and promote cartilage tissue formation in vitro and in vivo [41,42]. Other widely used natural polymers as scaffolds for musculoskeletal tissue engineering include fibrin, chitosan, and alginate [43−46].

Mesenchymal condensation is a critical process that precedes cartilage tissue formation during development. In vitro culture conditions, such as micromass and pellet cultures, promote cell−cell interactions that facilitate condensation-like cellular processes and have been shown to promote chondrogenesis. Athanasiou and colleagues have extended the ability of stem cells to form three-dimensional aggregates through cell−cell interactions toward scaffold-free cartilage tissue engineering [47]. The authors have used such a scaffold-free approach to create ∼2 mm in diameter cartilage tissues enriched with collagens and glycosaminoglycans (GAGs) from human embryonic stem cells (hESCs). A similar approach of using cell aggregates to create cartilage tissue was used by Bhumiratana et al. in assembling osteochondral tissue [48]. Here, the authors fused cell aggregates generated from human mesenchymal stem cells (hMSCs) with a decellularized bone matrix destined to support bone tissue formation. Although the above approaches use high cell density−mediated aggregation of cells to mimic mesenchymal condensation, biomaterials designs have also been used to promote the condensation of encapsulated MSCs [49]. For instance, HA hydrogels modified with N-cadherin mimetic peptides were used to promote condensation of hMSCs, which underwent early chondrogenesis and produced cartilage-specific matrices [50]. Polyethylene glycol-co-chondroitin sulfate hydrogels have also been shown to promote chondrogenic differentiation of stem cells and progenitor cells and in some cases even assisted aggregation of encapsulated cells similar to that observed during early stages of chondrogenesis-like mesenchymal condensation [51].

Articular cartilage is a highly organized tissue with depth-varying properties and functions. Composition and organization of ECM play a key role in maintaining the zonal architecture of cartilage with distinct cellular functions. Nguyen et al. have used multilayered hydrogels with varying ECM composition to direct zonal-specific differentiation of hMSCs toward cartilage tissue [52]. In a recent study, Peng et al. have shown enhanced boundary lubrication of engineered cartilage tissues exposed to medium conditions supplemented with ECM extracts from the superficial zone of native articular cartilage [53]. In addition to ECM components, chondrocyte-secreted factors also promote chondrogenic differentiation of stem cells [54,55].

Another class of materials that has been extensively studied as scaffolds is synthetic polymers and ceramics. Some examples of synthetic biomaterials that have extensively used as scaffolds for musculoskeletal tissue engineering are polyethylene glycol (PEG), poly-L-lactic acid (PLLA), polyglycolic acid (PGA), poly-lactic-co-glycolic acid (PLGA), and silk [56−58]. One of the unique characteristics of synthetic matrices is that they can be processed and modified easily. For example, synthetic biomaterials such as PEG can be easily functionalized with peptides, proteins, small molecules, and/or growth factors to modulate various cellular activities such as attachment, growth, and musculoskeletal differentiation of stem cells [59,60]. Besides prolonging their functions, tethering growth factors with the scaffolds can also be used to develop microenvironments with spatiotemporal control and specificity. Growth factors or morphogens can be loaded onto scaffolds through conjugation (covalent) or through noncovalent interactions. The latter can be achieved through electrostatic interactions between the scaffold and growth factors or through other secondary interactions [61]. Incorporation of heparins to scaffolds has been used for immobilization of growth factors, which harnesses electrostatic affinity of polysaccharides (e.g., heparin) with growth factors [62]. Some synthetic polymers can mimic certain functions of polysaccharides. For example, poly(styrene sulfonate) (PSS) molecules can bind to basic fibroblast growth factors (bFGFs) similar to heparin and regulate their signaling [63]. Cultures containing PSS moieties have been shown to support both self-renewal of hPSCs [64] and myogenic differentiation of progenitor cells [63].

Similarly, scaffolds containing calcium phosphate (CaP) minerals have been demonstrated to bind to various growth factors such as BMP, making such scaffolds highly osteoinductive and a potential alternative to demineralized bone matrix (DBM) or BMP in bone tissue engineering [65]. CaP is a major inorganic component of the bone ECM. Most CaP-based scaffolds are thought to be dynamic as the CaP minerals can undergo dissolution, precipitation, and even stimulate formation of CaP minerals. In the case of composite scaffolds (organic and inorganic components), the minerals can be incorporated through physical interactions or through chemical bonds [66,67]. We have recently shown that biomaterials containing CaP minerals will not only promote osteogenic differentiation of stem cells and promote bone tissue formation in vivo [9] but will also prevent adipogenic differentiation of hMSCs in adipogenic medium [68]. Using these matrices, we have also identified a previously unknown mechanism, phosphate-ATP-adenosine metabolic signaling, through which mineral phase (i.e., CaP) of bone ECM promotes osteogenesis while inhibiting adipogenesis [69].

Although both degradable and nondegradable scaffolds support biosynthesis and deposition of ECM, degradation of the scaffold as new tissue forms is often necessary to create functional tissues because scaffold degradation and its rate have a significant impact on the formation and function of engineered tissues. Thus, synthetic biomaterials are being heavily investigated as they can also be modified to achieve desired degradation. By using PEG hydrogels with varying levels of degradation, Bryant et al. have shown the importance of scaffold degradation on accumulation and distribution of cartilage ECM [70]. Similarly, studies by Park et al. have shown the influence of scaffold degradation on osteogenic differentiation of hMSCs in 3D silk scaffolds [71]. Scaffolds with degradation responding to various external and internal stimuli such as water [72], enzymes [73], matrix metalloproteinase [74], and cell-secreted molecules such as glutathione [75] have been developed and used for musculoskeletal tissue engineering (Fig. 9.2). Degradable biomaterials have also been used to assist cell transplantation [76,77]. These studies suggest that using biomaterials as delivery vehicles could improve viability and engraftment of transplanted cells [78].

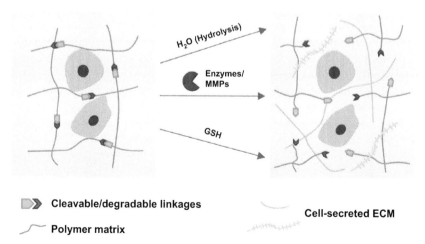

Cleavable/degradable linkages

Polymer matrix

Cell-secreted ECM

FIGURE 9.2 Schematic of degradable tissue engineering scaffolds. Scaffolds with cleavable/degradable linkages can undergo degradation to free up space for the deposition of cell-secreted extracellular matrix (ECM). These linkages include ester bonds, matrix metalloproteinase (MMP)–sensitive peptides, and disulfide bonds, which can be cleaved by water, MMPs, and glutathione (GSH), respectively.

Engineering Biomechanical and Biophysical Environment

The mechanical environment plays a key role in maintaining the health and function of load-bearing tissues such as cartilage and bone [79]. Bioreactors that provide hydrostatic pressure, compression, or shear forces have been extensively used to incorporate biomechanical environment in musculoskeletal tissue engineering. All of these different forms of mechanical cues have been shown to play a key role in stem cell differentiation and tissue formation [80,81]. Studies over the past years have shown the pivotal role of dynamic mechanical environment on cartilage tissue engineering. For instance, studies applying mechanical compression showed that the tissue constructs exposed to dynamic mechanical loading promoted chondrogenic differentiation of stem cells and cartilage tissue formation [82,83]. However, it is important to keep in mind that the dynamic mechanical loading also promotes diffusion, which will also have a positive effect on tissue formation. Exposure to cyclic mechanical loading has been shown to result in increased hypertrophy, myotube alignment, and increase in overall mass of the engineered skeletal muscle tissues from primary human muscle precursor cells [84].

In addition to aforementioned active mechanical cues, ECM-based mechanical and topographical cues also play a key role toward the engineering cartilage, bone, and skeletal muscle tissues from stem cells. Biomaterial designs have been used to mimic such biophysical environment. These include development of scaffolds with tissue-relevant Young's modulus (or commonly termed as matrix stiffness), topographical cues, or fibrillar structures [85]. Mechanical properties of the scaffold can be controlled through varying crosslinking density, concentration and/or molecular weight of the polymer precursor, and polymer composition [72,86]. Mostly, hMSCs on soft matrices with less cell spreading (maintaining ~ spherical shape) undergo chondrogenic (or adipogenic) differentiation while those on stiff matrices with cell significantly spreading show osteogenic differentiation [2,50] (Fig. 9.3). Matrix stiffness has also been shown to influence growth factor—mediated stem cell commitment. For example, studies by Park et al. have used biomaterials with varying mechanical properties (Young's modulus) to tease out the differential effect of TGF-β on differentiation commitment of hMSCs [87]. According to the authors, hMSCs on soft matrices displayed upregulation of chondrogenic marker in the presence of exogenous TGF-β compared with those on stiff matrices. The cells on stiff matrices showed upregulation of smooth muscle cell markers. Matrix stiffness—mediated changes in cell shape (less cell spreading on soft matrices compared with stiff matrices) could also be contributing to the differential outcome observed. Indeed, in another study, Chen and colleagues have attributed the differential outcome of TGF-β on hMSC differentiation to cell shape where the authors have used micropatterning to control the cell shape [88]. The authors have reported chondrogenic differentiation of hMSCs on micropatterned surfaces that prevented

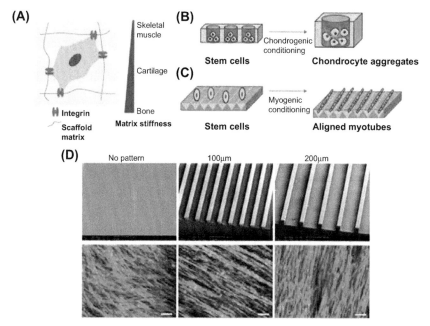

FIGURE 9.3 Matrix stiffness and topography affect stem cell differentiation in musculoskeletal tissue engineering. (A) Schematic of cells "feeling" the matrix stiffness through integrin-mediated cell—matrix interaction, which dictates cell shape and function. (B) Schematic of stem cells densely seeded in wells forming chondrocyte aggregates. (C) Schematic of stem cells seeded on groove-patterned surface forming aligned myotubes. (D) Substrate topography—guided cellular alignment. Top panel: SEM images of nonpatterned and patterned (microgrooves of width 100 or 200 μm and height 50 μm) substrates. Bottom panel: F-actin staining of human embryonic stem cell—derived progenitor cells cultured on micropatterned substrates, aligned along the groove [97].

cell spreading while those exhibiting well-spread shape (again controlled by micropatterned surface) underwent smooth muscle cell differentiation responding to exogenous TGF-β3. The shape-dependent chondrogenesis is concurrent with many original studies in the field [89,90]. Although most of these biomaterials design offers static matrices (or scaffold), Kloxin et al. have developed photodegradable hydrogels to modulate the physical and chemical properties of the extracellular environment [91]. Using such systems, the authors have achieved temporal variation in the biochemical composition and mechanical properties of the scaffolds subsequently influencing chondrogenic differentiation of encapsulated hMSCs. Lim et al. have developed electric field stimuli hydrogels and used such hydrogels to promote chondrogenic differentiation of hMSCs [92]. Such stimuli-responsive hydrogels have found a myriad of applications in cell culture, medical devices, and drug delivery [93,94].

Another key aspect of ECM architecture is their topographical features; these nano- and microscale features of various ECM proteins play an equally important role regulating cellular fate. To mimic the ECM topographical features, different topographically defined cell cultures have been developed. These involve grooved surfaces, nanotubes/nanopillars, aligned fibers [95—97] (Fig. 9.3). When compared with nonpatterned surfaces, surfaces providing topographical cues promote chondrogenic and osteogenic differentiation of stem cells [16,98]. Furthermore, the topography-mediated differentiation was found to be sensitive to the specific features of the topography [16]. Substrates with topographical cues that promote alignment of adhered cells have been shown to support their myogenic commitment and cell fusion to form multi-nucleated myotubes [99]. Although most of these approaches provide 2D or pseudo-2D cultures, electrospinning was used to create 3D scaffold with nano- and microfibers. There are a number of reviews describing the application of such scaffolds in musculoskeletal tissue engineering [100].

CONCLUSIONS AND FUTURE CONSIDERATIONS

The advent and development of biomaterials have greatly advanced the field of biomimetic engineering. The field of musculoskeletal tissue engineering has greatly benefited from these developments. Although the research highlighted above is promising, challenges persist, because of limited understanding of or ability to recapitulate the key aspects of the human development process that are necessary for successful development of tissue surrogates with functional and biological features similar to that of native tissues. The first major hurdle is establishing a reliable, high-volume cell source. Even though pluripotent cells are being extensively researched and their differentiation yield and purity can be improved by cell sorting, there is still small risk of impure populations causing teratoma formation after implantation [1]. More importantly, implanted cells still have low engraftment rates and viability. Without enough functional cells, this intervention may be futile. Moreover, biomimetic tissues engineered from pluripotent stem cells often suffer from poor maturation as seen in many of the reported cases where biomimetic constructs are unable to withstand the same compressive forces as native bone and cartilage.

Over the years, significant advancements have been made to create bio-mimetic materials recapitulating tissue-specific biochemical, mechanical, and topographical properties. Although the studies described in this review showed great progress in using these biomaterials to control cellular fates, methods should be further developed to direct cellular functions in a spatiotemporal manner. Another key feature of native tissue is diverse cell populations with distinct functions. The tissue-specific cell—cell communication plays a key role in tissue homeostasis and function. Thus, in an effort to create more physiologically relevant tissue surrogates, researchers are incorporating more of the cell types present in native tissue into their engineered constructs. Although all the cells may not be needed to mimic a desired function, because it is still not fully elucidated how each of the endogenous cells interact with each other and their environments in response to different stimuli, key aspects of disease models may be overlooked or neglected by not including all cell types. As researchers continue to elucidate the complex processes involved in human development, these principles will continue to be applied to enhance the functionality of engineered constructs and further the field of biomimetic tissue engineering.

LIST OF ACRONYMS AND ABBREVIATIONS

bFGF Basic fibroblast growth factor
BMP Bone morphogenetic protein
CaP Calcium phosphate
DBM Demineralized bone matrix
ECM Extracellular matrix
ESCs Embryonic stem cells
GAGs Glycosaminoglycans
HA Hyaluronic acid
iPSCs Induced pluripotent stem cells
MMP Matrix metalloproteinase
MSCs Mesenchymal stem cells
PCL Polycaprolactone
PEG Polyethylene glycol
PGA Polyglycolic acid
PLGA Poly-lactic-co-glycolic acid
PLLA Poly-L-lactic acid
PSCs Pluripotent stem cells
PSS Poly(styrene sulfonate)
RGD Arg−Gly−Asp tripeptide
TGF-β Transforming growth factor beta

GLOSSARY

Biomimetic tissue engineering A tissue engineering effort that closely recapitulates tissue-specific microenvironments with the knowledge of biological development to foster functional tissue formation.
Induced pluripotent stem cells (iPSCs) Pluripotent stem cells reprogrammed from adult stromal cells by introducing a set of four critical gene switches (Oct4 and Sox2 with either c-Myc and Klf4 or Nanog and Lin28).
Tissue engineering An interdisciplinary approach that aims to create functional tissue replacements for clinical applications.

REFERENCES

[1] Seale NM, Varghese S. Biomaterials for pluripotent stem cell engineering: from fate determination to vascularization. J Mater Chem B 2016;4(20):3454−63.
[2] Discher DE, Mooney DJ, Zandstra PW. Growth factors, matrices, and forces combine and control stem cells. Science 2009;324(5935):1673−7.
[3] Thomson JA, Itskovitz-Eldor J, Shapiro SS, Waknitz MA, Swiergiel JJ, Marshall VS, et al. Embryonic stem cell lines derived from human blastocysts. Science 1998;282(5391):1145−7.

[4] Robinton DA, Daley GQ. The promise of induced pluripotent stem cells in research and therapy. Nature 2012;481(7381):295−305.

[5] Takebe T, Sekine K, Enomura M, Koike H, Kimura M, Ogaeri T, et al. Vascularized and functional human liver from an iPSC-derived organ bud transplant. Nature 2013;499(7459):481−4.

[6] Leyton-Mange J, Milan D. Pluripotent stem cells as a platform for cardiac arrhythmia drug screening. Curr Treat Options Cardio Med 2014;16(9):1−18.

[7] Mullen CA, Haugh MG, Schaffler MB, Majeska RJ, McNamara LM. Osteocyte differentiation is regulated by extracellular matrix stiffness and intercellular separation. J Mech Behav Biomed Mater 2013;28:183−94.

[8] Levenberg S, Huang NF, Lavik E, Rogers AB, Itskovitz-Eldor J, Langer R. Differentiation of human embryonic stem cells on three-dimensional polymer scaffolds. Proc Natl Acad Sci USA 2003;100(22):12741−6.

[9] Kang H, Shih Y-RV, Hwang Y, Wen C, Rao V, Seo T, et al. Mineralized gelatin methacrylate-based matrices induce osteogenic differentiation of human induced pluripotent stem cells. Acta Biomater 2014;10(12):4961−70.

[10] Wen C, Kang H, Shih YV, Hwang Y, Varghese S. In vivo comparison of biomineralized scaffold-directed osteogenic differentiation of human embryonic and mesenchymal stem cells. Drug Deliv Transl Res 2015.

[11] Nakasaki M, Hwang Y, Xie Y, Kataria S, Gund R, Hajam EY, et al. The matrix protein Fibulin-5 is at the interface of tissue stiffness and inflammation in fibrosis. Nat Commun 2015;6:8574.

[12] Engler AJ, Sen S, Sweeney HL, Discher DE. Matrix elasticity directs stem cell lineage specification. Cell 2006;126(4):677−89.

[13] Gurdon JB, Bourillot PY. Morphogen gradient interpretation. Nature 2001;413(6858):797−803.

[14] Ayala R, Zhang C, Yang D, Hwang Y, Aung A, Shroff SS, et al. Engineering the cell−material interface for controlling stem cell adhesion, migration, and differentiation. Biomaterials 2011;32(15):3700−11.

[15] Lin S, Sangaj N, Razafiarison T, Zhang C, Varghese S. Influence of physical properties of biomaterials on cellular behavior. Pharm Res (NY) 2011;28(6):1422−30.

[16] Dalby MJ, Gadegaard N, Tare R, Andar A, Riehle MO, Herzyk P, et al. The control of human mesenchymal cell differentiation using nanoscale symmetry and disorder. Nat Mater 2007;6(12):997−1003.

[17] Huebsch N, Arany PR, Mao AS, Shvartsman D, Ali OA, Bencherif SA, et al. Harnessing traction-mediated manipulation of the cell/matrix interface to control stem-cell fate. Nat Mater 2010;9(6):518−26.

[18] Burdick JA, Vunjak-Novakovic G. Engineered microenvironments for controlled stem cell differentiation. Tissue Eng 2008;15(2):205−19.

[19] Beck EC, Barragan M, Libeer TB, Kieweg SL, Converse GL, Hopkins RA, et al. Chondroinduction from naturally derived cartilage matrix: a comparison between devitalized and decellularized cartilage encapsulated in hydrogel pastes. Tissue Eng 2016;22(7−8):665−79.

[20] Mehta M, Schmidt-Bleek K, Duda GN, Mooney DJ. Biomaterial delivery of morphogens to mimic the natural healing cascade in bone. Adv Drug Deliv Rev 2012;64(12):1257−76.

[21] Martino MM, Briquez PS, Maruyama K, Hubbell JA. Extracellular matrix-inspired growth factor delivery systems for bone regeneration. Adv Drug Deliv Rev 2015;94:41−52.

[22] Benders KE, van Weeren PR, Badylak SF, Saris DB, Dhert WJ, Malda J. Extracellular matrix scaffolds for cartilage and bone regeneration. Trends Biotechnol 2013;31(3):169−76.

[23] Crapo PM, Gilbert TW, Badylak SF. An overview of tissue and whole organ decellularization processes. Biomaterials 2011;32(12):3233−43.

[24] Sicari BM, Rubin JP, Dearth CL, Wolf MT, Ambrosio F, Boninger M, et al. An acellular biologic scaffold promotes skeletal muscle formation in mice and humans with volumetric muscle loss. Sci Transl Med 2014;6(234):234ra58.

[25] Yang Q, Peng J, Guo Q, Huang J, Zhang L, Yao J, et al. A cartilage ECM-derived 3-D porous acellular matrix scaffold for in vivo cartilage tissue engineering with PKH26-labeled chondrogenic bone marrow-derived mesenchymal stem cells. Biomaterials 2008;29(15):2378−87.

[26] Grayson WL, Fröhlich M, Yeager K, Bhumiratana S, Chan ME, Cannizzaro C, et al. Engineering anatomically shaped human bone grafts. Proc Natl Acad Sci USA 2010;107(8):3299−304.

[27] Sutherland AJ, Beck EC, Dennis SC, Converse GL, Hopkins RA, Berkland CJ, et al. Decellularized cartilage may be a chondroinductive material for osteochondral tissue engineering. PLoS One 2015;10(5):e0121966.

[28] Londono R, Badylak SF. Biologic scaffolds for regenerative medicine: mechanisms of in vivo remodeling. Ann Biomed Eng 2015;43(3):577−92.

[29] Chen C-C, Liao C-H, Wang Y-H, Hsu Y-M, Huang S-H, Chang C-H, et al. Cartilage fragments from osteoarthritic knee promote chondrogenesis of mesenchymal stem cells without exogenous growth factor induction. J Orthop Res 2012;30(3):393−400.

[30] Glowacki J, Mizuno S. Collagen scaffolds for tissue engineering. Biopolymers 2008;89(5):338−44.

[31] Vandenburgh HH, Karlisch P, Farr L. Maintenance of highly contractile tissue-cultured avian skeletal myotubes in collagen gel. In Vitro Cell Dev Biol 1988;24(3):166−74.

[32] Okano T, Matsuda T. Tissue engineered skeletal muscle: preparation of highly dense, highly oriented hybrid muscular tissues. Cell Transplant 1998;7(1):71−82.

[33] Lee PH, Vandenburgh HH. Skeletal muscle atrophy in bioengineered skeletal muscle: a new model system. Tissue Eng Part A 2013;19(19−20):2147−55.

[34] Chiron S, Tomczak C, Duperray A, Lainé J, Bonne G, Eder A, et al. Complex interactions between human myoblasts and the surrounding 3D fibrin-based matrix. PLoS One 2012;7(4):e36173.

[35] Juhas M, Engelmayr Jr GC, Fontanella AN, Palmer GM, Bursac N. Biomimetic engineered muscle with capacity for vascular integration and functional maturation in vivo. Proc Natl Acad Sci USA 2014;111(15):5508−13.

[36] Madden L, Juhas M, Kraus WE, Truskey GA, Bursac N. Bioengineered human myobundles mimic clinical responses of skeletal muscle to drugs. eLife 2015;4:e04885.

[37] Melkoumian Z, Weber JL, Weber DM, Fadeev AG, Zhou Y, Dolley-Sonneville P, et al. Synthetic peptide-acrylate surfaces for long-term self-renewal and cardiomyocyte differentiation of human embryonic stem cells. Nat Biotech 2010;28(6):606−10.

[38] Jin S, Yao H, Weber JL, Melkoumian ZK, Ye K. A synthetic, xeno-free peptide surface for expansion and directed differentiation of human induced pluripotent stem cells. PLoS One 2012;7(11):e50880.

[39] Lee HJ, Yu C, Chansakul T, Hwang NS, Varghese S, Yu SM, et al. Enhanced chondrogenesis of mesenchymal stem cells in collagen mimetic peptide-mediated microenvironment. Tissue Eng Part A 2008;14(11):1843−51.

[40] Hwang NS, Varghese S, Zhang Z, Elisseeff J. Chondrogenic differentiation of human embryonic stem cell-derived cells in arginine-glycine-aspartate-modified hydrogels. Tissue Eng 2006;12(9):2695−706.

[41] Kim IL, Khetan S, Baker BM, Chen CS, Burdick JA. Fibrous hyaluronic acid hydrogels that direct MSC chondrogenesis through mechanical and adhesive cues. Biomaterials 2013;34(22):5571−80.

[42] Highley CB, Prestwich GD, Burdick JA. Recent advances in hyaluronic acid hydrogels for biomedical applications. Curr Opin Biotechnol 2016;40:35−40.

[43] Griffon DJ, Sedighi MR, Schaeffer DV, Eurell JA, Johnson AL. Chitosan scaffolds: interconnective pore size and cartilage engineering. Acta Biomater 2006;2(3):313−20.

[44] Ahmed TA, Giulivi A, Griffith M, Hincke M. Fibrin glues in combination with mesenchymal stem cells to develop a tissue-engineered cartilage substitute. Tissue Eng 2010;17(3−4):323−35.

[45] Chaudhuri O, Gu L, Klumpers D, Darnell M, Bencherif SA, Weaver JC, et al. Hydrogels with tunable stress relaxation regulate stem cell fate and activity. Nat Mater 2016;15(3):326−34.

[46] Radhakrishnan J, Subramanian A, Krishnan UM, Sethuraman S. Injectable and 3D bioprinted polysaccharide hydrogels: from cartilage to osteochondral tissue engineering. Biomacromolecules 2016.

[47] Hu JC, Athanasiou KA. A self-assembling process in articular cartilage tissue engineering. Tissue Eng 2006;12(4):969−79.

[48] Bhumiratana S, Eton RE, Oungoulian SR, Wan LQ, Ateshian GA, Vunjak-Novakovic G. Large, stratified, and mechanically functional human cartilage grown in vitro by mesenchymal condensation. Proc Natl Acad Sci USA 2014;111(19):6940−5.

[49] Sekiya I, Vuoristo JT, Larson BL, Prockop DJ. In vitro cartilage formation by human adult stem cells from bone marrow stroma defines the sequence of cellular and molecular events during chondrogenesis. Proc Natl Acad Sci USA 2002;99(7):4397−402.

[50] Bian L, Guvendiren M, Mauck RL, Burdick JA. Hydrogels that mimic developmentally relevant matrix and N-cadherin interactions enhance MSC chondrogenesis. Proc Natl Acad Sci USA 2013;110(25):10117−22.

[51] Varghese S, Hwang NS, Canver AC, Theprungsirikul P, Lin DW, Elisseeff J. Chondroitin sulfate based niches for chondrogenic differentiation of mesenchymal stem cells. Matrix Biol 2008;27(1):12−21.

[52] Nguyen LH, Kudva AK, Saxena NS, Roy K. Engineering articular cartilage with spatially-varying matrix composition and mechanical properties from a single stem cell population using a multi-layered hydrogel. Biomaterials 2011;32(29):6946−52.

[53] Peng G, McNary SM, Athanasiou KA, Reddi AH. Superficial zone extracellular matrix extracts enhance boundary lubrication of self-assembled articular cartilage. Cartilage 2016;7(3):256−64.

[54] Aung A, Gupta G, Majid G, Varghese S. Osteoarthritic chondrocyte-secreted morphogens induce chondrogenic differentiation of human mesenchymal stem cells. Arthritis Rheum 2011;63(1):148−58.

[55] Varghese S, Hwang NS, Ferran A, Hillel A, Theprungsirikul P, Canver AC, et al. Engineering musculoskeletal tissues with human embryonic germ cell derivatives. Stem Cell 2010;28(4):765−74.

[56] Scaffaro R, Lopresti F, Botta L, Rigogliuso S, Ghersi G. Melt processed PCL/PEG scaffold with discrete pore size gradient for selective cellular infiltration. Macromol Mater Eng 2016;301(2):182−90.

[57] Jakus AE, Rutz AL, Jordan SW, Kannan A, Mitchell SM, Yun C, et al. Hyperelastic "bone": a highly versatile, growth factor−free, osteoregenerative, scalable, and surgically friendly biomaterial. Sci Transl Med 2016;8(358):358ra127.

[58] Wang Y, Kim H-J, Vunjak-Novakovic G, Kaplan DL. Stem cell-based tissue engineering with silk biomaterials. Biomaterials 2006;27(36):6064−82.

[59] Rosales AM, Anseth KS. The design of reversible hydrogels to capture extracellular matrix dynamics. Nature Rev Mater 2016;1:15012.

[60] Green JJ, Elisseeff JH. Mimicking biological functionality with polymers for biomedical applications. Nature 2016;540(7633):386−94.

[61] Macdonald ML, Samuel RE, Shah NJ, Padera RF, Beben YM, Hammond PT. Tissue integration of growth factor-eluting layer-by-layer polyelectrolyte multilayer coated implants. Biomaterials 2011;32(5):1446−53.

[62] Jeon O, Powell C, Solorio LD, Krebs MD, Alsberg E. Affinity-based growth factor delivery using biodegradable, photocrosslinked heparin-alginate hydrogels. J Contr Release 2011;154(3):258−66.

[63] Sangaj N, Kyriakakis P, Yang D, Chang C-W, Arya G, Varghese S. Heparin mimicking polymer promotes myogenic differentiation of muscle progenitor cells. Biomacromolecules 2010;11(12):3294−300.

[64] Varghese S, Chang C-W, Hwang Y. Synthetic matrices for self-renewal and expansion of stem cells. 2012. US Patent App. 13/706,900.

[65] Bose S, Roy M, Bandyopadhyay A. Recent advances in bone tissue engineering scaffolds. Trends Biotechnol 2012;30(10):546−54.

[66] Phadke A, Zhang C, Hwang Y, Vecchio K, Varghese S. Templated mineralization of synthetic hydrogels for bone-like composite materials: role of matrix hydrophobicity. Biomacromolecules 2010;11(8):2060−8.

[67] Leonardi E, Ciapetti G, Baldini N, Novajra G, Verné E, Baino F, et al. Response of human bone marrow stromal cells to a resorbable P_2O_5−SiO_2−CaO−MgO−Na_2O−K_2O phosphate glass ceramic for tissue engineering applications. Acta Biomater 2010;6(2):598−606.

[68] Kang H, Shih YR, Varghese S. Biomineralized matrices dominate soluble cues to direct osteogenic differentiation of human mesenchymal stem cells through adenosine signaling. Biomacromolecules 2015;16(3):1050−61.

[69] Shih Y-RV, Hwang Y, Phadke A, Kang H, Hwang NS, Caro EJ, et al. Calcium phosphate-bearing matrices induce osteogenic differentiation of stem cells through adenosine signaling. Proc Natl Acad Sci USA 2014;111(3):990−5.

[70] Bryant SJ, Anseth KS. Controlling the spatial distribution of ECM components in degradable PEG hydrogels for tissue engineering cartilage. J Biomed Mater Res 2003;64(1):70−9.

[71] Park SY, Ki CS, Park YH, Jung HM, Woo KM, Kim HJ. Electrospun silk fibroin scaffolds with macropores for bone regeneration: an in vitro and in vivo study. Tissue Eng 2010;16(4):1271−9.

[72] Zhang C, Aung A, Liao L, Varghese S. A novel single precursor-based biodegradable hydrogel with enhanced mechanical properties. Soft Matter 2009;5(20):3831−4.

[73] Perrone GS, Leisk GG, Lo TJ, Moreau JE, Haas DS, Papenburg BJ, et al. The use of silk-based devices for fracture fixation. Nat Commun 2014;5.

[74] Sridhar BV, Brock JL, Silver JS, Leight JL, Randolph MA, Anseth KS. Tissue engineering: development of a cellularly degradable PEG hydrogel to promote articular cartilage extracellular matrix deposition. Adv Healthc Mater 2015;4(5):635.

[75] Kar M, Shih Y-RV, Velez DO, Cabrales P, Varghese S. Poly(ethylene glycol) hydrogels with cell cleavable groups for autonomous cell delivery. Biomaterials 2016;77:186−97.

[76] Kabra H, Hwang Y, Lim HL, Kar M, Arya G, Varghese S. Biomimetic material-assisted delivery of human embryonic stem cell derivatives for enhanced in vivo survival and engraftment. ACS Biomater Sci Eng 2015;1(1):7−12.

[77]   Huebsch N, Lippens E, Lee K, Mehta M, Koshy S, Darnell M, et al. Matrix elasticity of void-forming hydrogels controls matrix elasticity of void-forming hydrogels controls transplanted stem cell-mediated bone. Nat Mater 2015;14:1–19.

[78]   Burdick Jason A, Mauck Robert L, Gerecht S. To serve and protect: hydrogels to improve stem cell-based therapies. Cell Stem Cell 2016;18(1):13–5.

[79]   Yokota H, Leong DJ, Sun HB. Mechanical loading: bone remodeling and cartilage maintenance. Curr Osteoporos Rep 2011;9(4):237.

[80]   Rauh J, Milan F, Günther K-P, Stiehler M. Bioreactor systems for bone tissue engineering. Tissue Eng B Rev 2011;17(4):263–80.

[81]   Elder BD, Athanasiou KA. Hydrostatic pressure in articular cartilage tissue engineering: from chondrocytes to tissue regeneration. Tissue Eng B Rev 2009;15(1):43–53.

[82]   Huang AH, Farrell MJ, Mauck RL. Mechanics and mechanobiology of mesenchymal stem cell-based engineered cartilage. J Biomech 2010;43(1):128–36.

[83]   Bian L, Zhai DY, Mauck RL, Burdick JA. Coculture of human mesenchymal stem cells and articular chondrocytes reduces hypertrophy and enhances functional properties of engineered cartilage. Tissue Eng 2011;17(7–8):1137–45.

[84]   Moon DG, Christ G, Stitzel JD, Atala A, Yoo JJ. Cyclic mechanical preconditioning improves engineered muscle contraction. Tissue Eng 2008;14(4):473–82.

[85]   Guilak F, Cohen DM, Estes BT, Gimble JM, Liedtke W, Chen CS. Control of stem cell fate by physical interactions with the extracellular matrix. Cell Stem Cell 2009;5(1):17–26.

[86]   Nguyen QT, Hwang Y, Chen AC, Varghese S, Sah RL. Cartilage-like mechanical properties of poly(ethylene glycol)-diacrylate hydrogels. Biomaterials 2012;33(28):6682–90.

[87]   Park JS, Chu JS, Tsou AD, Diop R, Tang Z, Wang A, et al. The effect of matrix stiffness on the differentiation of mesenchymal stem cells in response to TGF-β. Biomaterials 2011;32(16):3921–30.

[88]   Gao L, McBeath R, Chen CS. Stem cell shape regulates a chondrogenic versus myogenic fate through Rac1 and N-Cadherin. Stem Cells 2010;28(3):564–72.

[89]   Archer CW, Rooney P, Wolpert L. Cell shape and cartilage differentiation of early chick limb bud cells in culture. Cell Differentiation 1982;11(4):245–51.

[90]   Zanetti NC, Solursh M. Induction of chondrogenesis in limb mesenchymal cultures by disruption of the actin cytoskeleton. J Cell Biol 1984;99(1):115–23.

[91]   Kloxin AM, Kasko AM, Salinas CN, Anseth KS. Photodegradable hydrogels for dynamic tuning of physical and chemical properties. Science 2009;324(5923):59–63.

[92]   Lim HL, Chuang JC, Tran T, Aung A, Arya G, Varghese S. Dynamic electromechanical hydrogel matrices for stem cell culture. Adv Funct Mater 2011;21(1).

[93]   Matsuda N, Shimizu T, Yamato M, Okano T. Tissue engineering based on cell sheet technology. Adv Mater 2007;19(20):3089–99.

[94]   Koetting MC, Guido JF, Gupta M, Zhang A, Peppas NA. pH-responsive and enzymatically-responsive hydrogel microparticles for the oral delivery of therapeutic proteins: effects of protein size, crosslinking density, and hydrogel degradation on protein delivery. J Contr Release 2016;221:18–25.

[95]   Brammer KS, Oh S, Frandsen CJ, Varghese S, Jin S. Nanotube surface triggers increased chondrocyte extracellular matrix production. Mater Sci Eng C 2010;30(4):518–25.

[96]   Nikkhah M, Edalat F, Manoucheri S, Khademhosseini A. Engineering microscale topographies to control the cell–substrate interface. Biomaterials 2012;33(21):5230–46.

[97] Hwang Y, Seo YN, Hariri S, Choi C, Varghese S. Matrix topographical cue-mediated myogenic differentiation of human embryonic stem cell derivatives. Polymers 2017:580.

[98] Baker BM, Nathan AS, Gee AO, Mauck RL. The influence of an aligned nanofibrous topography on human mesenchymal stem cell fibrochondrogenesis. Biomaterials 2010;31(24):6190−200.

[99] Zhao Y, Zeng H, Nam J, Agarwal S. Fabrication of skeletal muscle constructs by topographic activation of cell alignment. Biotechnol Bioeng 2009;102(2):624−31.

[100] Murphy SV, Atala A. 3D bioprinting of tissues and organs. Nat Biotechnol 2014;32(8):773−85.

Chapter 10

Clinical Translation of Cartilage Tissue Engineering, From Embryonic Development to a Promising Long-Term Solution

Diego Correa, Annie C. Bowles
University of Miami, Miami, FL, United States

INTRODUCTION

Tissue engineering (TE) is an interdisciplinary field that seeks to fabricate constructs, mostly in vitro, to restore the structure and function of diseased and/or injured tissues and organs once implanted, helping to compensate for their universal donor shortage [1]. The central theme of TE-derived approaches is to recapitulate the structure of a tissue by recreating its components, namely, the parenchyma and the surrounding supporting stroma (extracellular matrix, vasculature, nerves, and related cellular types), while promoting their relationship through an inductive environment/milieu. The resulting structure requires the reestablishment of the functional tissue unit (the parenchymal cells centered around a capillary within diffusion limits of $\sim 100 \, \mu m$) [2], for the engineered implant to be therapeutically viable. Furthermore, for multitissue organs (most cases), the integration between the different tissues, including structural epithelia, [3] is of critical importance.

Based on all those requirements and characteristics, TE predicted to be more straightforward and effective when dealing with tissues with homogeneous and thin structures, composed by single (or few) cell types and not dependent on vascular supply for its well-being [4]. Articular cartilage, closely matching all those criteria, constituted an "ideal" candidate. Bone, on the other hand, with a complex structure, multiple resident cell types and well vascularized was not initially considered promising. However, the prediction fell short, as clinical solutions to treat articular cartilage injuries based on TE have been more elusive than their counterpart for bone [4,5].

Developmental Biology and Musculoskeletal Tissue Engineering. https://doi.org/10.1016/B978-0-12-811467-4.00010-3
225

As we discuss throughout this chapter, multiple variables account for the delayed clinical translation of viable, reproducible, and long-term efficacious therapeutic solutions based on an engineered implant to treat articular cartilage defects. On a positive note, the redefinition of manufacturing protocols, now incorporating concepts derived from embryonic development such as the ones comprehensively elaborated in previous chapters, have significantly impacted the field offering potential solutions. Therefore, we foresee the advent of novel alternatives in the coming years that will offer consistent and durable clinical solutions.

EMBRYONIC DEVELOPMENT AS A SOURCE OF INFORMATION

Elucidating the chemical and mechanical processes at precise spatial and temporal stages during embryonic development renders fundamental mechanisms to test for the possibility of recapitulation (i.e., regeneration) in adulthood. Substantial evidence has identified the robust regulators of development including Bone Morphogenetic Proteins (BMPs), transforming growth factor β, WNTs, fibroblast growth factors [6–10], that are significant candidates leading the way to target cellular pathways and phenotypes in adult cells that possess intrinsic differentiation programs, otherwise referred to as developmental plasticity, for recapitulation of desired tissues. These approaches are the basis for "developmental engineering," which is newly termed yet conceptually a long-standing field, that aims to mimic developmental processes for TE applications [11,12]. Using pluripotent, multipotent, or progenitor cells directed under various stimuli to generate stable and defined tissue constructs in vitro are key endeavors proposed to achieve clinically translational success in vivo.

Translating Concepts Into Possible Solutions

Embryonic studies reveal the synchronization of concerted events originating from the earliest, most unspecialized cells that ultimately generate an autonomous complex being equipped with specialized tissues and organs. Although developmental research conceptually focuses on embryonic stem cell biology (e.g., proliferation, differentiation, etc.), because of ethical limitations and teratoma development, these cells are not commonly used for TE or therapeutic applications, rather adult stem cells are. Nevertheless, adult stem cells (e.g., mesenchymal stem cells—MSCs) [13] can be coopted to mimic key features of embryonic stem cells during development to recapitulate processes underlying the generation of specialized tissues that we, as adults, can no longer (re)generate or repair. This is in fact the central theme of developmental engineering.

Technological advances bring immense value to identifying the exact contributions of each cell toward creating specialized tissues with biological, anatomical, and physiological functions. Lineage-tracing techniques to genetically engineering knockout mice strains have devised approaches to resolving long unanswered questions and validating, or invalidating, proposed dogmas. For instance, we now know that there are multiple mesenchymal cells sources that make up the stromal compartment of the bone marrow during development. These cells were identified in the adjacent perichondrium, the hypertrophic layer of the growth plate, and the "borderland" between the growth plate and perichondrium [14,15]. Furthermore, researchers have identified the superficial layer of articular cartilage to contain a finite quantity of chondrocytes showing stem cell features and capable of constituting cells of each layer and potentially repair its damage [16,17]. The identification of a "true" embryonic-like chondroprogenitor, present in articular cartilage or other neighboring adult tissues [18], may thus alleviate one of the toughest challenges reiterated from several of the previous chapters and other bodies of literature and related with the ultimate phenotypic fate of adult MSCs when forced to sustain chondrogenesis.

Notwithstanding these biological barriers, the promise to 1 day regenerate hyaline cartilage as a treatment modality for one of the largest, unmet clinical challenges perseveres with each attempt. Insights into development will amass the foundation for preclinical approaches that will lead to clinical success.

Challenges Ahead, the Special Case of the Cell Phenotype

Several challenges still remain ahead before accepting TE-based technologies as frontline solutions to treat musculoskeletal entities, especially for articular cartilage injuries [19]. In fact, each one of the main components of TE (i.e., cells, scaffolds, morphogenetic inducers, and bioreactors) have their own limitations and difficulties, extensively reviewed elsewhere [19–21]. Here, we focus only on the cellular phenotype as a key determinant of the potential success of articular cartilage TE, exposing some of the challenges and the developing solutions (Fig. 10.1).

A conspicuous choice for cell source when engineering implants would be to obtain them from the tissue itself, as they naturally build it; although for articular cartilage, this poses an even greater challenge. In addition to harvesting-associated morbidity, isolated chondrocytes require prolonged and costly ex vivo expansion exhibiting a fragile phenotypic maintenance leading to dedifferentiation and survival compromise before and after implantation [22,23]. Despite its positive yet variable clinical results [24], autologous chondrocyte implantation (ACI) technique still presents limitations, especially with cost-effectiveness [25]. Therefore, alternative cell types with stable phenotypes need to be proposed and studied. Multiple slow-cycling adult chondroprogenitors have been identified within the joint tissues, all with

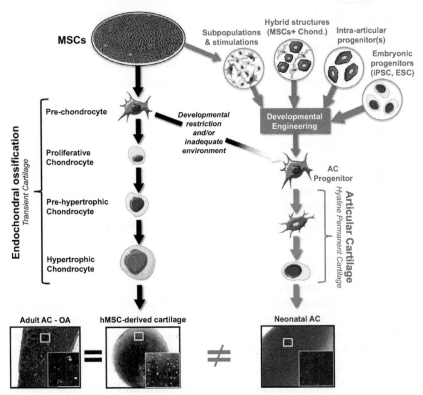

FIGURE 10.1 Phenotypic fate of mesenchymal stem cells (MSCs) undergoing chondrogenesis. MSCs follow an intrinsic endochondral bone formation program (black arrows), culminating as hypertrophic chondrocytes fabricating a transient cartilaginous matrix reminiscent of adult articular cartilage (AC) with signs of osteoarthritis (OA). Developmental restrictions and/or an inadequate instructive environment prevent differentiating MSCs to acquire a hyaline permanent AC progenitor phenotype (green arrows). Various approaches (blue arrows), incorporating developmental engineering concepts have been proposed to circumvent this restriction, offering the possibility of generating a chondrocyte capable of producing permanent AC type of matrix.

features of stem cells. Those progenitor cells seem to be retained in various tissues inside the joint such as synovium, synovial fluid, fat pad, perichondrium/periosteum, ligaments, and Prg4$^+$ chondrocytes at the superficial layer of the cartilage, exhibiting enhanced chondrogenic potential [16,17,26,27]. These progenitors have been identified also in human articular cartilage [28], play a role in endogenous repair [29], and can be used to induce the fabrication of a hyaline type of tissue, especially those derived from the cartilage itself [18,30].

Alternatively, stem cells have been proposed based on a high expansion capacity and multipotential differentiation including chondrogenesis [31]. Although there are several types of stem cells (e.g., embryonic, adult, induced pluripotent, etc.) [32], we will focus on the most common adult type, the MSCs, because of the large body of evidence that has demonstrated robust modalities for their use. MSCs have been found in numerous tissues of the body, where they can be easily procured, residing in the tissue stroma strategically situated at the perivascular space [33−37]. Regardless of tissue source (e.g., bone, fat, muscle, liver), MSCs possess unique characteristics that have been well described including the abilities of self-renewal, multi-differentiation, and adherence to plastic [38]. Much effort has contributed to the current protocols used to differentiate human MSCs (hMSCs) into mature cells types; however, most fall short of desired stable terminal phenotypes, especially during chondrogenesis [39]. In general, differentiating MSCs follow an intrinsic endochondral ossification program culminating in terminal hypertrophic chondrocytes [40−42]. This poses significant limitations to their in vivo use as they fabricate a temporal cartilaginous matrix distant from native permanent articular cartilage required for successful repair [22]. As we elaborate in later sections, recent efforts have addressed this issue with very promising results. For instance, providing the cells with a proper molecular and physical environment during expansion and differentiation, it is possible to deviate the resulting chondrocytes from a hypertrophic fate [9,43−46]. New approaches may help overcome the existing phenotype-related issue for both autologous chondrocytes and MSCs, thus providing a combinatorial solution. Assembling both cell types in hybrid structures has shown that the chondrogenic phenotype of expanding chondrocytes is maintained while the differentiation of MSCs reaching a prehypertrophic stage is promoted [47−51]. The benefits of the intercellular cross talk were further mediated by combining hMSCs with chondrocytes embedded in their own territorial matrix (i.e., chondrons), showing highly promising preclinical and clinical results [52,53].

Although a vast body of evidence characterizes and defines MSCs as a particular cellular entity [38], they exhibit a large heterogeneity regardless of the tissue source from which they were derived. In other words, MSCs obtained by plastic adherence have been described as "bulk" or "crude" MSCs, which constitute multiple subphenotypes that may reflect in vivo−distinct functions, as evident in the bone marrow [54,55]. This phenotypic discrimination has been translated into potential ways to overcome differentiation limitations by preselecting cells with special chondrogenic traits, via the expression of specific cell surface markers [60,61].

FROM PRECLINICAL MODELS TO THE CLINIC

Preclinical Experimental Design With the Clinical Research in Mind

The target of clinical research should build on the current "status quo" while contributing to the advancement for the greater good of humanity. Generating evidence with sound approaches, feasibility, reproducibility, and expected outcomes for human patients indicate high translational impacts. However, to do so, it requires a great deal of effort when designing the experiment, which relies on many contributing factors. We here present a few of those aspects, which become pivotal when translating preclinical evidence into clinical protocols.

Animal Model Selection

Selection of an animal model suitable for your experiments requires many considerations. The conflict arises when basic and preclinical research relies on reproducibility, which depends on the uniformity of the subjects (e.g., genetic strain, age, sex, tissue-specific maturation times), sufficient subject numbers and appropriate controls while selecting a model with outcomes translatable to humans. All the while, researchers must consider affordability, time, equipment, and housing. For musculoskeletal indications, there are thorough reviews that detail the applicability of models [56–58]. Here, we will focus on the parameters that will greatly impact the considerations for choosing a model (Table 10.1).

Some of the parameters listed in Table 10.1 are less objective than others; nevertheless, the overall relativity among the models is indicative of consensual evidence. Genetic variability is generally pertinent to studies that elucidate specific mechanistic dissections of the causes of a specific condition/disease, by using transgenic, knockout, or athymic strains that permit tailored experimental designs but are limited to the smaller animal models [56]. Depending on the aims of the research project, the anatomical, physiological, and biological resemblance of the model to the human structures is very important. For instance, porcine, caprine, and equine models have joints with high anatomical and biomechanical similarities with their human counterpart, permitting outcomes correlations and similar surgical procedures (e.g., arthroscopy, synovial fluid easy harvesting, etc.). However, the caprine model presents some biological differences when healing postprocedure because of a tendency to develop cysts in the operative space, which may confound results [56]. Although larger animals closely resemble features of humans, suggesting reduced subject numbers are necessary, the indirect specifications (e.g., equipment, husbandry, staff) needed for their maintenance, the type, and length of the procedures involved markedly increase the cost of the proposed experiments [56–58]. This is especially relevant for caprine and equine models. The list of factors has been reduced to obvious considerations when designing experiments; thus, understanding the resources and attainability for selected models that surround you will provide additional criteria for making the best choice. In summary, we could say that for articular cartilage TE, small animal models (i.e., rodents and rabbits) are good to establish proof of concepts when standardized procedures are applied and used in large numbers.

TABLE 10.1 Parameters for Selection of Preclinical Animal Models

Selection Factors	Rodent	Rabbit	Canine	Caprine/Ovine	Porcine	Equine
Controlled genetic variability	✓✓✓✓	✓✓	✓	✓	✓	✓✓✓
Anatomically comparable with humans	✓	✓	✓✓	✓✓✓	✓✓✓✓	✓✓✓
Physiologically comparable with humans	✓	✓	✓✓	✓✓✓	✓✓✓✓	✓✓✓
Biologically comparable with humans	✓	✓✓	✓✓		✓✓✓✓	✓✓✓
Rapid study period	✓✓✓	✓✓	✓✓	✓	✓✓	✓
Affordable	✓✓✓✓	✓✓✓	✓✓	✓	✓✓	✓
Preferential size (e.g., tissues, defects)	✓✓✓✓	✓	✓✓✓	✓✓✓	✓✓✓	✓✓✓✓
Ease of procedural requisites	✓✓✓	✓✓✓	✓✓	✗	✓	✗
Ease of maintenance, housing, husbandry	✓✓✓✓	✓✓✓	✓✓	✗	✓	✗
Attainability	✓✓✓	✓✓✓	✓✓	✓✓	✓✓	✓
Uniform and sufficient subject numbers	✓✓✓✓	✓✓	✓✓	✓✓	✓✓	✓✓
Ethically acceptable	✓✓✓✓	✓✓✓	✓	✓✓	✓✓	✓

Then, to move toward a clinical phase, a large animal model is required for validation. Among the available ones, the porcine constitutes the most cost-effective one, becoming the nonrodent model of preference [59,60].

General Considerations, Controls, and Variability

Animal studies, just as clinical trials, seek to establish initial safety profiles and efficacy of a product. The clinical indication directly drives the animal study design. Various considerations need to be incorporated, such as the following: (1) including both genders and different age ranges when feasible; (2) recapitulation of the human disease as close as possible to allow the extrapolation of outcomes; (3) determination of defined end points that can demonstrate beneficial effects; (4) specification of the timing of intervention, as sometimes "preventive" schemes in animals not necessarily translates into human efficient therapies; (5) dosing regimens, which in this case would deal with cell densities for the most part; and (6) an optimized time course of observations and follow-ups that permits establish true therapeutic effects.

Setting up the right controls will always be a critical piece of any study. There are various types of controls that can be included, depending on the study design. Noninjured and nontreated controls should always be included, as they provide the comparable boundaries necessary to establish an effect. The former shows how intact structures look like, whereas the latter provides information regarding potential innate/endogenous healing contributions. A placebo-treated group/extremity (e.g., saline or a carrier alone) should be involved at a minimum to determine an effect. Alternatively, established procedures (e.g., microfracture) can be used to ascertain therapeutic superiority. When directly comparing two techniques or products (e.g., a scaffold with and without cells), they should be administered side-by-side (e.g., contralateral extremities) as much as possible to reduce interanimal variability. This is not always possible and greatly depends on the animal model used.

Variability is intrinsic to biology and medicine and constitutes the foundation of how species evolve. When conducting in vivo animal studies, variability among the subjects used is the norm. Animals, just as us humans, have different injury and healing responses, more evident in some species than others. For instance, the rabbit model is one that presents high variability in terms of cartilage repair responses, typically requiring large number of animals to justify meaningful conclusions. At this point, the guidance of a biostatistician is key to determine the minimal animal number required to achieve significant differences. This largely depends on several factors such as the type of data gathered, whether they can be quantified (e.g., clinical and histological scores) and how the comparisons will be paired. On the other hand, variability also touches the product being tested, with cell-based formulations exhibiting special challenges, mainly applicable to allogeneic and even xenogeneic schemes using hMSCs (see below). Decades of research has now described differences among hMSCs from various donors according to age, sex, ethnicity, and health status, all directly affecting their therapeutic potential [61,62]. This issue fuels the question of whether using a single, standardized, homogeneous, "off-the-shelf" cell-based product obtained from a healthy/ screened donor would eliminate or reduce the product variability.

Types and Sources of Cells

In section Challenges Ahead, the Special Case of the Cell Phenotype, we introduced the concept of cell phenotype as a key parameter determining the outcomes of articular cartilage TE. Here, we review a couple technical aspects critical when defining both preclinical and clinical therapeutic modalities.

Studies involving articular chondrocytes (with or without scaffolds) typically follow an autologous model in which cells are isolated from a non−weight-bearing surface of the cartilage from the same subject and serially expanded before implantation in a second procedure. Articular cartilage has been historically seen as an "immunoprivileged" tissue, based on its avascular nature−limiting immune cells access. Therefore, allogeneic and even xenogeneic models have been explored as well. Nevertheless, according to Arzi et al., this "stealth" behavior does not apply for all models, as they observed signs of rejection with xenogeneic implants, whereas allogeneic structures are tolerated when implanted in the trochlea and not in the patella, as it is distant from the immune active synovium [63]. In sharp contrast, other reports have shown full acceptance and no signs of immune rejection when human-based cartilage constructs were implanted in goats [52,64]. Bekkers et al. used xenogeneic-recycled human chondrons along with syngeneic MSCs evidencing no signs of rejection and good results overall [52], suggesting that MSCs could have exerted an immunomodulatory paracrine role as largely described for cell-based therapy protocols [65,66]. Moreover, human juvenile articular chondrocytes harvested from cadaveric donors have been tested in animal models [67] and are the basis of advanced clinical trials to evaluate clinical efficacy with no signs of rejection reported [68].

We will leave chondrocytes aside for a moment to discuss MSCs as the elected cell type. Bornes et al. reviewed in detail the advantages and disadvantages with MSCs derived from various tissues, in terms of the following: (1) ease of harvesting (i.e., access and yields); (2) chondrogenic potential and propensity to form bone tissue; and (3) evidence available [69]. Bone marrow−derived mesenchymal stem cells (BM-MSCs) are by far the most rigorously investigated with strong supporting evidence both at the preclinical and clinical levels. Adipose tissue−derived stem cells (ADSCs) are easy to harvest, present in highest numbers per tissue volumetric unit [70], but their chondrogenic potential is reduced compared with BM-MSCs, and the clinical investigation is still lacking. Synovial-derived cells seem to outperform other MSC sources in terms of their chondrogenic differentiation potential and stability; however, their limited abundance, difficulties associated with their harvesting, and the lack of clinical support remain concerning. In animal studies, both autologous and allogeneic (primarily) schemes have been used successfully.

Preclinical and Clinical Evidence, From Elusive Success to a Promising Long-Term Solution

Current surgical alternatives to induce cartilage repair (e.g., chondroplasty, mosaicplasty, microfracture) present significant clinical limitations related to their inconsistency in inducing hyaline cartilage formation, filling the defect completely and integrating with the adjacent tissue. As an alternative, cartilage TE approaches have been studied for decades, designed primarily to repair chondral (partial and full thickness) and osteochondral focal defects, with the femoral condyles and trochlea, tibial plateaus, and the talus as main locations. Various recent reviews comprehensively cover the existing literature studying cartilage TE in traumatic articular cartilage defects, at both preclinical and clinical phases [69,71,72]. We encourage the reader to access those publications for highly detailed information. Here are some relevant highlights and thoughts that deserve mention.

Several modalities and combinations have been tested using animal models, some of which have transcended to human clinical trials with variable degree of success. Each component of the TE "triad" (i.e., cells, scaffolds, and morphogens) can be used independently or through multiple combinations. This, in a very simplistic view, may discriminate two main approaches of TE-based protocols.

In an in vivo TE approach, each component is independently injected directly into the joint, attempting to promote endogenous repair/regeneration by different modes of action and in a progressive manner. Autologous chondrocytes were first introduced as a first-generation cell therapy in the early 90s as part of ACI [73], obtaining a wide range of results and with nontrivial biological and surgery-related shortcomings [74,75]. Several advancements have been introduced over the years to overcome some of the limitations ACI exhibit, with better success rates at both preclinical and clinical stages. These advancements relate to the use of a scaffold to guide the cells (i.e., Matrix-Associated Autologous Chondrocyte Implantation (MACI)), improvements to the culture conditions, cell handling (including allogeneic source) [67], and implanting techniques, and as elaborated below, the incorporation of an in vitro cell preconditioning period and the use of cocultures with adult stem cells. Despite these advancements, the associated morbidity of the donor site, the required dual intervention, the intrinsic phenotypic issues with expanded chondrocytes, and the failures observed preferentially within the first 5 years [76] have prompted researchers and clinicians to continue the search for better alternatives. In that regard, Kon et al. proposed to categorize the recent developments that may reach clinical phases into two major trends: identification of alternative cell sources (e.g., MSCs) and the use of cell-free scaffolds to guide regeneration [72].

Single intraarticular injections of MSCs alone have shown not to be able to rebuild hyaline cartilage, as structural and mechanical suboptimal fibrocartilage tissue tends to form at the surface of the defects [77]. Some improvements, related to structural parameters (proteoglycans and type II collagen presence) and integration with the host have been reached with the following adaptations: (1) serial cell injections; (2) implantation of the cells directly into the defect; and (3) combination of the cell injection with either viscosupplementation (e.g., hyaluronic acid—HA) or bone marrow stimulation (i.e., microfracture) [78–80]. Cell-free scaffold implantation, although sometimes excluded as a true TE approach [71], has also received attention given its simplicity, the fact that expensive and long cell cultures are obviated thus overcoming key regulatory barriers and the development of novel biomaterials capable of potentially exploiting endogenous regenerative capacities [74,81]. Various preclinical models have shown that scaffolds with cells (primarily bone marrow—derived and adipose tissue—derived MSCs) receive more attention and in fact seem to outperform scaffolds alone, of course taking into account the inherent great heterogeneity of both the cell-based products used and the study designs.

Another promising approach is the strategic combination of articular cartilage chondrocytes with expanded MSCs. This combination has shown in vitro benefits for both cell types including proliferative responses and mutual phenotypic stabilization [47,49,50]. When using freshly isolated autologous chondrons (chondrocytes with their pericellular matrix) in combination with allogeneic (i.e., "off-the-shelf") MSCs, preclinical [52] and clinical [53] studies have shown encouraging results, allowing a single-stage procedure instead of a two-step approach. Interestingly, it is becoming clear that MSCs provide a supportive effect to chondrocytes rather than a progenitor one, as the cells present in biopsy-derived tissue are exclusively from a host origin.

Contrary to the in vivo approach, the *in vitro—enhanced* TE approach combines TE components to fabricate a full chondral or osteochondral construct in the laboratory before its implantation. Within this approach, cells are seeded in biological (e.g., collagen gels or membranes, Hyaluronic Acid (HA), fibrin-rich gels or glues), synthetic (e.g., Polyglycolic Acid (PGA), Polylactic Acid (PLA), etc.,) or hybrid matrices [82]. The resulting TE construct (TEC) is grown, preconditioned (using bioreactors), characterized (ideally in a nondestructive/noninvasive manner as in Ref. [83]), and then implanted into the defect expecting integration and functional adaptation [21,84,85]. Huang et al. extensively reviewed the current available products for cartilage TE, defined either as second- and third-generation ACI or first-generation TE solutions [71]. The preconditioning period (few days to several weeks), in which the growing structure is exposed to biochemical and biophysical environments, seems to have a positive effect on the performance of the implanted tissue as better resurfacing, integration of the implant with the surrounding native tissue, and underlying bone regeneration in osteochondral defects is observed in animal models [86–88]. Interestingly, in those models predifferentiated MSCs showed better results than articular chondrocytes as part of the growing TEC [86]. At the clinical stage, scaffolds with cells are commonly used, with chondrocytes as the preferred cell type and bone marrow–derived MSCs as very promising and gaining popularity [72,89,90]. Nevertheless, the use of MSCs as a cell of choice to treat focal cartilage defects is in its infancy, with high-quality clinical studies still pending [91].

In general, positive results have been reported up to 24 months based on clinical parameters (i.e., functional scores and imaging—MRI), and direct observation when performed (i.e., second-look arthroscopy evaluating integration, surface, and filling) [69]. Histologically, also when available, a hyalinelike tissue rich in type II collagen and proteoglycans has been typically observed, sometimes mixed with circumscribed hypertrophic cells [92,93]. Currently, several groups are engaged in evaluating longer follow-up periods to solidly establish efficacy beyond safety.

When Is "Enough Evidence" to Move on

Now that the basic research has elucidated developmental processes, from which key information can be borrowed and exploited to improve regenerative medicine techniques, who or what determines the quantity and quality of the preclinical evidence as "enough" to move on to clinical phases? Let us first understand the basic conditions for the submission of an Investigational New Drug (IND) and/or an Investigational Device Exemption (IDE) for a cell-based TE product (biologic, device, or their combination), termed "investigational product," intended for articular cartilage repair or regeneration, as it sets up the "minimal" requirements for moving along.

In addition to specific information regarding the manufacturing of a device and the chemical, manufacturing, and control (CMC) of a biological product, thorough reports of any and all preclinical studies performed using the investigation product are requisite for IND/IDE submission, and it is of utmost importance that they convey the *safety* of its use, especially when cells are part of it. Moreover, the procedures conducted in these studies must comply with good laboratory practice regulations. The *efficacy* of the product is specific to the intended purpose to determine the relevant structural and biological characteristics of the product that will be transferred to the tissue defect. Data acquired from animal studies and mechanical testing (if required) should be sufficient to assess the biological response to product, durability of response, toxicology, and dose response, as supportive information to establish a scientific rationale for clinical investigation and "first use in humans." These stipulations are governed by the severity of the damage and length of time for preclinical trials, which is specific to the used models.

As discussed earlier in this chapter, choosing an animal model is critical. It is important, yet not always such an obvious goal, to collect evidence that is analogous or closely resembles what is to be expected in a human patient and not an artifact of the model. For instance, intrinsic species-derived differences in life span, skeletal development, and healing properties. Furthermore, clearly demonstrating that the investigational product is safe (i.e., no adverse effects) and effective with all assessments adequately addressed, and that these stipulations can be reproducibly generated in a sufficient number of animals to determine an overarching outcome, should deem sufficient for IND approval and rapid advancement into human clinical trials.

Finally, optimizing all TE components (i.e., cells, scaffolds, stimulatory molecular regimes, and bioreactors) to generate the "perfect" implant is far from reality. There are far too many variables to consider, that the resulting testing matrix would be unmanageable. Perhaps, as scientists and clinicians, the underlying problem stems from our never-ending temptation to improve the "good." Here, we would like to quote Dr. Chris Evans, who elegantly summarized this into a couple provoking sentences [94]: "However, at some point, it is necessary to stop tread-milling. Instead, it is better to identify the best available components at the time, and move forward into the next phase of development, even if it will lead to transitional technology to be superseded by later generation products..." So, the idea would be to reach a midpoint between "perpetual advancement" and "rushing to human use," to maximize the chances of seeing well-designed, long-term, and functional solutions supported by solid science.

CONSIDERATIONS TO DESIGN EARLY-PHASE CLINICAL TRIALS

Tissue engineering and regenerative medicine (TERM)−based products have very special features (e.g., multiple components, living cells, etc.), significantly impacting the bench-to-bedside transition. As a consequence, regulatory aspects have to always be kept in mind when embarking in projects with high chances of clinical translation. Depending on the final product's composition and complexity, the FDA evaluates it as a device, a biologic, a drug, or a combination, all with different demands and challenges. Here, we elaborate on simple concepts based on guidelines and recommendations setup by the FDA and international scientific societies such as the International Cannabinoid Research Society (ICRS) [95].

Clinical studies need to comply with IND/IDE regulations in addition to informed consent and IRB regulations. Two main types of clinical trials (exploratory and confirmatory) are typically conducted to establish safety, efficacy, and to address general considerations (e.g., need of special procedures, tissue integrity, appropriateness of the target population, product activity, etc.). Exploratory trials are intended to support the design of the confirmatory, which are typically randomized, controlled, and blinded studies centered around gathering hypothesis-testing data to assess efficacy and safety. Additionally, various elements need special attention to maximize the information and conclusions drawn from the study. We will review the most important ones below.

- **Controls**: just as described for animal models, the establishment of a right control is critical to obtain the most informative clinical data possible. Concurrent controls include placebo (unpopular), sham surgery, active comparator, and a standard of care, of which the last two offer meaningful conclusions the most. Depending on the complexity of the lesion (duration, size, and localization), the type and extent of the procedure being tested, the standard treatment typically used and the subsequent rehabilitation scheme, good alternatives are as follows: microfracture, debridement, mosaicplasty, ACI, osteochondral allografts, and perichondral/periosteal flaps. Other important factors that both the experimental and control groups should have as comparable as possible, even in the presence of randomization, are as follows: patient demographics, body mass index, status of other joint tissues (especially menisci), leg alignment, and stability. Sham surgeries, despite their inherent flaws and ethical disputes, may constitute viable controls when subjective outcomes are established, as they can discriminate those generated by the tested therapy from other factors such as patient or observer expectations. However, they have to be carefully designed and be well justified. Studies can involve a head-to-head comparison between a comparator and the test therapy or involve more than two arms with more than one treatment arm.
- **Patient population and recruitment**: insufficient sample size is often found in trials for cartilage damage, due, in part, to the broad range of injuries presentation. The most important objective of clinical research is to recapitulate "real-life" scenarios and not "ideal" ones, arguably reproduced in randomized clinical trials (RCTs). Several patient selection characteristics need to be strictly defined and measured (when required) using standardized procedures or techniques to reduce—or even eliminate—ambiguity. Those characteristics are as follows: degree of pain; presence or absence of general cartilage damage (osteoarthritis) and how it was determined; degree of physical function and previous levels of activity; location, depth, size of the lesion; concomitant joint pathology (e.g., meniscal and/or ligament tears); and whether there have been any previous treatments (surgical and nonsurgical). When specific values are used, the rationale for their use needs to be well supported. Consequently, all these factors define a set of exclusion criteria that "homogenizes" the population sample, which in turn guarantees sound statistical comparisons. As an

alternative, and aiming at covering "real" clinical situations, longitudinal assessments of groups of treated patients "as they are" may provide clinical information that complements RCT's strictly obtained data. Of course, extreme care needs to be taken when drawing conclusions from such studies.

- **Study end points**: overall, primary and secondary end points are setup, depending on the characteristics of the investigational product and its way of administration. As for primary outcomes, improvement (i.e., degree of change) in pain and/or physical function of the involved joint should always be used. To have quantified and comparable data, validated questionnaires are used as patient-reported outcome measures (PROMs) to collect it (e.g., Knee Injury and Osteoarthritis Outcome Score (KOOS), Western Ontario and McMaster Universities Osteoarthritis Index (WOMAC), etc.). Secondary outcomes involve more objectives measurements, including the following:
  - Arthroscopic assessments of the surface, integrity, size, stiffness/firmness, binding to surrounding tissue, and general aspect of the resulting tissue, using established scales (e.g., ICRS).
  - Clinical evaluation of the involved joint (e.g., range of motion, effusion, muscle strength, alignment, crepitus, etc.).
  - Histological evaluation of a tissue biopsy to assess key structural parameters (e.g., zonal organization, cell density and morphology, collagen and proteoglycan deposition, inflammatory responses, etc.).
  - Serological and/or synovial fluid samples to assess for inflammatory markers and potential antibody formation.
  - Imaging-based assessment (e.g., MRI) of the entire joint, interpreted by blinded readers (minimum two). To the best of our knowledge, the T2 relaxation mapping protocol (i.e., CartiGram) constitutes the most sensitive imaging modality to assess structural changes of the articular cartilage.

Other elements include the administration of the investigational product, the follow-up schemes that may go from 12 months to several years depending on the type of product and therapy, and adverse events reporting. A final important parameter, when using MSCs, is the genetic relationship between donor and recipient. Autologous schemes have been historically the norm based on the long-standing notion of immune rejection of allotransplanted antigens. As adults, minute levels of MSCs reside in each of our tissues forcing culture expansion to obtain adequate numbers for a therapy. Even though very effective, expansion regimes can produce untoward effects on the cultured cells [19]. On the other hand, a very attractive quality of MSCs is their little-to-no expression of self-antigens (i.e., human leukocyte antigens—HLA), which allows them to evade immune responses when introduced into another host [38]. Because of the absence or low presence of those self-antigens on MSCs, allogeneic transplantation is popular for stem cell therapies as they offer screened/characterized "off-the-shelf" products that come from healthy young donors [96].

CONCLUSIONS

Articular cartilage TE has been an area with significant research over the years, yet with scarce clinical translation and approved products. Various factors, including the modulation of the chondrocytic phenotype of chondroprogenitors and harvested primary chondrocytes, and the resulting tissue functional behavior, have contributed to this slow bench-to-the-bedside jump. The incorporation of key concepts from emerging fields such as developmental engineering, borrowing from embryonic development precepts, along with advances in stem cell biology and continuous improvements in material science makes us to truly believe that a long-term solution for focal cartilage defects is not distant.

LIST OF ACRONYMS AND ABBREVIATIONS

ACI Autologous chondrocyte implantation
CMC Chemical, manufacturing, and control
HA Hyaluronic acid
IDE Investigational device exemption
IND Investigational new drug
MSCs Mesenchymal stem cells
PROMs Patient-reported outcome measures
RCTs Randomized clinical trials
TE Tissue engineering
TEC Tissue engineering construct

ACKNOWLEDGMENTS

We are in gratitude with the Soffer Family Foundation and the Diabetes Research Institute Foundation (DRIF) for their generous support.

REFERENCES

[1] Langer R, Vacanti JP. Tissue engineering. Science May 14, 1993;260(5110):920—6.

[2] de Bono B, Grenon P, Baldock R, Hunter P. Functional tissue units and their primary tissue motifs in multi-scale physiology. J Biomed Semantics October 8, 2013;4(1):22.

[3] Vrana NE, Lavalle P, Dokmeci MR, Dehghani F, Ghaemmaghami AM, Khademhosseini A. Engineering functional epithelium for regenerative medicine and in vitro organ models: a review. Tissue Eng Part B Rev December 2013;19(6):529—43.

[4] Huey DJ, Hu JC, Athanasiou KA. Unlike bone, cartilage regeneration remains elusive. Science November 16, 2012;338(6109):917—21.

[5]   O'Keefe RJ, Mao J. Bone tissue engineering and regeneration: from discovery to the clinic—an overview. Tissue Eng Part B Rev 2011:389—92.

[6]   Cleary MA, van Osch GJ, Brama PA, Hellingman CA, Narcisi R. FGF, TGFbeta and Wnt crosstalk: embryonic to in vitro cartilage development from mesenchymal stem cells. J Tissue Eng Regen Med April 2015;9(4):332—42. 2013 ed.

[7]   Salazar VS, Gamer LW, Rosen V. BMP signalling in skeletal development, disease and repair. Nat Rev Endocrinol February 19, 2016;12(4):203—21. Nature Publishing Group.

[8]   Narcisi R, Signorile L, Verhaar JAN, Giannoni P, van Osch GJVM. TGFβ inhibition during expansion phase increases the chondrogenic re-differentiation capacity of human articular chondrocytes. Osteoarthr Cartil October 2012;20(10):1152—60.

[9]   Correa D, Somoza RA, Lin P, Greenberg S, Rom E, Duesler L, et al. Sequential exposure to fibroblast growth factors (FGF) 2, 9 and 18 enhances hMSC chondrogenic differentiation. Osteoarthr Cartil March 2015;23(3):443—53. 2014 ed.

[10]   Lories RJ, Corr M, Lane NE. To Wnt or not to Wnt: the bone and joint health dilemma. Nat Rev Rheumatol March 5, 2013;9(6):328—39.

[11]   Lenas P, Moos M, Luyten FP. Developmental engineering: a new paradigm for the design and manufacturing of cell-based products. Part II: from genes to networks: tissue engineering from the viewpoint of systems biology and network science. Tissue Eng Part B Rev December 1, 2009;15(4):395—422.

[12]   Scotti C, Tonnarelli B, Papadimitropoulos A, Scherberich A, Schaeren S, Schauerte A, et al. Recapitulation of endochondral bone formation using human adult mesenchymal stem cells as a paradigm for developmental engineering. Proc Natl Acad Sci USA April 20, 2010;107(16):7251—6. 2010 ed.

[13]   Caplan AI. Mesenchymal stem cells. J Orthop Res September 1, 1991;9(5):641—50. Wiley Subscription Services, Inc., A Wiley Company.

[14]   Maes C, Kobayashi T, Selig MK, Torrekens S, Roth SI, Mackem S, et al. Osteoblast precursors, but not mature osteoblasts, move into developing and fractured bones along with invading blood vessels. Dev Cell August 17, 2010;19(2):329—44.

[15]   Bianco P, Cancedda FD, Riminucci M, Cancedda R. Bone formation via cartilage models: the "borderline" chondrocyte. Matrix Biol July 1998;17(3):185—92.

[16]   Kozhemyakina E, Zhang M, Ionescu A, Ayturk UM, Ono N, Kobayashi A, et al. Identification of a Prg4-expressing articular cartilage progenitor cell population in mice. Arthritis Rheumatol May 2015;67(5):1261—73.

[17]   Candela ME, Yasuhara R, Iwamoto M, Enomoto-Iwamoto M. Resident mesenchymal progenitors of articular cartilage. Matrix Biol October 2014;39:44—9.

[18]   O'Sullivan J, D'Arcy S, Barry FP, Murphy JM, Coleman CM. Mesenchymal chondroprogenitor cell origin and therapeutic potential. Stem Cell Res Ther 2011;2(1):8.

[19]   Somoza RA, Welter JF, Correa D, Caplan AI. Chondrogenic differentiation of mesenchymal stem cells: challenges and unfulfilled expectations. Tissue Eng Part B Rev December 2014;20(6):596—608.

[20]   Musumeci G, Castrogiovanni P, Leonardi R, Trovato FM, Szychlinska MA, Di Giunta A, et al. New perspectives for articular cartilage repair treatment through tissue engineering: a contemporary review. World J Orthop April 18, 2014;5(2):80—8.

[21]   Kock L, van Donkelaar CC, Ito K. Tissue engineering of functional articular cartilage: the current status. Cell Tissue Res March 2012;347(3):613—27. Springer-Verlag.

[22]   Correa D, Lietman SA. Articular cartilage repair: current needs, methods and research directions. Semin Cell Dev Biol July 2016.

[23] Pestka JM, Schmal H, Salzmann G, Hecky J, Südkamp NP, Niemeyer P. In vitro cell quality of articular chondrocytes assigned for autologous implantation in dependence of specific patient characteristics. Arch Orthop Trauma Surg December 17, 2010;131(6):779—89.

[24] Peterson L, Vasiliadis HS, Brittberg M, Lindahl A. Autologous chondrocyte implantation: a long-term follow-up. Am J Sports Med June 2010;38(6):1117—24. American Orthopaedic Society for Sports Medicine.

[25] Hinckel BB, Gomoll AH. Autologous chondrocytes and next-generation matrix-based autologous chondrocyte implantation. Clin Sports Med July 2017;36(3):525—48.

[26] Kurth TB, Dell'Accio F, Crouch V, Augello A, Sharpe PT, De Bari C. Functional mesenchymal stem cell niches in adult mouse knee joint synovium in vivo. Arthritis Rheum May 2011;63(5):1289—300. Wiley Subscription Services, Inc., A Wiley Company.

[27] Williams R, Khan IM, Richardson K, Nelson L, McCarthy HE, Analbelsi T, et al. Identification and clonal characterisation of a progenitor cell sub-population in normal human articular cartilageAgarwal S, editor. PLoS ONE 2010;5(10):e13246.

[28] Muiños-López E, Rendal-Vázquez ME, Hermida-Gómez T, Fuentes-Boquete I, Díaz-Prado S, Blanco FJ. Cryopreservation effect on proliferative and chondrogenic potential of human chondrocytes isolated from superficial and deep cartilage. Open Orthop J 2012;6(1):150—9.

[29] Seol D, McCabe DJ, Choe H, Zheng H, Yu Y, Jang K, et al. Chondrogenic progenitor cells respond to cartilage injury. Arthritis Rheum November 2012;64(11):3626—37. Wiley Subscription Services, Inc., A Wiley Company.

[30] Tuan RS, Chen AF, Klatt BA. Cartilage regeneration. J Am Acad Orthop Surg May 2013;21(5):303—11.

[31] Pittenger MF, Mackay AM, Beck SC, Jaiswal RK, Douglas R, Mosca JD, et al. Multilineage potential of adult human mesenchymal stem cells. Science April 2, 1999;284(5411):143—7. 1999 ed.

[32] Biehl JK, Russell B. Introduction to stem cell therapy. J Cardiovasc Nurs March 2009;24(2):98—103. quiz 104—5.

[33] Crisan M, Yap S, Casteilla L, Chen C-W, Corselli M, Park TS, et al. A perivascular origin for mesenchymal stem cells in multiple human organs. Cell Stem Cell September 11, 2008;3(3):301—13.

[34] Corselli M, Crisan M, Murray IR, West CC, Scholes J, Codrea F, et al. Identification of perivascular mesenchymal stromal/stem cells by flow cytometry. Cytometry August 2013;83(8):714—20.

[35] James AW, Zara JN, Corselli M, Askarinam A, Zhou AM, Hourfar A, et al. An abundant perivascular source of stem cells for bone tissue engineering. Stem Cells Transl Med September 2012;1(9):673—84. AlphaMed Press.

[36] Sacchetti B, Funari A, Michienzi S, Di Cesare S, Piersanti S, Saggio I, et al. Self-renewing osteoprogenitors in bone marrow sinusoids can organize a hematopoietic microenvironment. Cell October 19, 2007;131(2):324—36. 2007 ed.

[37] Caplan AI. All MSCs are pericytes? Cell Stem Cell September 11, 2008;3(3):229—30.

[38] Dominici M, Le Blanc K, Mueller I, Slaper-Cortenbach I, Marini F, Krause D, et al. Minimal criteria for defining multipotent mesenchymal stromal cells. The International Society for Cellular Therapy position statement. Cytotherapy 2006;8(4):315—7.

[39] Somoza RA, Correa D, Labat I, Sternberg H, Forrest ME, Khalil AM, et al. Transcriptome-wide analyses of human neonatal articular cartilage and human mesenchymal stem cell-derived cartilage provide a new molecular target for evaluating engineered cartilage. Tissue Eng Part A July 28, 2017. https://doi.org/10.1089/ten.TEA.2016.0559.

[40] Pelttari K, Winter A, Steck E, Goetzke K, Hennig T, Ochs BG, et al. Premature induction of hypertrophy during in vitro chondrogenesis of human mesenchymal stem cells correlates with calcification and vascular invasion after ectopic transplantation in SCID mice. Arthritis Rheum October 1, 2006;54(10):3254−66.

[41] Abrahamsson CK, Yang F, Park H, Brunger JM, Valonen PK, Langer R, et al. Chondrogenesis and mineralization during in vitro culture of human mesenchymal stem cells on three-dimensional woven scaffolds. Tissue Eng Part A December 2010;16(12):3709−18. Mary Ann Liebert, Inc. 140 Huguenot Street, 3rd Floor New Rochelle, NY 10801 USA.

[42] Mueller MB, Fischer M, Zellner J, Berner A, Dienstknecht T, Prantl L, et al. Hypertrophy in mesenchymal stem cell chondrogenesis: effect of TGF-β isoforms and chondrogenic conditioning. Cells Tissues Organs 2010;192(3):158−66.

[43] Marsano A, Medeiros da Cunha CM, Ghanaati S, Gueven S, Centola M, Tsaryk R, et al. Spontaneous in vivo chondrogenesis of bone marrow-derived mesenchymal progenitor cells by blocking vascular endothelial growth factor signaling. Stem Cells Transl Med July 26, 2016. AlphaMed Press.

[44] Narcisi R, Cleary MA, Brama PAJ, Hoogduijn MJ, Tüysüz N, Berge ten D, et al. Long-term expansion, enhanced chondrogenic potential, and suppression of endochondral ossification of adult human MSCs via WNT signaling modulation. Stem Cell Reports March 10, 2015;4(3):459−72. 2015 ed.

[45] Hellingman CA, Davidson ENB, Koevoet W, Vitters EL, van den Berg WB, van Osch GJVM, et al. Smad signaling determines chondrogenic differentiation of bone-marrow-derived mesenchymal stem cells: inhibition of Smad1/5/8P prevents terminal differentiation and calcification. Tissue Eng Part A April 2011;17(7−8):1157−67. 2010 ed.

[46] Panadero JA, Lanceros-Mendez S, Ribelles JLG. Differentiation of mesenchymal stem cells for cartilage tissue engineering: individual and synergetic effects of three-dimensional environment and mechanical loading. Acta Biomater March 2016;33:1−12. 2016 ed.

[47] Acharya C, Adesida A, Zajac P, Mumme M, Riesle J, Martin I, et al. Enhanced chondrocyte proliferation and mesenchymal stromal cells chondrogenesis in coculture pellets mediate improved cartilage formation. J Cell Physiol January 1, 2012;227(1):88−97. Wiley Subscription Services, Inc., A Wiley Company.

[48] Meretoja VV, Dahlin RL, Kasper FK, Mikos AG. Enhanced chondrogenesis in co-cultures with articular chondrocytes and mesenchymal stem cells. Biomaterials September 1, 2012;33(27):6362−9.

[49] Bian L, Zhai DY, Mauck RL, Burdick JA. Coculture of human mesenchymal stem cells and articular chondrocytes reduces hypertrophy and enhances functional properties of engineered cartilage. Tissue Eng Part A April 1, 2011;17(7−8):1137−45.

[50] Fischer J, Dickhut A, Rickert M, Richter W. Human articular chondrocytes secrete parathyroid hormone-related protein and inhibit hypertrophy of mesenchymal stem cells in coculture during chondrogenesis. Arthritis Rheum September 1, 2010;62(9):2696−706.

[51] Jikko A, Kato Y, Hiranuma H, Fuchihata H. Inhibition of chondrocyte terminal differentiation and matrix calcification by soluble factors released by articular chondrocytes. Calcif Tissue Int October 1, 1999;65(4):276−9.

[52] Bekkers JEJ, Tsuchida AI, van Rijen MHP, Vonk LA, Dhert WJA, Creemers LB, et al. Single-stage cell-based cartilage regeneration using a combination of chondrons and mesenchymal stromal cells: comparison with microfracture. Am J Sports Med September 2013;41(9):2158−66. American Orthopaedic Society for Sports Medicine.

[53] de Windt TS, Vonk LA, Slaper Cortenbach ICM, van den Broek MPH, Nizak R, van Rijen MHP, et al. Allogeneic mesenchymal stem cells stimulate cartilage regeneration and

are safe for single-stage cartilage repair in humans upon mixture with recycled autologous chondrons. Stem Cells January 1, 2017;35(1):256—64.

[54] Tormin A, Li O, Brune JC, Walsh S, Schütz B, Ehinger M, et al. CD146 expression on primary nonhematopoietic bone marrow stem cells is correlated with in situ localization. Blood May 12, 2011;117(19):5067—77.

[55] Méndez-Ferrer S, Michurina TV, Ferraro F, Mazloom AR, Macarthur BD, Lira SA, et al. Mesenchymal and haematopoietic stem cells form a unique bone marrow niche. Nature August 12, 2010;466(7308):829—34.

[56] Moran CJ, Ramesh A, Brama PAJ, O'Byrne JM, O'Brien FJ, Levingstone TJ. The benefits and limitations of animal models for translational research in cartilage repair. J Exp Orthop January 4, 2016;3(1):1—12. 3rd ed.

[57] Teeple E, Jay GD, Elsaid KA, Fleming BC. Animal models of osteoarthritis: challenges of model selection and analysis. AAPS J January 18, 2013;15(2):438—46.

[58] Chu CR, Szczodry M, Bruno S. Animal models for cartilage regeneration and repair. Tissue Eng Part B Rev February 2010;16(1):105—15. Mary Ann Liebert, Inc. 140 Huguenot Street, 3rd Floor New Rochelle, NY 10801 USA.

[59] Walters EM, Agca Y, Ganjam V, Evans T. Animal models got you puzzled?: think pig. Ann NY Acad Sci December 2011;1245(1):63—4. Blackwell Publishing Inc.

[60] Gregory MH, Capito N, Kuroki K, Sherman SL, Gregory MH, Capito N, et al. A review of translational animal models for knee osteoarthritis. Arthritis December 27, 2012;2012(7):1—14. Hindawi Publishing Corporation.

[61] Oh J, Lee YD, Wagers AJ. Stem cell aging: mechanisms, regulators and therapeutic opportunities. Nat Med August 2014;20(8):870—80.

[62] Deasy BM, Lu A, Tebbets JC, Feduska JM, Schugar RC, Pollett JB, et al. A role for cell sex in stem cell-mediated skeletal muscle regeneration: female cells have higher muscle regeneration efficiency. J Cell Biol April 9, 2007;177(1):73—86.

[63] Arzi B, Huey DJ, Borjesson DL, Murphy BG, Athanasiou KA. Cartilage immunoprivilege depends on donor source and lesion location. Acta Biomater September 2015;23:72—81.

[64] Lewis PB, McCarty LP, Yao JQ, Williams JM, Kang R, Cole BJ. Fixation of tissue-engineered human neocartilage constructs with human fibrin in a caprine model. J Knee Surg July 2009;22(3):196—204.

[65] Caplan AI, Correa D. The MSC: an injury drugstore. Cell Stem Cell July 8, 2011;9(1):11—5.

[66] Stagg J, Galipeau J. Mechanisms of immune modulation by mesenchymal stromal cells and clinical translation. Curr Mol Med June 2013;13(5):856—67.

[67] Adkisson HD, Martin JA, Amendola RL, Milliman C, Mauch KA, Katwal AB, et al. The potential of human allogeneic juvenile chondrocytes for restoration of articular cartilage. Am J Sports Med July 2010;38(7):1324—33. SAGE Publications Sage CA: Los Angeles, CA.

[68] McCormick F, Cole BJ, Nwachukwu B, Harris JD, Adkisson HDIV, Farr J. Treatment of focal cartilage defects with a juvenile allogeneic 3-dimensional articular cartilage graft. Operat Tech Sports Med June 2013;21(2):95—9.

[69] Bornes TD, Adesida AB, Jomha NM. Mesenchymal stem cells in the treatment of traumatic articular cartilage defects: a comprehensive review. Arthritis Res Ther September 26, 2014;16(5):432.

[70] Semon JA, Maness C, Zhang X, Sharkey SA, Beuttler MM, Shah FS, et al. Comparison of human adult stem cells from adipose tissue and bone marrow in the treatment of

experimental autoimmune encephalomyelitis. Stem Cell Res Ther January 9, 2014;5(1):2. BioMed Central.

[71] Huang BJ, Hu JC, Athanasiou KA. Cell-based tissue engineering strategies used in the clinical repair of articular cartilage. Biomaterials August 2016;98:1−22.

[72] Kon E, Roffi A, Filardo G, Tesei G, Marcacci M. Scaffold-based cartilage treatments: with or without cells? A systematic review of preclinical and clinical evidence. Arthroscopy April 2015;31(4):767−75.

[73] Brittberg M, Lindahl A, Nilsson A, Ohlsson C, Isaksson O, Peterson L. Treatment of deep cartilage defects in the knee with autologous chondrocyte transplantation. N Engl J Med October 6, 1994;331(14):889−95. Massachusetts Medical Society.

[74] Filardo G, Kon E, Roffi A, Di Martino A, Marcacci M. Scaffold-based repair for cartilage healing: a systematic review and technical note. Arthroscopy January 2013;29(1):174−86.

[75] McCormick F, Harris JD, Abrams GD, Frank R, Gupta A, Hussey K, et al. Trends in the surgical treatment of articular cartilage lesions in the United States: an analysis of a large private-payer database over a period of 8 years. Arthroscopy February 2014;30(2):222−6.

[76] Andriolo L, Merli G, Filardo G, Marcacci M, Kon E. Failure of autologous chondrocyte implantation. Sports Med Arthrosc March 2017;25(1):10−8.

[77] Koga H, Muneta T, Nagase T, Nimura A, Ju Y-J, Mochizuki T, et al. Comparison of mesenchymal tissues-derived stem cells for in vivo chondrogenesis: suitable conditions for cell therapy of cartilage defects in rabbit. Cell Tissue Res August 2008;333(2):207−15. Springer-Verlag.

[78] Nam HY, Karunanithi P, Loo WC, Naveen S, Chen H, Hussin P, et al. The effects of staged intra-articular injection of cultured autologous mesenchymal stromal cells on the repair of damaged cartilage: a pilot study in caprine model. Arthritis Res Ther 2013;15(5):R129. BioMed Central.

[79] Saw K-Y, Hussin P, Loke S-C, Azam M, Chen H-C, Tay Y-G, et al. Articular cartilage regeneration with autologous marrow aspirate and hyaluronic acid: an experimental study in a goat model. Arthroscopy December 2009;25(12):1391−400.

[80] Lee KBL, Hui JHP, Song IC, Ardany L, Lee EH. Injectable mesenchymal stem cell therapy for large cartilage defects−a porcine model. Stem Cells November 2007;25(11):2964−71. John Wiley & Sons, Ltd.

[81] Kon E, Filardo G, Roffi A, Andriolo L, Marcacci M. New trends for knee cartilage regeneration: from cell-free scaffolds to mesenchymal stem cells. Curr Rev Musculoskelet Med September 2012;5(3):236−43. Current Science Inc.

[82] Lee EJ, Kasper FK, Mikos AG. Biomaterials for tissue engineering. Ann Biomed Eng February 2014;42(2):323−37. 2nd ed. Springer US.

[83] Correa D, Somoza RA, Caplan AI. Non-destructive/non-invasive imaging evaluation of cellular differentiation progression during in vitro MSC-derived chondrogenesis. Tissue Eng Part A August 21, 2017. https://doi.org/10.1089/ten.TEA.2017.0125.

[84] Gadjanski I, Spiller K, Vunjak-Novakovic G. Time-dependent processes in stem cell-based tissue engineering of articular cartilage. Stem Cell Rev and Rep October 21, 2011;8(3):863−81.

[85] Grayson WL, Bhumiratana S, Grace Chao PH, Hung CT, Vunjak-Novakovic G. Spatial regulation of human mesenchymal stem cell differentiation in engineered osteochondral constructs: effects of pre-differentiation, soluble factors and medium perfusion. Osteoarthr Cartil May 2010;18(5):714−23.

[86] Marquass B, Schulz R, Hepp P, Zscharnack M, Aigner T, Schmidt S, et al. Matrix-associated implantation of predifferentiated mesenchymal stem cells versus articular chondrocytes:

in vivo results of cartilage repair after 1 year. Am J Sports Med July 2011;39(7):1401−12. American Orthopaedic Society for Sports Medicine.

[87] Zscharnack M, Hepp P, Richter R, Aigner T, Schulz R, Somerson J, et al. Repair of chronic osteochondral defects using predifferentiated mesenchymal stem cells in an ovine model. Am J Sports Med September 2010;38(9):1857−69. American Orthopaedic Society for Sports Medicine.

[88] Khozoee B, Mafi P, Mafi R, Khan W. Mechanical stimulation protocols of human derived cells in articular cartilage tissue engineering − a systematic review. Curr Stem Cell Res Ther June 13, 2016.

[89] Wakitani S, Nawata M, Tensho K, Okabe T, Machida H, Ohgushi H. Repair of articular cartilage defects in the patello-femoral joint with autologous bone marrow mesenchymal cell transplantation: three case reports involving nine defects in five knees. J Tissue Eng Regen Med January 2007;1(1):74−9. John Wiley & Sons, Ltd.

[90] Kuroda R, Ishida K, Matsumoto T, Akisue T, Fujioka H, Mizuno K, et al. Treatment of a full-thickness articular cartilage defect in the femoral condyle of an athlete with autologous bone-marrow stromal cells. Osteoarthr Cartil February 2007;15(2):226−31.

[91] Filardo G, Madry H, Jelic M, Roffi A, Cucchiarini M, Kon E. Mesenchymal stem cells for the treatment of cartilage lesions: from preclinical findings to clinical application in orthopaedics. Knee Surg Sports Traumatol Arthrosc August 2013;21(8):1717−29. Springer Berlin Heidelberg.

[92] Giannini S, Buda R, Vannini F, Cavallo M, Grigolo B. One-step bone marrow-derived cell transplantation in talar osteochondral lesions. Clin Orthop Relat Res December 2009;467(12):3307−20. Springer-Verlag.

[93] Giannini S, Buda R, Cavallo M, Ruffilli A, Cenacchi A, Cavallo C, et al. Cartilage repair evolution in post-traumatic osteochondral lesions of the talus: from open field autologous chondrocyte to bone-marrow-derived cells transplantation. Injury November 2010;41(11):1196−203.

[94] Evans CH. Barriers to the clinical translation of orthopedic tissue engineering. Tissue Eng Part B Rev December 2011;17(6):437−41.

[95] Mithoefer K, Saris DBF, Farr J, Kon E, Zaslav K, Cole BJ, et al. Guidelines for the design and conduct of clinical studies in knee articular cartilage repair. Cartilage March 23, 2011;2(2):100−21.

[96] Ryan JM, Barry FP, Murphy JM, Mahon BP. Mesenchymal stem cells avoid allogeneic rejection. J Inflamm July 26, 2005;2:8.

Index

Printed and bound by CPI Group (UK) Ltd, Croydon, CR0 4YY

08/05/2025

01865001-0001